Development of Analytical Methods in the Field of Food Analysis

Development of Analytical Methods in the Field of Food Analysis

Editor

Gianfranco Picone

MDPI

Basel • Beijing • Wuhan • Barcelona • Belgrade • Novi Sad • Cluj • Manchester

Editor
Gianfranco Picone
Department of Agricultural
and Food Sciences
University of Bologna
Cesena, Italy

Editorial Office
MDPI
St. Alban-Anlage 66
4052 Basel, Switzerland

This is a reprint of articles from the Special Issue published online in the open access journal *Foods* (ISSN 2304-8158) (available at: https://www.mdpi.com/journal/foods/special_issues/Development_Field_Food_Analysis).

For citation purposes, cite each article independently as indicated on the article page online and as indicated below:

Lastname, A.A.; Lastname, B.B. Article Title. *Journal Name* **Year**, *Volume Number*, Page Range.

ISBN 978-3-0365-9810-9 (Hbk)
ISBN 978-3-0365-9811-6 (PDF)
doi.org/10.3390/books978-3-0365-9811-6

Contents

About the Editor

Gianfranco Picone

Gianfranco Picone achieved in 2009 a European Ph.D. in Food Science, Technology, and Biotechnology at the University of Bologna. Since 2010, he has been involved in several projects concerning the metabolomics evaluation of the effects of enriched food on patients affected by chronic diseases such as Crohn's disease, Diverticulitis, and Constipation, using HF 1H-NMR. These studies are based on the principles of metabolomics and nutrimetabolomics. During these years, he acquired broad competence in food analysis, working also on the assessment of in vitro digested food, specifically focusing the research on food rich in protein such as dairy products, meat, and fish. He participates in different national and international projects; he has worked on the project CHANCE, PATHWAY, ECOPROLIVE, FOODBALL, FutureEUAqua, NewTechAqua, and BIOZOOSTAIN. He is coordinating the Project (SOE_0000116) "Metabolomics Approach for the assessment of Baby-Mother Enteric microbiota Legacy (MABEL)" funded under the National Recovery and Resilience Plan (PNRR) of the Italian Ministry of Education, University and Research (MIUR), designed to support researchers who obtain a Seal of Excellence for a Mary Curie Project. He achieved scientific results, as displayed in more than 50 papers, through an effective collaboration with several research groups.

Preface

Any product intended for human consumption needs to be rigorously tested, and because foodstuffs are ingested, the testing is often especially crucial to prevent any potential health issues. Thus, food analysis is a very important branch of analytical chemistry, as it provides information about the chemical composition, processing, quality control, and contamination of foodstuffs, ensuring compliance with food and trade laws. For this reason, it impacts both the economic and medical aspects of modern societies. The development of analytical methods in food matrices, as well as the understanding of how the food matrix influences nutrients' bioaccessibility, has always been difficult due to the large variety of their physicochemical properties. In terms of the techniques that are used to name a few common examples, the range of instruments includes nuclear magnetic resonance (NMR) spectroscopy, gas chromatography (GC), atomic absorption spectroscopy (AAS), and high-performance liquid chromatography (HPLC). The choice of technique(s) depends on what foodstuff is being analyzed, what is being analyzed within the foodstuff, and the reasons for the analysis, which can change analyte structure and extraction efficiencies because of different processing procedures throughout preparation and distribution. Furthermore, the multiplexing trend, which means the simultaneous detection of multiple analytes at the same time, requires the development of more appropriate sample preparation methods, as well as instruments that take into account such factors as sustainability, green chemistry, and operator intervention.

Gianfranco Picone
Editor

foods

Article

Effect of High Hydrostatic Pressure on the Metabolite Profile of Striped Prawn (*Melicertus kerathurus*) during Chilled Storage

Qiuyu Lan [1], Silvia Tappi [1], Giacomo Braschi [1], Gianfranco Picone [1], Pietro Rocculi [1,2] and Luca Laghi [1,2,*]

[1] Department of Agricultural and Food Sciences, Alma Mater Studiorum, University of Bologna, 47521 Cesena, Italy
[2] Interdepartmental Centre for Industrial Agrofood Research, Alma Mater Studiorum, University of Bologna, 47521 Cesena, Italy
* Correspondence: l.laghi@unibo.it; Tel.: +39-0547-338-105

Abstract: A variety of metabolites contribute to the freshness and taste characteristics of seafood. This study investigated the effects of high hydrostatic pressure (HHP; 400, 500, and 600 MPa) for 10 min) on the metabolome of striped prawn during chilled storage, in relation to microorganisms' development. All treated samples showed lower viable counts throughout storage compared to the untreated counterparts. The limit of acceptability from a microbiological point of view was extended from 9 to as many as 35 days by 600 MPa treatment. Metabolites were quantified by ^{1}H-NMR through a targeted-untargeted metabolomic approach. Molecules linked to nucleotides' degradation and amines' anabolism suggested an overall freshness improvement granted by HHP. Notably, putrescine and cadaverine were detected only in untreated prawn samples, suggesting the inactivation of degradative enzymes by HHP. The concentration of molecules that influence umami perception was significantly elevated by HHP, while in untreated samples, the concentration of molecules contributing to a sour taste gradually increased during storage. As metabolomics was applied in its untargeted form, it allowed us to follow the overall set of metabolites related to HHP processing and storage, thus providing novel insights into the freshness and taste quality of striped prawn as affected by high hydrostatic pressure.

Keywords: fish freshness; high hydrostatic pressure; fish storage; spoilage; metabolomics; ^{1}H-NMR

Citation: Lan, Q.; Tappi, S.; Braschi, G.; Picone, G.; Rocculi, P.; Laghi, L. Effect of High Hydrostatic Pressure on the Metabolite Profile of Striped Prawn (*Melicertus kerathurus*) during Chilled Storage. *Foods* **2022**, *11*, 3677. https://doi.org/10.3390/foods11223677

Academic Editor: Rosario Ramirez

Received: 16 October 2022
Accepted: 14 November 2022
Published: 17 November 2022

1. Introduction

The striped prawn (*Melicertus kerathurus*) belongs to a species of prawn autochthonous of the Mediterranean Sea, appreciated and traded for its delicate flavor and pleasant aroma. Striped prawn catches in the Mediterranean Sea recorded more than 6900 tons in 2016, with an additional 300 tons in the Central-Eastern Atlantic [1]. Even if catching practices are evolving toward freezing of the products on the fishing boat, nearly half of the catch is still consumed in its "live, fresh, or chilled" forms [1], prone to the logistic problems connected to high perishability. The primary causes of spoilage, in addition to autolysis, are microorganisms, whose growth negatively impacts shelf life and overall product quality [2].

High hydrostatic pressure (HHP) is a non-thermal technology that has been effectively used on perishable seafoods to reduce the growth of undesirable spoilage microorganisms, thereby substantially extending their shelf life and improving their safety [3]. For these purposes, pressures of 100–600 MPa for 2–10 min at 2–25 °C are generally applied [4,5]. Reyes et al. [6] found that treatment at 450 MPa for 3 min increases the microbiological shelf life of Chilean Jack mackerel from 6 to 29 days, while treatment at 550 MPa for 4 min increases it to 40 d. The extension of the microbiological shelf life is mainly attributed to a reduced initial load of spoilage microflora in the samples. As a consequence, HHP application contributes to delaying the formation of nitrogen-based volatile compounds, mainly trimethylamine (TMA), related to reduced freshness [7].

Microbial growth, and many of its consequences on the sensory characteristics associated by consumers with freshness and high quality, can be conveniently followed by observing the evolution of the fish flesh metabolome, the ensemble of the low-weight molecules it harbors [8]. For example, bacterial growth leads to inosine (Ino) and hypoxanthine (Hx) from adenosine-5′-triphosphate (ATP), adenosine-5′-diphosphate (ADP), adenosine-5′-monophosphate (AMP), and inosine-5′-monophosphate (IMP) [9,10]. The relative amounts of these molecules have been therefore combined for decades in freshness scores, such as the K-value [11]. Moreover, in addition to TMA, the microbial degradation of fish constituents generates biogenic amines and organic acids associated with spoilage [12], as well as unpleasant, bitter-taste-related compounds, including arginine, valine, and methionine. Moreover, some water-soluble, low-weight components are known as taste-active components because they contribute to the specific aspects of seafood's organoleptic characteristics linked to sweetness, sourness, or umami taste [13].

Unfortunately, all the above-mentioned compounds are characterized by a high variety of functional groups that hinder their rapid and simultaneous determination using a single analytical platform. In this respect, proton nuclear magnetic resonance (^{1}H-NMR) can be a convenient tool because the physical principles it relies on may make it intrinsically quantitative regardless of the chemistry of the molecules observed. In fact, numerous works have been devoted to the evaluation of seafood freshness by ^{1}H-NMR [8,14–17].

It is also important to notice that ^{1}H-NMR spectra used for analyses targeted toward specific molecules of interest can be simultaneously used for untargeted observations. This is particularly appealing as the consequences of HHP on the fish metabolome are still largely unknown. Therefore, with this work, we sought to perform a combined targeted and untargeted study on the metabolome of prawn flesh treated with HHP during storage, a topic at present limitedly covered by the published literature.

2. Materials and Methods

2.1. Sample Preparation

The striped prawns (*Melicertus kerathurus*) used in this study were fished in the Adriatic Sea. The company Economia del Mare (Cesenatico, Italy) fast-froze them at −30 °C for 24 h until sample preparation, as typically performed at the industrial level for products intended for raw consumption. For sample preparation, thawing at 4 °C for 16 h was followed by mechanical deboning and removal of the shell. Finally, the striped prawns were manually finely diced, similarly to a fish tartare, into portions (15–20 g) and vacuum-packed in a polypropylene (PP) tray with PP film. A total of 21 packages were created, each containing 6 portions, for subsequent analysis.

2.2. HHP Treatment

Vacuum-packaged samples were placed in a 350 L chamber (HPP Italia s.r.l, Parma, Italy) filled with water at 5 °C. The samples were then subjected to 0 (used as a control), 400, 500, and 600 MPa pressure for 10 min. During the rise in pressure, lasting approximately 1 min, adiabatic heating of 3.3 °C every 100 MPa was registered.

2.3. Storage

After HHP treatment, all samples were stored at 2 ± 1 °C, and analytical determinations were planned at 1, 6, 9, 14, 21, 28, and 35 days. For each HHP treatment, the effective storage duration was based on the results of microbiological analysis, considering the reaching of a microbial load of 6 log CFU/g as the end of the shelf life. For each treatment, and at each storage time, 2 portions were used for metabolomics analysis by ^{1}H-NMR and 3 portions were used for microbiological analysis. Each portion for the same analysis was obtained from different packages.

2.4. Microbiological Analysis

By following the rationale of previous works [18], to assess the microbiological quality of packaged striped prawns in relation to the HHP treatment applied, a few selected microbial groups were considered by giving priority to those with the highest expected impact on the metabolome. These were total mesophilic bacteria (TMB; ISO 4833), *Lactobacillus* spp. (ISO 15214), *Pseudomonas* spp. (ISO 13720), *Clostridia* (ISO 7937), total Coliform and *E. coli* (ISO 16649-2), *Staphylococcus aureus* and coagulase positive staphylococci (ISO 6888-1), *Salmonella* spp. (ISO 6579-1:2017/A1:2020), and *Listeria monocytogenes* (ISO 11290-1:2017). The microbial groups were determined according to the corresponding ISO protocols.

2.5. Metabolomics Analysis by ^{1}H-NMR

A ^{1}H-NMR analysis solution was prepared, comprising a 10 mM D_2O solution of 3-(trimethylsilyl)-propionic-2,2,3,3-d_4 acid sodium salt (TSP) as a chemical-shift reference (δ −0.017). A 1 M phosphate buffer granted a pH of 7.00 ± 0.02, while 10 µL of NaN3 2 mM avoided microbial proliferation.

By modifying the procedure set up by Ciampa et al. [15], trichloroacetic acid (TCA) extraction was performed by adding 0.5 g of fish muscle to 3 mL of 7% (*w*/*w*) TCA, followed by homogenization by Ultra-Turrax (IKA, Germany) at 14,000 rpm for 20 s. The homogenate was centrifuged at 18,630× *g* for 10 min at 4 °C, and then, 0.7 mL of the supernatant was added with 0.100 mL of a D_2O solution of 10 mmol/L of 3-(trimethylsilyl)-propionic-2,2,3,3-d4 acid sodium salt (TSP). The pH was adjusted to 7.00 ± 0.02 using 9 mol/L of KOH in an Eppendorf microfuge tube. After centrifuging once more under the above-mentioned conditions, 0.65 mL of the supernatant was transferred to an NMR tube for analysis.

To register ^{1}H-NMR spectra, an AVANCE III spectrometer (Bruker, Milan, Italy) was used, operating at a frequency of 600.13 MHz and equipped with Topspin (ver. 3.5) software. ^{1}H-NMR spectra were acquired at 298 K using a CPMG pulse sequence, with suppression of the solvent signal, by setting the key parameters as follows: fid constituted by 32 k points for an acquisition time of 2.28 s, 256 scans, 16 dummy scans, 12 ppm of spectral width, and relaxation delay of 5 s. The NMR spectra were pre-processed using Topspin.

Signal assignment to compounds was performed using Chenomx ver 8.3 software (Chenomx Inc., Edmonton, AB, Canada) through comparison with Chenomx (ver. 10) HMDB (release 2) databases. Absolute quantification of molecules was performed in one reference sample by adding 100 µL of maleic acid (9.26 mM) as an internal standard. Spectra from any other sample were adjusted toward the reference by probabilistic quotient normalization (PQN) [19] to compensate for differences in water content. The concentration of each molecule was calculated from the area of one of its signals, measured by global spectra deconvolution, implemented in MestReNova (ver. 14.2.0-26256) software (Mestrelab research S.L. Santiago De Compostela, Spain), by considering a limit of quantification (LOQ) of 5. This was carried out after applying a line broadening of 0.3 and a baseline adjustment by the Whittaker Smoother procedure.

2.6. Statistical Analysis

The significance of the differences among groups was tested with SPSS Statistics (ver. 8.0) software by analysis of variance (ANOVA), followed by Tukey's post hoc test ($p < 0.05$). In R computational language (ver. 4.0.2), Spearman's correlation tests were performed to determine the relationships between microorganisms and metabolites. For this purpose, *p*-values were adjusted for multiple comparisons according to Benjamini and Hochberg by relying on the R package "*p*.adjust."

To express the concentration of taste-active molecules in terms of their contribution to sensorial characteristics, their taste-active value (TAV) was calculated as the ratio of the concentration of an individual compound in the striped prawns and its corresponding taste recognition threshold. In this respect, compounds were considered to contribute to the certain taste characteristic of striped prawns when its TAV was higher than 1. Taste thresholds were obtained from the literature [20,21]. The TAVs of individual compounds were then summed up to estimate their overall contribution to the striped prawn taste component (umami, sweet, bitter, sour).

3. Results

3.1. Characterization of the Striped Prawn Flesh Metabolome

A ^{1}H-NMR spectrum representative of the samples observed in the experiment is shown in Figure 1. In total, 53 metabolites were identified in the aqueous extract of striped prawns. The metabolite groups mostly characterizing these spectra (Table S1) included amino acids, amines, carbohydrates, nucleotides, and organic acids. The most concentrated metabolites were glycine and betaine in each sample. Signals from specific molecules, such as spoilage-related putrescine and cadaverine, were detected only on the last day of storage. The intensities of other signals, such as ethanol, acetate, or ornithine, showed marked differences between treated and untreated samples, especially at the end of the storage period.

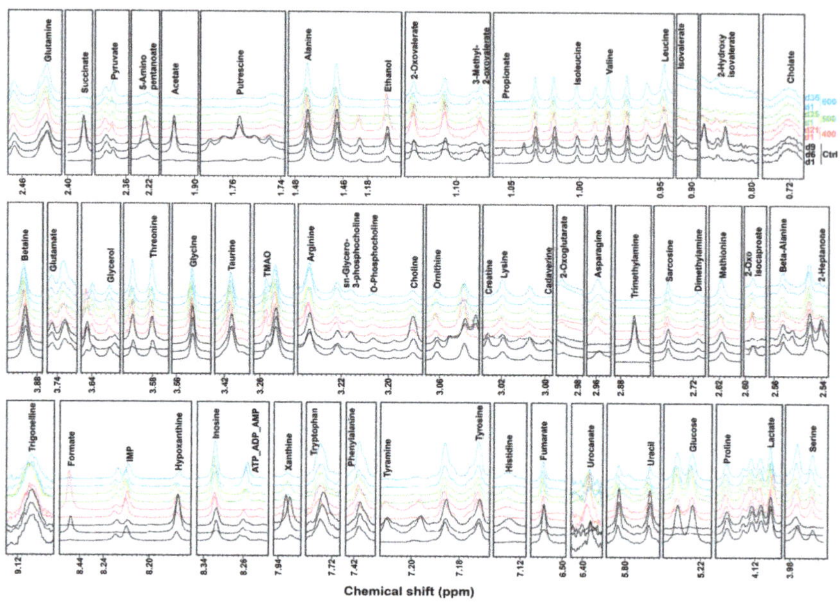

Figure 1. Examples of ^{1}H-NMR spectra registered on striped prawn samples analyzed for the investigation.

To highlight the main trends in the patterns of metabolites, their concentrations were expressed as a distance matrix and then grouped by means of hierarchical clustering. Two heat maps were then generated to show the overall changes in the metabolites due to different processing pressures on day 1 and due to storage time (Figure 2). A visual inspection allowed us to appreciate two main trends. On day 1 (Figure 2a), untreated samples could be easily distinguished from samples treated with any level of pressure. Among samples observed along with storage (Figure 2b), untreated samples collected on days 9 and 14 differed substantially from any other, witnessing a compromised microbial profile.

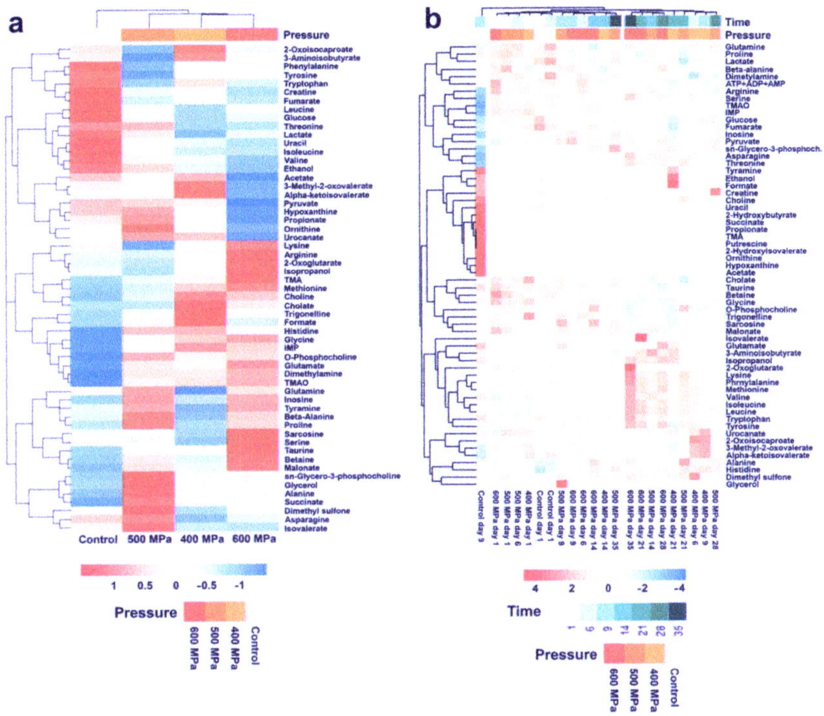

Figure 2. Heatmap visualization of the differential concentrations of metabolites among striped prawn samples subjected to different treatments on day 1 (**a**) and during storage (**b**). Red and green squares highlight metabolites more and less concentrated, respectively, than the average, with the depth of colors indicating greater differences.

3.2. Effect of HHP on Freshness-Related Metabolites during Storage

3.2.1. Nucleotide Degradation

Among the 53 molecules that we were able to observe by ^{1}H-NMR, it was possible to notice the complete set of molecules from which freshness is typically estimated, under the parameter K-value, namely ATP, ADP, AMP, IMP, inosine, and hypoxanthine. As Table 1 shows, there was a gradual increase in the K-value with storage time for all treatments. In detail, the K-value increased for the control group from 39.93% on day 1 to 93.52% on day 9, while it increased from 31.83% on day 1 to 44.13% on day 9 for the samples treated with a pressure of 600 MPa. Intermediate pressures led, as a trend, to intermediate K-values. Focusing on the last day of storage for each treatment, the K-values of the treated samples were all at least 55% lower than the K-values of control samples.

Trends during storage in the nucleotide concentration for each treatment are detailed in Figure 3. Unexpectedly, the total content of ATP, ADP, and AMP in samples treated with a pressure of 400 MPa was systematically lower than in any other and compared to the control group. Moreover, the total concentration of ATP, ADP, and AMP remained low (<1.4 μmol/g) in all samples during the whole storage period. The main differences between control and treated samples was found in IMP, inosine, and hypoxanthine. In control samples, IMP and inosine showed the lowest concentrations, particularly on days 9 and 14, while the hypoxanthine content increased sharply after day 6. Finally, as a general trend, the rate of change in IMP, inosine, and hypoxanthine levels during storage was higher in the control group.

Table 1. K-values measured during chilled storage in striped prawns untreated (control) or treated by HHP at 400, 500, and 600 MPa.

Time (Days)	Treatment			
	Control	400 MPa	500 MPa	600 MPa
1	39.93 ± 8.51 AB *	33.62 ± 1.06 AB	40.11 ± 1.91 A	31.83 ± 0.95 B
6	42.24 ± 0.66 A	38.68 ± 0.29 B	37.13 ± 0.44 C	37.91 ± 1.97 AC
9	93.52 ± 1.23 A	50.28 ± 3.57 B	45.95 ± 2.63 BC	44.13 ± 0.38 C
14		40.36 ± 0.21	41.12 ± 0.03	40.48 ± 1.42
21		50.3 ± 0.83	44.33 ± 0.41	44.78 ± 0.48
28			45.45 ± 1.24	45.56 ± 0.6
35			52.12 ± 1.15	46.32 ± 0.14

* Different capital letters in the same row indicate significant differences ($p < 0.05$).

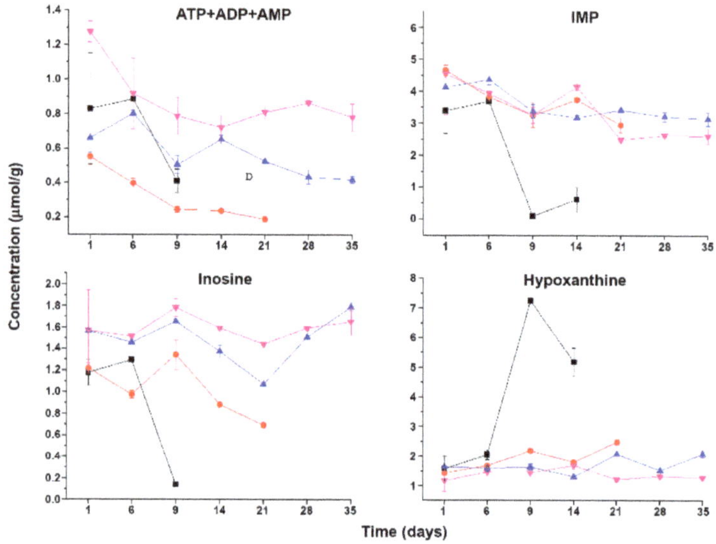

Figure 3. Changes during storage in nucleotide concentrations in striped prawn flesh control samples (black squares) and in samples treated with HHP at 400 (red circles), 500 (blue upward triangles), and 600 MPa (purple downward triangles).

3.2.2. TMAO, TMA, and DMA

Molecules largely quantified for their ability to provide information about fish freshness connected to microbial spoilage are trimethylamine-N-oxide (TMAO) and its volatile catabolites trimethylamine (TMA) and dimethylamine (DMA) [7].

The evolution of TMAO, TMA, and DMA content during storage is shown in Table 2. The TMAO concentration decreased sharply on day 9 in the control group, reaching a concentration of 17.47 mg/100 g of flesh. The same molecule changed moderately with storage in HHP-treated samples, so its levels were all greater than 115 mg/100 g of flesh during storage. A similar but opposite trend was found for TMA. In detail, TMA reached a concentration of 139.1 mg/100 g of flesh in untreated samples after 9 days, while in treated samples, its concentration was under 3 mg/100 g of flesh during the entire storage period. In addition, at each storage time, TMA concentrations in treated samples were always lower than those in the untreated counterparts. DMA was detected in each sample, but no significant differences were highlighted between treated and untreated samples.

Table 2. Amine concentrations (mg/100 g of flesh) during chilled storage of striped prawn untreated (control) or treated with HHP.

Treatments	Time (Days)	TMAO	TMA	DMA	Cadaverine	Putrescine	Tyramine
Control	1	137.8 ± 3.13 ^{Ab} *	0.48 ± 0.04 ^{Bc} **	0.1 ± 0 ^{Aa}	- ***	-	-
	6	178.64 ± 0.71 ^{Aa}	1.09 ± 0.18 ^{Ac}	0.15 ± 0.02 ^{Aa}	-	-	-
	9	17.47 ± 2.43 ^{Bc}	139.1 ± 0.89 ^{Aa}	0.1 ± 0.01 ^{Aa}	2.78 ± 0.36	102 ± 13.5	14.19 ± 1.18
400 MPa	1	183.33 ± 14.1 ^{Aa}	0.7 ± 0.04 ^{ABa}	0.12 ± 0 ^{Aa}	-	-	-
	6	141.7 ± 4.1 ^{Abc}	0.7 ± 0.05 ^{ABa}	0.06 ± 0 ^{Bc}	-	-	-
	9	135.93 ± 4.87 ^{Abc}	2.55 ± 1.08 ^{Ba}	0.1 ± 0 ^{Ab}	-	-	-
	14	156.45 ± 5.34 ^{Aab}	1.82 ± 0.48 ^{Ba}	0.09 ± 0 ^{Ab}	-	-	0.69 ± 0.15
	21	125.8 ± 4.61 ^{Bc}	0.96 ± 0.01 ^{Aa}	0.09 ± 0 ^{Ab}	-	-	8.2 ± 0.36
500 MPa	1	173.62 ± 15.48 ^{Aa}	0.56 ± 0.05 ^{Bb}	0.12 ± 0 ^{Aab}	-	-	-
	6	157.63 ± 3.17 ^{Aa}	0.52 ± 0.03 ^{Bb}	0.12 ± 0 ^{ABa}	-	-	-
	9	159.13 ± 2.67 ^{Aa}	1.41 ± 0.22 ^{Ba}	0.1 ± 0.01 ^{Aabc}	-	-	-
	14	150.26 ± 2.72 ^{Aa}	0.71 ± 0 ^{Bb}	0.1 ± 0 ^{Acd}	-	-	-
	21	148.7 ± 4.65 ^{Aa}	0.79 ± 0.03 ^{Ab}	0.1 ± 0 ^{Abc}	-	-	-
	28	138.74 ± 8.26 ^{a}	0.4 ± 0.01 ^{b}	0.08 ± 0 ^{d}	-	-	-
	35	156.72 ± 0.35 ^{a}	0.47 ± 0.03 ^{b}	0.11 ± 0 ^{ab}	-	-	-
600 MPa	1	186.41 ± 18.67 ^{Aa}	0.87 ± 0.01 ^{Aa}	0.12 ± 0.02 ^{Aa}	-	-	-
	6	161.97 ± 15.03 ^{Aabc}	0.5 ± 0.02 ^{Bd}	0.13 ± 0.02 ^{ABa}	-	-	-
	9	151.44 ± 6.29 ^{Aabc}	0.8 ± 0.03 ^{Bab}	0.11 ± 0 ^{Aa}	-	-	-
	14	173.41 ± 4.35 ^{Aab}	0.69 ± 0.02 ^{Bbc}	0.1 ± 0 ^{Aa}	-	-	-
	21	115.76 ± 6.1 ^{Bc}	0.59 ± 0.03 ^{Bcd}	0.1 ± 0.01 ^{Aa}	-	-	-
	28	120.67 ± 9.25 ^{c}	0.5 ± 0.05 ^{d}	0.11 ± 0 ^{a}	-	-	-
	35	130.21 ± 3.59 ^{bc}	0.64 ± 0.02 ^{d}	0.1 ± 0 ^{a}	-	-	-

* Different capital letters (A, B, C) in the same column indicate significant differences on the same day among treatments ($p < 0.05$). ** Different lowercase letters (a, b, c) in the same column indicate significant differences among different storage times for the same treatment ($p < 0.05$). ***—below the LOQ.

3.2.3. Biogenic Amines

Biogenic amines testify to the ongoing proteolysis or amino acid decarboxylase activity of microorganisms during storage and are therefore considered biomarkers of spoilage [12]. The evolution of putrescine, cadaverine, and tyramine content during storage is given in Table 2. Putrescine and cadaverine were detected only in untreated samples on days 9 and 14. The tyramine level could be determined at the end of the storage period in untreated samples and samples treated with 400 MPa HHP, while it was under the limit of detection in samples treated with 500 and 600 MPa HHP.

3.3. *Effect of HHP on Taste-Active Metabolites during Storage*

A part of the metabolites that contribute to the four tastes umami, sweet, bitter, and sour could be detected and quantified by ^{1}H-NMR. As shown in Table S2, on day 1, lactate, IMP, and arginine showed a concentration above the threshold for sour, umami, and bitter tastes, respectively, while glycine, lysine, and alanine exceeded the threshold for sweet taste. Both HHP treatment and storage had an impact on the concentrations of taste-active compounds.

The TAV of lactate systematically decreased below the threshold on the last day of storage in all groups, while IMP and arginine decreased too but never below the threshold. Interestingly, glycine followed a trend like IMP and arginine only in treated samples, while it tended to increase in the control group. The TAV of valine and methionine tended to reach levels above 1 at the end of the storage period for all the treatments tested, while acetate and succinate did so only for untreated samples or for those treated with the lowest pressures.

To get an overview of the effects of these molecules on umami, sweet, bitter, and sour tastes, their sum is shown as radar graphs in Figure 4. The overall umami TAV on day 1, as estimated by metabolomics, was higher in treated than in untreated samples. On the following days, the difference faded away toward values around 6 for each treatment. From the metabolomic perspective, a sour TAV was below the threshold on day 1, while it tended toward values above 1 in untreated samples. Bitterness, as assessed by the combined quantification of nine molecules through ^{1}H-NMR, tended to decrease with time as a consequence of each treatment, particularly in untreated or less treated samples.

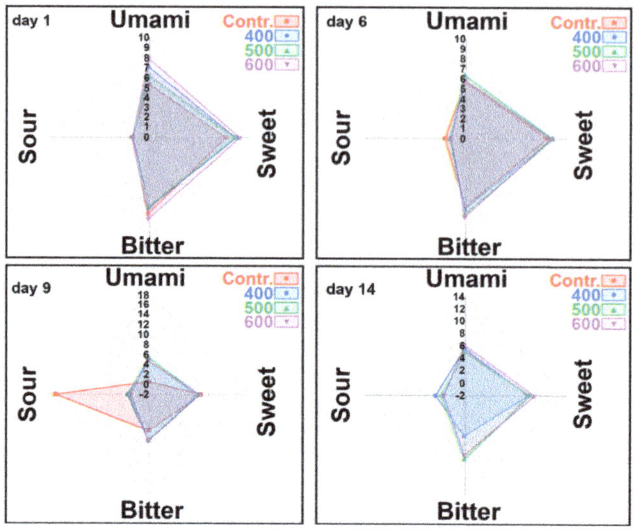

Figure 4. Overall umami, sour, sweet, and bitter taste activity by compounds detected by ^{1}H-NMR in control and treated (400, 500, 600 MPa) striped prawns' samples on days 1, 6, 9, and 14.

3.4. *Effect of HHP Treatment on Other Metabolites*

Key metabolites that could be quantified by an untargeted approach are presented in Figure 5. Pyruvate in seafood is mainly produced through pyruvate from glucose, in turn mainly derived from glycogen. Control samples showed a steady decrease, while the opposite was found in samples treated with HHP at 500 and 600 MPa, and an intermediate trend was observed in samples treated with HHP at 400 MPa.

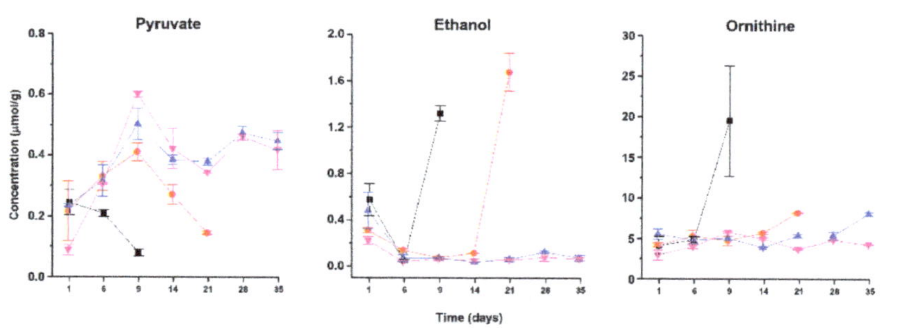

Figure 5. Changes during storage in pyruvate, ethanol, and ornithine concentrations in striped prawn flesh control samples (black squares) and in samples treated with HHP at 400 (red circles), 500 (blue upward triangles), and 600 MPa (purple downward triangles).

Ethanol decreased from day 1 to day 6 in both treated and untreated samples, and then, it spiked on the last day of storage in the control group and in samples treated with HHP at 400 MPa, reaching, respectively, a concentration 3 and 4 times higher compared to day 1.

Ornithine increased 2-fold on the last day of storage, compared to the corresponding fresh samples, in all treated samples, while it showed a 4-fold increase in the control group.

3.5. Effect of HHP on Microorganisms

The effects of HHP treatment on the microbiological quality of packaged striped prawns are reported in Figure 6. In all samples, *Salmonella* spp., *Listeria monocytogenes*, or coagulase-positive staphylococci were never detected during the microbial shelf life of striped prawns. In general, the application of HHP treatment increased the microbiological shelf life of the considered products, and the inactivation effect was more severe at higher pressures.

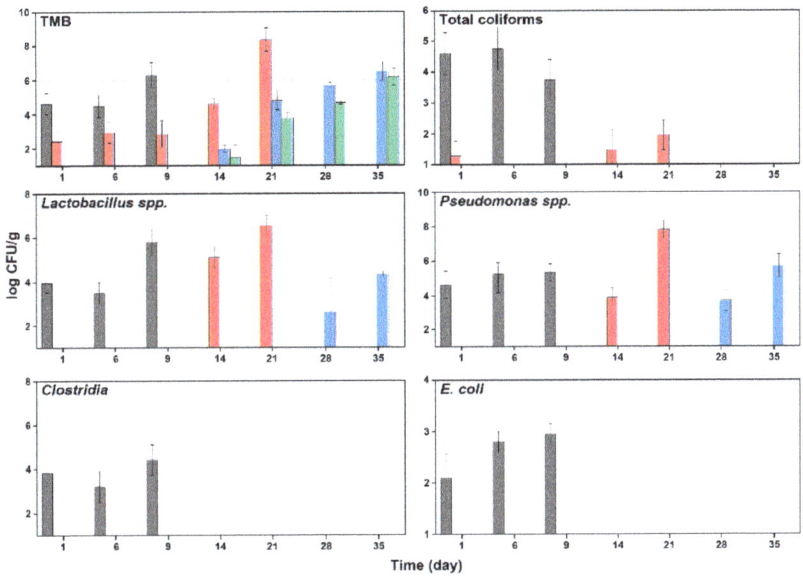

Figure 6. Changes in microbial cell loads (log CFU/g) of total mesophilic bacteria, total Coliforms, *Lactobacillus* spp., *Pseudomonas* spp., *Clostridia*, and *E. coli* during chilled storage of packaged striped prawn untreated (gray) or after treatment with HHP at 400 (red), 500 (blue), and 600 MPa (green).

Regarding untreated striped prawns, initial TMB loads (T1) of control and samples treated at 400 MPa showed values of 4.65 and 2.45 log CFU/g, respectively, while for samples treated at 500 MPa and 600 MPa, mesophilic bacterial counts were below the detection limit (<1 log CFU/g); see Figure 6. Considering a threshold level for acceptability of 6.0 log CFU/g, the TMB in samples treated at 400, 500, and 600 MPa reached such a limit in 21, 28, and 35 days of storage, respectively, compared to 9 days of the control group. A marked difference was evident also in the trends in single species. *E. coli* and *Clostridia* were always under the detection limit for any treatment. In samples treated at 600 MPa, this was true also for total *Coliforms*, *Lactobacilli*, and *Pseudomonas* spp., while intermediate trends could be observed for samples treated at 400 and 500 MPa (Figure 6).

3.6. Correlations between Microorganisms and Metabolites

To evaluate the relationship between microorganisms and metabolites, the results of Spearman's correlation tests between total mesophilic bacteria (TMB) cell loads (log CFU/g) and metabolites are listed in Table 3. TMAO appeared as being negatively correlated with TMB, with a correlation coefficient of −0.66. It is worth noting that TMAO, which produces TMA by bacterial activity, was significantly correlated ($p < 0.05$) with microorganisms. This suggests that the consumption of TMAO is associated with bacteria during the chilled storage of striped prawns. The amino acids could be divided into two groups according to the trend of their correlation. Nine amino acids (tryptophan, phenylalanine, threonine, ornithine, creatine, methionine, isoleucine, valine, and leucine) were positively correlated with TMB, while the opposite could be observed for betaine, glycine, arginine, and sarcosine. All nucleotides (AMP, Ino, IMP) were negatively correlated with TMB, with IMP showing a higher negative correlation of −0.69. This result was consistent with the idea that nucleotides are related to the quality of fish. In addition, significant positive correlations were found between TMB and succinate, acetate, ethanol, and propionate, while opposite correlations were found between TMB and alpha-ketoisovalerate, glycerol, and fumarate.

Table 3. Correlations between total mesophilic bacteria (TMB) and the metabolite concentrations in striped prawns during chilled storage.

	Metabolite	Correlation	Adjusted *p*-Value
Amines			
	Trigonelline	−0.41	1.46×10^{-2}
	TMAO	−0.66	8.72×10^{-6}
Amino Acids			
	Arginine	−0.59	1.21×10^{-4}
	Glycine	−0.51	1.29×10^{-3}
	Betaine	−0.47	4.77×10^{-3}
	Sarcosine	−0.45	5.64×10^{-3}
	Ornithine	0.34	4.87×10^{-2}
	Creatine	0.36	3.33×10^{-2}
	Tryptophan	0.38	2.38×10^{-2}
	Threonine	0.43	8.70×10^{-3}
	Methionine	0.46	5.64×10^{-3}
	Phenylalanine	0.52	1.17×10^{-3}
	Leucine	0.59	1.21×10^{-4}
	Isoleucine	0.63	3.49×10^{-5}
	Valine	0.73	4.87×10^{-7}
Organic acids			
	Fumarate	−0.51	1.29×10^{-3}
	Succinate	0.59	1.21×10^{-4}
	Acetate	0.56	3.38×10^{-4}
Nucleotides			
	AMP	−0.5	2.01×10^{-3}
	Inosine	−0.45	5.64×10^{-3}
	IMP	−0.69	2.40×10^{-6}
Other			
	Glycerol	−0.4	1.74×10^{-2}
	Alpha-ketoisovaleric acid	−0.4	1.59×10^{-2}
	Ethanol	0.34	4.87×10^{-2}
	Propionate	0.44	8.12×10^{-3}

4. Discussion

The effects of HHP treatment on preservation have been observed in a variety of seafood products, with results varying considerably depending on process parameters and also on the seafood species [22–24].

Setting up non-thermal methods for the microbiological extension of the shelf life of a fresh food product is a challenging task because it implies a never-perfect compromise between the need for safety and the need to maintain, or even improve, the original organoleptic characteristics. This is especially true when dealing with the application of high hydrostatic pressures to control microorganisms' proliferation in a product based on fresh striped prawns. Fresh seafood is one of the most prone to rapid microorganisms' growth. The application of HHP to hinder such growth is still in its infancy, with key pieces of information for optimal application still lacking. This is especially true for the consequences on the food's molecular profile, as the microorganisms' metabolism has a direct effect on a wide range of molecules both easily perceived by the consumer and traditionally associated with freshness.

It seems therefore a natural choice to investigate the effects of HHP treatment on seafood by combining microbiological measurements with the quantification of the widest possible set of molecules. In this context, it seems wise to flank observations targeted toward specific molecules, known to have an impact on already established quality indexes, with untargeted metabolomics in order to evidence potential biomarkers or unexpected switches of the microorganisms' metabolism caused by sub-lethal treatments.

Metabolomics by ^{1}H-NMR allowed us to follow the entire set of molecules used to calculate the nucleotide breakdown freshness index K-value, which shows the extent of enzymatic spoilage in hypoxanthine-forming seafood, characterizing those species that contain a low inosine concentration [25–28]. The differences between untreated and treated samples were striking, with control samples showing 6.7 points of increase per day against 0.35 and 0.43 caused by 500 and 600 MPa treatments, respectively. It is anyway interesting that the 400 MPa treatment caused the K-value to increase by 0.8 points per day. This number is far lower than the one obtained for control samples but anyway double that obtained with stronger treatments, suggesting that the K-value can finely discriminate among treatments of different intensity. Such sensitivity is likely granted by the fact that the K-value combines the concentration of several molecules, so spikes in one are compensated by more stable values of others. A visual impression that the combination of molecules gives finer information than single molecules can be obtained by comparing Table 1 with the trends in single molecules in Figure 3. This figure shows also that the decrease in ATP, ADP, AMP, and IMP does not lead to a corresponding increase in inosine and hypoxanthine, witnessing the evolution of the latter molecules to further degradation products.

The convenience of ^{1}H-NMR as an analytical platform for our case is reinforced by the possibility to quantify from the same spectrum used for the K-value also the freshness indicators TMAO and its catabolites, despite the different chemical structure. It is established that microbial activity has the greatest influence on the concentrations of these molecules. Examples can be traced in the work of Prabhakar et al. [29], who found that in Rohu fish stored at 5 °C, TMA passes from values around 0.1 mg/100 g of flesh to approximately 2 mg/100 mg of flesh after 20 days. Our data, anyway, suggest that HHP treatment itself affects, even if at a far lower extent, the concentration of TMAO and TMA. In detail, although control samples show a concentration of TMA of 0.48 mg/100 g of flesh on day 1, treated samples show concentrations from 0.56 to 0.87 mg/100 g of flesh. In parallel, the concentration of TMAO passed from 137.8 mg/100 g of flesh in the control group to values higher than 170 mg/100 g of flesh in treated samples. Changes in the TMA concentration with HHP are well documented in the literature, but they seem to be matrix and temperature related, likely determined by the procedure applied for the extraction from fish flesh. Consistent with our results, Briones-Labarca et al. (2012) found that the application of HHP at 550 MPa for 5 min increases the TMA-N concentration to 0.81 mg/100 g of flesh compared to 0.08 of the control. Contrary to our results, Erkan et al. [22], focusing on horse mackerel, found that applying HHP at 0, 220, 250, and 330 MPa at 7 °C for 5 min leads to TMA-N concentrations of 2.58, 2.45, 1.98, and 1.90, respectively. Intriguingly, Senturk and Alpas [30] found that treating Atlantic mackerel for 5 min with increasing pressures decrease TMA-N at 5 °C but increases it at 15 °C.

Despite the described peculiarities that can be noticed right after HHP treatment, the main differences in TMA and TMAO concentrations can be undoubtedly noticed throughout storage, related to microbial activity. In fact, although in treated samples, TMAO decreased at most by 30% and the TMA concentration did not show appreciable modifications, TMAO decreased and TMA increased by 1 order of magnitude in untreated samples. The complexity of the relationship between treatment and matrix can be observed in this context, too. In fact, although in samples treated at 500 MPa, TMAO decreased by almost 10%, in samples treated at 600 MPa, it showed a decrease close to 30%.

Correlations with total mesophilic bacteria seem to separate amino acids into two groups, with arginine, glycine, betaine, and sarcosine being negatively correlated and nine others showing a positive trend. As many metabolic pathways regulate the concentration of amino acids, it is hard to give a single explanation of the biological reasons. Anyway, Kegg metabolic pathway databases (https://www.genome.jp/kegg/pathway.html, accessed on 31 October 2022) show that sarcosine, betaine, and glycine are doubly linked to glycine through the action of several enzymes, among which are sarcosine dehydrogenase and glycine-N-methyl transferase.

It is intriguing to notice that a single ^{1}H spectrum obtained by NMR makes it possible not only to calculate objective indexes of freshness but also to measure molecules with a known direct effect on the consumer's gustatory perception. The clearest trend that could be observed concerned umami taste on day 1, with pressure increasing the IMP concentration, thus emphasizing umami perception, as observed from a metabolomics point of view. This observation could be rationalized by noticing that high-pressure treatments have been demonstrated to reduce the activity of 5′-nucleotidase (5′-NT) enzyme and nucleoside phosphorylase (NP), which carry out the dephosphorylation of IMP to form Ino and Hx, respectively. Karim et al. [27] reported that high-pressure processing results in significantly lower ($p < 0.05$) mean 5′-NT and NP activity in pressure-treated haddock (*Melanogrammus aeglefinus*) and herring (*Clupea harengus*) compared to the controls. They speculated that high-pressure treatment effectively slows down the conversion of inosine to hypoxanthine.

The possibility to obtain instrumental data on taste preliminary to a proper taste investigation through panelists is compelling. Anyway, it is necessary to emphasize the main limit of our application—that molecules known to have a conspicuous impact on umami taste could not be quantified. This is particularly true for AMP, which could amplify umami taste up to 8-fold [31]. In perspective, the fact that we could actually quantify AMP, but only together with ATP and ADP, allows us to foresee a convenient space for improvements.

Biogenic amines are related to fish spoilage since they accumulate through ongoing proteolysis and amino acid decarboxylase activity of microorganisms in storage [32]. Biogenic amines (putrescine, cadaverine, tyramine, and histamine) have been used as a quality indicator for fish freshness [33,34]. In this study, putrescine, cadaverine, and tyramine were only detected at the end of storage in untreated samples, while putrescine and cadaverine were not detected in treated samples. Our results suggest that the freshness evaluation using biogenic amines as indicators was not relevant for striped prawns because of the low accumulation rate detected. A possible explanation for this might be the low levels of tyrosine, ornithine, and lysine in striped prawns, which degrade to arginine, tyramine, putrescine, and cadaverine, respectively [35]. Another possible explanation for this is the inactivation of microorganisms by HHP, such as *Pseudomonas* spp. and *E. coli*. As reported by previous studies, *Pseudomonas* spp., *E. coli*, *Photobacterium*, *Shewanella* spp., and *Psychrobacter* are responsible for the formation of putrescine and cadaverine in fish and shrimp [35–37]. Combined with the observations on microorganisms, this suggests that HHP could reduce the accumulation of biogenic amines and the microbial load. A similar result was reported by Gou et al. in semi-dried squid [38].

The possibility to quantify molecules besides those targeted in connection to previous knowledge, through an untargeted investigation, represents a positive characteristic of ^{1}H-NMR. This is particularly evident with ethanol, which showed a marked increase at the end of the storage time in control samples and in samples treated at 400 MPa, thus appearing as a potential convenient biomarker of microbial spoilage. In fact, ethanol is mainly a result of microorganisms' activity, as documented, among others, by Xu et al. [39]. Another example of potential biomarkers, in line with the observations of Lou et al. [8], is represented by acetate, which increased by 40 and 7 times in samples untreated and treated at 400 MPa but by less than 2 times in the other samples.

5. Conclusions

Fish freshness is the result of a complex interaction among a plethora of molecules. Any analytical technique aiming at exploring the topic should, likewise, give the widest possible picture of the fish flesh molecular profile. In this investigation, we observed the consequences of HHP treatments at 400–600 MPa on the metabolome of striped prawn flesh during storage. This was carried out through the calculation of objective indexes of freshness, such as the K-value and the absolute quantification of the amines, key molecules in this respect. From the same ^{1}H-NMR spectra, therefore with no extra effort needed, we estimated characteristics linked to perceived freshness, namely umami, sweet, bitter, and sour tastes. More molecules, quantified simultaneously to the so-outlined indexes, were found to be related to total mesophilic bacteria, setting the stage for indexes allowing a direct estimation of microbial spoilage and for a deeper understanding of the underlying mechanisms. Comparisons with literature evidence that the effects of this non-thermal preservation treatment are deeply matrix related, so investigations of this kind should be extended to other kinds of fish and emerging processing technologies.

Supplementary Materials: The following supporting information can be downloaded at: https://www.mdpi.com/article/10.3390/foods11223677/s1, Table S1: Metabolites of striped prawn samples untreated or treated with HHP at 400, 500, and 600 MPa during chilled storage; Table S2: The taste, taste threshold, and TAVs of taste-active molecules in untreated and treated (400, 500, 600 MPa) striped prawns during storage [20,21,40].

Author Contributions: Conceptualization, L.L. and Q.L.; methodology, L.L. and G.P.; software, Q.L. and G.P.; validation, L.L.; formal analysis, Q.L. and L.L.; investigation, S.T., G.B. and P.R.; resources, S.T., L.L., G.B. and P.R.; data curation, Q.L. and G.B.; writing—original draft preparation, Q.L. and L.L.; writing—review and editing, G.P., S.T., G.B. and P.R.; visualization, Q.L. and L.L.; supervision, L.L. All authors have read and agreed to the published version of the manuscript.

Funding: This research received no external funding.

Data Availability Statement: Data in contained within the article.

Acknowledgments: Q.L. gratefully acknowledges financial support from the Chinese Scholarship Council.

Conflicts of Interest: The authors declare no conflict of interest.

References

1. FAO. The State of World Fisheries and Aquaculture 2020. In *Sustainability in Action*; FAO: Rome, Italy, 2020. [CrossRef]
2. Dalgaard, P.; Madsen, H.L.; Samieian, N.; Emborg, J. Biogenic Amine Formation and Microbial Spoilage in Chilled Garfish (*Belone belone belone*)—Effect of Modified Atmosphere Packaging and Previous Frozen Storage. *J. Appl. Microbiol.* **2006**, *101*, 80–95. [CrossRef] [PubMed]
3. Ekonomou, S.I.; Boziaris, I.S. Non-Thermal Methods for Ensuring the Microbiological Quality and Safety of Seafood. *Appl. Sci.* **2021**, *11*, 833. [CrossRef]
4. Banerjee, R.; Verma, A.K. Minimally Processed Meat and Fish Products. In *Food Engineering Series*; Springer: Berlin/Heidelberg, Germany, 2015.
5. Roobab, U.; Fidalgo, L.G.; Arshad, R.N.; Khan, A.W.; Zeng, X.; Bhat, Z.F.; Bekhit, A.E.A.; Batool, Z.; Aadil, R.M. High-pressure Processing of Fish and Shellfish Products: Safety, Quality, and Research Prospects. *Compr. Rev. Food Sci. Food Saf.* **2022**, *21*, 3297–3325. [CrossRef]
6. Reyes, J.E.; Tabilo-Munizaga, G.; Pérez-Won, M.; Maluenda, D.; Roco, T. Effect of High Hydrostatic Pressure (HHP) Treatments on Microbiological Shelf-Life of Chilled Chilean Jack Mackerel (*Trachurus murphyi*). *Innov. Food Sci. Emerg. Technol.* **2015**, *29*, 107–112. [CrossRef]
7. Prabhakar, P.K.; Vatsa, S.; Srivastav, P.P.; Pathak, S.S. A Comprehensive Review on Freshness of Fish and Assessment: Analytical Methods and Recent Innovations. *Food Res. Int.* **2020**, *133*, 109157. [CrossRef]
8. Lou, X.; Zhai, D.; Yang, H. Changes of Metabolite Profiles of Fish Models Inoculated with *Shewanella baltica* during Spoilage. *Food Control* **2021**, *123*, 107697. [CrossRef]
9. Márquez-Ríos, E.; Moraán-Palacio, E.F.; Lugo-Saánchez, M.E.; Ocano-Higuera, V.M.; Pacheco-Aguilar, R. Postmortem Biochemical Behavior of Giant Squid (*Dosidicus gigas*) Mantle Muscle Stored in Ice and Its Relation with Quality Parameters. *J. Food Sci.* **2007**, *72*, C356–C362. [CrossRef] [PubMed]

10. Mendes, R.; Quinta, R.; Nunes, M.L. Changes in Baseline Levels of Nucleotides during Ice Storage of Fish and Crustaceans from the Portuguese Coast. *Eur. Food Res. Technol.* **2001**, *212*, 141–146. [CrossRef]
11. Karube, I.; Matsuoka, H.; Suzuki, S.; Watanabe, E.; Toyama, K. Determination of Fish Freshness with an Enzyme Sensor System. *J. Agric. Food Chem.* **1984**, *32*, 314–319. [CrossRef]
12. Visciano, P.; Schirone, M.; Paparella, A. An Overview of Histamine and Other Biogenic Amines in Fish and Fish Products. *Foods* **2020**, *9*, 1795. [CrossRef]
13. Sarower, M.G.; Farah, A.; Hasanuzzaman, M.; Biswas, B.; Abe, H. Taste Producing Components in Fish and Fisheries Products: A Review. *Int. J. Food Ferment. Technol.* **2012**, *2*, 113–121.
14. Shumilina, E.; Ciampa, A.; Capozzi, F.; Rustad, T.; Dikiy, A. NMR Approach for Monitoring Post-Mortem Changes in Atlantic Salmon Fillets Stored at 0 and 4 °C. *Food Chem.* **2015**, *184*, 12–22. [CrossRef]
15. Ciampa, A.; Picone, G.; Laghi, L.; Nikzad, H.; Capozzi, F. Changes in the Amino Acid Composition of Bogue (*Boops boops*) Fish during Storage at Different Temperatures by 1H-NMR Spectroscopy. *Nutrients* **2012**, *4*, 542–553. [CrossRef]
16. Lu, Z.; Wang, S.; Ji, C.; Shan, X.; Wu, H. Evaluation of Metal Pollution-Induced Biological Effects in Chinese Shrimp *Fenneropenaeus chinensis* by NMR-Based Metabolomics. *Mar. Pollut. Bull.* **2020**, *150*, 110688. [CrossRef]
17. Cappello, T.; Giannetto, A.; Parrino, V.; de Marco, G.; Mauceri, A.; Maisano, M. Food Safety Using NMR-Based Metabolomics: Assessment of the Atlantic Bluefin Tuna, Thunnus Thynnus, from the Mediterranean Sea. *Food Chem. Toxicol.* **2018**, *115*, 391–397. [CrossRef]
18. Iacumin, L.; Pellegrini, M.; Sist, A.; Tabanelli, G.; Montanari, C.; Bernardi, C.; Comi, G. Improving the Shelf-Life of Fish Burgers Made with a Mix of Sea Bass and Sea Bream Meat by Bioprotective Cultures. *Microorganisms* **2022**, *10*, 1786. [CrossRef]
19. Dieterle, F.; Ross, A.; Schlotterbeck, G.; Senn, H. Probabilistic Quotient Normalization as Robust Method to Account for Dilution of Complex Biological Mixtures. Application In1H NMR Metabonomics. *Anal. Chem* **2006**, *78*, 4281–4290. [CrossRef]
20. Rotzoll, N.; Dunkel, A.; Hofmann, T. Quantitative Studies, Taste Reconstitution, and Omission Experiments on the Key Taste Compounds in Morel Mushrooms (*Morchella deliciosa* Fr.). *J. Agric. Food Chem.* **2006**, *54*, 2705–2711. [CrossRef]
21. Yamaguchi, S.; Yoshikawa, T.; Ikeda, S.; Ninomiya, T. Measurement of the Relative Taste Intensity of Some L-α-Amino Acids and 5′-Nucleotides. *J. Food Sci.* **1971**, *36*, 846–849. [CrossRef]
22. Erkan, N.; Üretener, G.; Alpas, H.; Selçuk, A.; Özden, Ö.; Buzrul, S. Effect of High Hydrostatic Pressure (HHP) Treatment on Physicochemical Properties of Horse Mackerel (*Trachurus trachurus*). *Food Bioproc. Tech.* **2011**, *4*, 1322–1329. [CrossRef]
23. de Alba, M.; Pérez-Andrés, J.M.; Harrison, S.M.; Brunton, N.P.; Burgess, C.M.; Tiwari, B.K. High Pressure Processing on Microbial Inactivation, Quality Parameters and Nutritional Quality Indices of Mackerel Fillets. *Innov. Food Sci. Emerg. Technol.* **2019**, *55*, 80–87. [CrossRef]
24. Kaur, B.P.; Kaushik, N.; Rao, P.S.; Chauhan, O.P. Effect of High-Pressure Processing on Physical, Biochemical, and Microbiological Characteristics of Black Tiger Shrimp (*Penaeus monodon*): High-Pressure Processing of Shrimp. *Food Bioproc. Tech.* **2013**, *6*, 1390–1400. [CrossRef]
25. Ghosh, S.; Sarker, D.; Misra, T.N. Development of an Amperometric Enzyme Electrode Biosensor for Fish Freshness Detection. *Sens. Actuators B Chem.* **1998**, *53*, 58–62. [CrossRef]
26. Özogul, Y.; Ahmad, J.I.; Hole, M.; Özogul, F.; Deguara, S. The Effects of Partial Replacement of Fish Meal by Vegetable Protein Sources in the Diet of Rainbow Trout (*Onchorynchus mykiss*) on Post Mortem Spoilage of Fillets. *Food Chem.* **2006**, *96*, 549–561. [CrossRef]
27. Karim, N.U.; Kennedy, J.T.; Linton, M.; Patterson, M.; Watson, S.; Gault, N. Determination of Nucleotide and Enzyme Degradation in Haddock (*Melanogrammus aeglefinus*) and Herring (*Clupea harengus*) after High Pressure Processing. *Peer J.* **2019**, *2019*, e7527. [CrossRef] [PubMed]
28. Alasalvar, C.; Taylor, K.D.A.; Öksüz, A.; Garthwaite, T.; Alexis, M.N.; Grigorakis, K. Freshness Assessment of Cultured Sea Bream (*Sparus aurata*) by Chemical, Physical and Sensory Methods. *Food Chem.* **2001**, *72*, 33–40. [CrossRef]
29. Prabhakar, P.K.; Srivastav, P.P.; Pathak, S.S. Kinetics of Total Volatile Basic Nitrogen and Trimethylamine Formation in Stored Rohu (*Labeo rohita*) Fish. *J. Aquat. Food Prod. Technol.* **2019**, *28*, 452–464. [CrossRef]
30. Senturk, T.; Alpas, H. Effect of High Hydrostatic Pressure Treatment (HHPT) on Quality and Shelf Life of Atlantic Mackerel (*Scomber scombrus*). *Food Bioproc. Tech.* **2013**, *6*, 2306–2318. [CrossRef]
31. Johnson, R.J.; Nakagawa, T.; Sánchez-Lozada, L.G.; Lanaspa, M.A.; Tamura, Y.; Tanabe, K.; Ishimoto, T.; Thomas, J.; Inaba, S.; Kitagawa, W.; et al. Umami: The Taste That Drives Purine Intake. *J. Rheumatol.* **2013**, *40*, 1794–1796. [CrossRef]
32. Visciano, P.; Schirone, M.; Tofalo, R.; Suzzi, G. Biogenic Amines in Raw and Processed Seafood. *Front. Microbiol.* **2012**, *3*, 188. [CrossRef]
33. Benner, R.A.; Staruszkiewicz, W.F.; Rogers, P.L.; Otwell, W.S. Evaluation of Putrescine, Cadaverine, and Indole as Chemical Indicators of Decomposition in Penaeid Shrimp. *J. Food Sci.* **2003**, *68*, 2178–2185. [CrossRef]
34. Baixas-Nogueras, S.; Bover-Cid, S.; Veciana-Nogués, M.T.; Mariné-Font, A.; Vidal-Carou, M.C. Biogenic Amine Index for Freshness Evaluation in Iced Mediterranean Hake (*Merluccius merluccius*). *J. Food Prot.* **2005**, *68*, 2433–2438. [CrossRef] [PubMed]
35. Wunderlichová, L.; Buňková, L.; Koutný, M.; Jančová, P.; Buňka, F. Formation, Degradation, and Detoxification of Putrescine by Foodborne Bacteria: A Review. *Compr. Rev. Food Sci. Food Saf.* **2014**, *13*, 1012–1030. [CrossRef]

36. Jia, S.; Li, Y.; Zhuang, S.; Sun, X.; Zhang, L.; Shi, J.; Hong, H.; Luo, Y. Biochemical Changes Induced by Dominant Bacteria in Chill-Stored Silver Carp (*Hypophthalmichthys molitrix*) and GC-IMS Identification of Volatile Organic Compounds. *Food Microbiol.* **2019**, *84*, 103248. [CrossRef]
37. Liu, X.; Huang, Z.; Jia, S.; Zhang, J.; Li, K.; Luo, Y. The Roles of Bacteria in the Biochemical Changes of Chill-Stored Bighead Carp (*Aristichthys nobilis*): Proteins Degradation, Biogenic Amines Accumulation, Volatiles Production, and Nucleotides Catabolism. *Food Chem.* **2018**, *255*, 174–181. [CrossRef]
38. Gou, J.; Choi, G.P.; Ahn, J. Biochemical Quality Assessment of Semi-Dried Squid (*Todarodes pacificus*) Treated with High Hydrostatic Pressure. *J. Food Biochem.* **2012**, *36*, 171–178. [CrossRef]
39. Xu, Y.; Li, L.; Mac Regenstein, J.; Gao, P.; Zang, J.; Xia, W.; Jiang, Q. The Contribution of Autochthonous Microflora on Free Fatty Acids Release and Flavor Development in Low-Salt Fermented Fish. *Food Chem.* **2018**, *256*, 259–267. [CrossRef]
40. Kato, M.; Ra Rhue, M.; Nishimura, T. Role of Free Amino Acids and Peptides in Food Taste. In *Flavor Chemistry*; American Chemical Society: Washington, DC, USA, 1989; Chapter 13, pp. 158–174. [CrossRef]

foods

MDPI

Article

A New Electrochemical Method to Determine Tryptophan in Fruit Juices: Development and Validation

Assefa Takele [1,2], José María Palacios-Santander [1,*] and Miguel Palma [2,*]

[1] Institute of Research on Electron Microscopy and Materials (IMEYMAT), Campus de Excelencia Internacional del Mar (CEIMAR), Department of Analytical Chemistry, Faculty of Sciences, University of Cadiz, Campus Universitario de Puerto Real, Polígono del Río San Pedro S/N, Puerto Real, 11510 Cádiz, Spain; asetak@yahoo.com

[2] Wine and Agrifood Research Institute (IVAGRO), Campus de Excelencia Internacional Agroalimentario (ceiA3), Department of Analytical Chemistry, Faculty of Sciences, University of Cadiz, Campus Universitario de Puerto Real, Polígono del Río San Pedro S/N, Puerto Real, 11510 Cádiz, Spain

* Correspondence: josem.palacios@uca.es (J.M.P.-S.); miguel.palma@uca.es (M.P.)

Abstract: Tryptophan (Trp) is an essential amino acid usually found in fruit juices. Its determination is necessary for food companies because of its relation to human health. In this work, a new electrochemical method based on sonogel–carbon electrodes (SNGCEs) was developed and validated using an ultra performance liquid chromatography (UPLC) method as a reference method for the determination of Trp in fruit juices. Cyclic voltammetry (CV), chronoamperometry, and differential pulse voltammetry (DPV) techniques were applied to investigate the oxidation of Trp on a previously polarized SNGCE surface in a Britton–Robinson (BR) buffer solution at pH 3.6. The operating conditions for electroanalysis were optimized using a Box–Behnken design (BBD), obtaining an oxidation peak for Trp at 0.749 V. The linear range for this method was from 0.1 to 5 mg/L. The intraday and interday precision, expressed as a relative standard deviation (RSD), were 3.1% and 2.7%, respectively. The average recovery was 99.01%, and the limit of detection and quantitation were 0.33 and 1.09 mg/L, respectively. Therefore, from the quality analytical parameters obtained, it can be concluded that the new electrochemical method can be successfully used for the routine analysis of Trp in fruit juices. As far as we are concerned, this is the first time that a methodology for Trp determination was performed in this kind of real food matrices.

Keywords: tryptophan; ultra performance liquid chromatography; differential pulse voltammetry; sonogel–carbon electrodes; fruit juice

Citation: Takele, A.; Palacios-Santander, J.M.; Palma, M. A New Electrochemical Method to Determine Tryptophan in Fruit Juices: Development and Validation. *Foods* **2022**, *11*, 2149. https://doi.org/10.3390/foods11142149

Academic Editor: Gianfranco Picone

Received: 3 June 2022
Accepted: 18 July 2022
Published: 20 July 2022

1. Introduction

Tryptophan (Trp) is an essential amino acid [1,2] with various physiological roles: it can function independently or by incorporation into the structure of larger molecules or polymers, such as proteins [1,3]. Trp is required for the biosynthesis of proteins, and it is important in nitrogen balance and the maintenance of muscle mass and body weight in humans [4]. It is also a precursor of many biologically active substances, such as serotonin, 5-hydroxytryptamine (5-HT, a monoamine neurotransmitter), melatonin (a neurohormone), vitamin B3 (niacin) [4,5], and kynurenine (a pathway in which 95% of Trp is catabolized) [6]. If there is a deficiency of vitamin B6 in the body, Trp may be transformed into vitamin B3. The production of vitamin B3 is very important since the body considers its production to be more important than that of serotonin. Therefore, if the body is not getting enough vitamin B3 from the diet, it uses up its available Trp, which will lead to a deficiency of serotonin and melatonin [5]. An abnormal level of serotonin can cause depression, while melatonin is associated with sleeping disorders and Alzheimer's and Parkinson's diseases [7]. Therefore, the daily intake of Trp is an important matter for human health.

As mentioned before, Trp is an essential amino acid, so it is commonly added as a food fortifier in dietary food products and in pharmaceutical products because it is scarcely present in vegetables and cannot be directly synthesized in the human body [1,4,8]. Trp is essential for humans and acts predominantly as the precursor of serotonin, which is an important neurotransmitter with physiological activity [3,9] for the transduction of a neurochemical signal between neurons. The synthesis of serotonin in the brain is dependent on the availability of its precursor, the Trp amino acid. Serotonin is considered essential in the modulation of several behavioral and physiological functions, such as mood, sleep and wakefulness, sexual behavior, cognition, appetite, impulsivity, aggression, neurodevelopment, circadian rhythms, body temperature, and neuroendocrine function [10]. Therefore, Trp plays also an important role in physiological processes, such as nerve transmission and immune response, and pathological processes, such as stress, depression, sleep, and appetite disorders, because of its relationship with serotonin [3]. However, improper metabolism of Trp results in toxic products that can be accumulated in the brain, and which can then cause problems, such as hallucinations, delusions, and schizophrenia. If Trp is taken in high doses, it may show side effects, such as agitation, confusion, diarrhea, fever, and nausea [1].

In the past, various methods were developed for the determination of Trp in food samples, such as spectroscopy, high performance liquid chromatography (HPLC), fluorimetric methods, capillary electrophoresis and electroanalysis [11], flow injection chemiluminescence [9,12], and colorimetric methods [13]. Some of the previously reported methods for the determination of Trp in different samples are shown in Table S1.

The electrochemical detection method was developed as a potentially useful technique for pharmaceutical applications [1]. It has been found to be a more attractive technique for the determination of electroactive compounds in biological samples and foodstuffs because of its sensitivity, fast operation, reproducibility, accuracy, low cost, very small sample, and solvent consumption. In fact, modern electrochemical methods are more regularly used for the study of industrial and environmental applications and drug analysis in dosage forms and biological samples [14].

As shown above in Table S1 and other reported methods, Trp was determined in different food samples, different kinds of biological samples, and pharmaceutical preparations using various analytical methods. However, there few applications related to the determination of Trp in fruit juices that have been reported.

Fruit juices play an important role worldwide in the human diet, and the industry has become large and profitable [15]. When fruit juices are consumed in moderation as part of a balanced diet, they have a positive effect on promoting health and reducing disease risk [16]. As fruit juices are healthier choices among consumers, the quality and the safety of juice products are always a worry, and they are always subjected to very detailed legislation, ensuring all the necessary information on their nutritional benefits and consumptions [17]. Therefore, the development of a simple, sensitive, and less expensive detection method for Trp is of great significance and interesting to public health and food companies. Thus, the aim of this study was to develop an electrochemical method using sonogel–carbon electrodes for the direct determination of Trp in different fruit juices (apple; pineapple; tropical; five fruits; mixture of pineapple and grape; mixture of peach and grape; mixture of pineapple, apple, and grape; and mixture of peach, apple, and grape) and to validate the methodology using ultra performance liquid chromatography (UPLC) as a reference method. To the extent of our knowledge, this is the first time that Trp determination has been accomplished in fruit juices using this analytical technique.

2. Materials and Methods

2.1. Chemicals and Reagents

Tryptophan standard (T0254-5G, reagent grade, purity $\geq$ 98%) was obtained from Sigma-Aldrich (Sigma-Aldrich Co., St. Louis, MO, USA). HPLC-grade methanol, acetonitrile, acetic acid, trimethoxymethylsilane, hydrochloric acid, sulfuric acid, phosphoric acid, and boric acid were purchased from Merck KGaA (Darmstadt, Germany); sodium hydroxide from Panreac Quimica SAU (Barcelona, Spain); graphite powder from Alfa Aesar (Kandel, Germany); and gold (III) chloride trihydrate ($HAuCl_4$) from Sigma Aldrich (St. Louis, MO, USA). All the reagents were of analytical grade and used without further purification. Purified water was obtained from the Milli-Q System (Billerica, MA, USA). Samples of fruit juices were purchased from local supermarkets (Mercadona and Supersol) in Puerto Real, Cadiz, Spain.

2.2. Preparation of Buffer, Tryptophan Standard Solution, and Real Samples

An amount of 0.04 M Britton–Robinson (BR) buffer solution was used as an electrolyte and placed in the electrochemical cell. The pH of the solution was adjusted using a pH meter (Crison GLP21). The buffer solution at pH 3.6 was prepared from a mixture of 2.3 mL of acetic acid (CH_3COOH), 2.7 mL of phosphoric acid (H_3PO_4), and 2.48 g of boric acid (H_3BO_4) in a 1 L volumetric flask and by diluting to a volume with Milli-Q water. Then the pH of the buffer solution was adjusted to 3.6 using sodium hydroxide (NaOH). The prepared BR buffer solution was used as a blank in the determination of Trp. The pH of the buffer solution was adjusted to 3.6, considering the pH's of the fruit juices to be determined, which were in the range of 3.3 to 3.8.

A Trp standard stock solution (100 mg/L) was prepared in methanol. Working standard solutions were prepared from stock solution at concentration levels ranging from 0.1 to 5 mg/L in a 50:50 (v/v) Milli-Q water/methanol mixture.

Different types of fruit juice samples were purchased from local supermarkets, Mercadona (apple, pineapple, tropical, five fruits, mixture of pineapple and grape, and mixture of peach and grape), and Supersol (apple; mixture of pineapple, apple, and grape; and mixture of peach, apple, and grape). Apple; pineapple; mixture of pineapple and grape; and mixture of pineapple, apple, and grape were prepared directly by filtering through a 0.45 μm filter, then filtered through a 0.22 μm filter, transferred to a vial, and injected into the UPLC system. Other fruit juice samples—tropical; five fruits; peach; and mixture of peach, apple, and grape—were first clarified by centrifugation (4000× g, 5 min), filtered through a 0.45 μm filter and then through a 0.22 μm filter, transferred to a vial, and injected to the UPLC system. As it is not required by law, no information about the levels of Trp appeared in the commercial labels.

2.3. Preparation of Sonogel–Carbon Electrodes

Sonogel–carbon was synthesized by mixing 500 μL of trimethoxymethylsilane and 100 μL of HCl 0.2 M, as reported in [18]. The mixture was homogenized using a high-power ultrasound generator, Sonicator 3000, from Misonix (Misonix, Inc., Farmingdale, NY, USA) (equipped with a 13 mm titanium tip), which provides a maximum power of 600 W. In this case, the energy applied was between 90 and 100 joules. After sonication, 0.5 g of graphite powder was added to the mixture and well homogenized. Then the SNGCEs were prepared by inserting the material obtained into glass capillary tubes (internal diameter of 1.15 and external diameter of 1.55 mm) and connecting the ceramic material to a copper wire for electric contact.

2.4. Polarization and Modification of Sonogel–Carbon Electrodes

The measurements of standard solutions and quantitation of tryptophan in fruit juice samples were carried out using a SNGCE. Before conducting the measurements, SNGCEs were well polished with a waterproof silicon carbide paper (FEPA P#1200, Struers, Germany) and white paper until the surface become shiny. Then the polished electrodes were polarized amperometrically in 25 mL of 0.1 M H_2SO_4 in an electrochemical cell. The polarization was performed as follows: The working electrode potential was poised at −0.7 V for 10 s, and then stepped at +1.8 V for 10 s, the procedure being repeated 8 times, as reported previously [19,20]. After that, the SNGCEs were conditioned in a 4 mL blank solution from −0.5 to 1.2 V, with a step potential of 0.005 V, at a scan rate of 0.05 V/s, using cyclic voltammetry technique, being then ready for measurements.

The SNGCEs were modified using two approaches: electrodeposition from a gold solution (0.05 mM of $HAuCl_4$) and drop casting of gold nanoparticles (AuNPs, 4 μL of AuNP colloidal solution) onto the surface of SNGCE [21]. A gold solution (0.05 mM of $HAuCl_4$) was prepared by dissolving 0.005 g $HAuCl_4$ in a H_2SO_4 0.5 M solution and diluting to a volume with the same solution in a 25 mL volumetric flask. Then the electrodeposition was performed by polarizing the SNGCE in 25 mL of the previous solution using amperometry technique. Additionally, the effect of duration time on the electrodeposition of a gold solution on the electrode surface was studied by setting the duration time at 200, 600, and 1000 s under the amperometric conditions.

2.5. Chromatographic Conditions

An ACQUITY UPLC® H-Class system coupled to an Acquity UPLC® Fluorescence Detector and the system was controlled by Empower™ 3 Chromatography Data Software (Waters Corporation, Milford, MA, USA). The Acquity UPLC system was equipped with a binary solvent manager, a sample manager including the column heater, an optional sample manager, pumps, and the detector.

Analysis of tryptophan standard solution and real samples (fruit juice samples) was performed using an Acquity UPLC® H-Class system coupled to an Acquity UPLC® Fluorescence Detector, and the system was controlled by Empower™ 3 Chromatography Data Software (Waters Corporation, Milford, MA, USA). Separations were performed using a reverse-phase RP C18 Cortecs UPLC® Solid-Core-Based (SCB) Column (silica-based solid-core particle, 100 mm length, 2.1 mm ID, 1.6 μm particle size) obtained from Waters. Identification of Trp was carried out by setting the photodiode array detector in the wavelength range of 210–400 nm for the 3D scan and set at 278 nm for peak integration. Quantitation of real samples was carried out, setting the fluorescence detector (FD) at an excitation wavelength, λ_{ex} = 280 nm, and an emission wavelength, λ_{em} = 325 nm. The FD sensitivity for the 3D scan was set at PMT gain 1000. The mobile phase was a binary solvent system consisting of phase A (water with 2% acetic acid) and phase B (acetonitrile with 2% acetic acid). The injection volume of the sample was 1.5 μL, and the flow rate was 0.6 mL/min. The analyses were performed at 47 °C (column temperature) for 4 min using a gradient elution (Table 1). The column was equilibrated for 3 min before starting injections. The samples were filtered before injections using a 0.45 μm filter and then filtered through 0.22 μm syringe-driven filter. Under these conditions, the resulting retention time for Trp was 2.64 min for 11 concentration levels ranging from 0.1 to 5 mg/L.

Table 1. Elution gradient for tryptophan determination by UPLC (phase A: water with 2% acetic acid; phase B: acetonitrile with 2% acetic acid).

	Time (min)	Flow Rate (mL/min)	Phase A (%)	Phase B (%)	Curve
1	Initial	0.6	100.0	0.0	Initial
2	1.00	0.6	100.0	0.0	6
3	3.00	0.6	95.0	5.0	6
4	4.00	0.6	90.0	10.0	6

2.6. Electrochemical Measurements

Differential pulse voltammetry (DPV) and cyclic voltammetry (CV) experiments were carried out using an Autolab PGSTAT128N (Ecochemie, Utrecht, The Netherlands) potentiostat/galvanostat connected to 663 VA stand (Metrohm, Swiss) and a personal computer. The GPES program version 4.9 software was used to generate signals, data analysis, and storage. A three-electrode electrochemical system composed of a platinum wire as the auxiliary (counter) electrode, Ag/AgCl (KCl 3 M) electrode as the reference electrode, and unmodified sonogel–carbon electrode (SNGCE) as the working electrode was employed.

The calibration curve for Trp standard solution and the quantitation of real samples were performed using differential pulse voltammetry. The DPV measurements were performed in Britton–Robinson buffer by setting the interval time at 0.6 V, step potential at 0.016 V, and modulation amplitude at 0.1 V; the solutions were magnetically stirred for 20 s.

2.7. Electrochemical Analysis of Tryptophan

The calibration curve for Trp standard solution and the quantitation of real samples were performed using differential pulse voltammetry. After the polishing, amperometrically polarizing, and CV conditioning of the SNGCE, the DPV measurements were performed by adding Trp standard solution into an electrochemical cell containing 4 mL of Britton–Robinson buffer and stirring the solution for 20 s with a magnetic stirrer. Real samples were determined, first, running the blank. Subsequently, the blank was discarded, and 4 mL of a real sample was added into the electrochemical cell and stirred for 20 s with a magnetic stirrer. After each measurement of the real sample, the SNGCE was repolished, repolarized, and reconditioned. This methodology ensured the reproducibility and repeatability of the measurements, avoiding fouling of the electrode surface.

2.8. Experimental Design and Statistical Analysis

A Box-Behnken design (BBD) is a kind of experimental design commonly used to optimize successfully experimental variables [22,23]. In this work, it was employed to measure the effect of three independent variables: modulation amplitude (×1), step potential (×2), and interval time (×3) on the determination of Trp using DPV. It is supposed that the independent variables are continuous and controllable during the experiments. Three levels were selected for each factor, being coded as −1, 0, and +1 (low, central, and high value), respectively. The values considered for each variable were: modulation amplitude at 0.01 (−1), 0.055 (0), and 0.1 (+1) V; step potential at 0.004 (−1), 0.010 (0), and 0.016 (+1) V; and interval time at 0.2 (−1), 0.4 (0), and 0.6 (+1) s.

The matrix experiment based on BBD was established by combining the treatment variables at the midpoints of the edges of the process space and the center, thus avoiding extreme treatment combinations, which are the advantage of the design. The number of experiments required was calculated by applying the particular BBD equation [24]: $N = 2k(k - 1) + C_0$, where N is the total number of experiments required, k is the number of factors (×1, ×2, ×3), and C_0 is the number of central points. Accordingly, $N = 2 \times 3 (3 - 1) + 3 = 15$, the design consisting of 15 experimental points, including 3 central points, was used to assess the effects of three variables and the interaction effects between variables on responses by fitting the data to a polynomial model [24–26].

A Trp standard solution of 0.5 mg/L was measured by performing 15 experimental runs. Statgraphics® Centurion software was used to design the experiment and to calculate the model from the experimental data that best describes the variation of the signals. For this purpose, analysis of variance (ANOVA) was performed at a significance level of 95% (p-value = 0.05). A t-value of 2.571 was established to evaluate which independent variable was statistically significant.

2.9. Method Validation

A variety of general validation protocols have been recommended by organizations, such as the Food and Drug Administration (FDA) and the International Conference on Harmonization (ICH). Consequently, the validation of the developed electrochemical method and the proposed UPLC method for the analysis of Trp was carried out as outlined by FDA [27] and ICH [28,29] guidelines. The validation parameters used were linearity, precision, accuracy, limit of detection, and limit of quantitation.

Linearity and range

The linearity of Trp was determined by addition method. A working solution at 11 concentration levels ranging from 0.1 to 5 mg/L was added from a standard stock solution of Trp (100 mg/L) into the electrochemical cell containing 4 mL of the blank (BR buffer solution) for DPV measurements.

Precision

Repeatability, as recommended by the ICH guideline, is determined from a minimum of 9 determinations covering the specified range of the procedure (e.g., 3 concentrations, 3 replicates each) or from a minimum of 6 determinations at 100% of the test or target concentration [28,29].

The precision of the proposed method was determined as repeatability (intraday precision) and intermediate precision (interday) and expressed in terms of percentage relative standard deviation (%RSD) of the peak current. Repeatability was determined from 10 replicate measurements ($n = 10$) of Trp standard solution (2.5 mg/L), and intermediate precision was analyzed at different concentration levels in triplicate ($n = 3$) on 3 separate days.

Accuracy and recovery

The accuracy of the method was determined by standard addition. It was determined by spiking the real sample with standard Trp solution. Then the recovery was calculated as %RSD from the peak current.

Limit of detection and limit of quantitation

Limit of detection (LOD) and limit of quantitation (LOQ) for Trp were calculated from the equation of regression analysis obtained from the calibration curve as follows: LOD = $3\sigma/S$ and LOQ = $10\sigma/S$, where σ is the standard deviation of the intercept and S is the sensitivity expressed by the slope of the calibration curve [30,31].

3. Results

3.1. Optimization of the Electrochemical Method

3.1.1. Modification of Sonogel–Carbon Electrodes

As it is well known, the modification of the electrode surface with some kind of nanomaterials, for instance, gold nanoparticles, affects positively the analytical signal [32]. Therefore, different procedures were evaluated for enhancing the Trp signal and sensitivity of SNGCEs. This includes electrodeposition of a gold solution and drop casting of AuNPs onto the surface of the electrodes. Then the modified electrodes and bare electrodes were used to determine the Trp standard solution at concentration levels of 0.1 to 10 mg/L. Calibration curves were constructed from DPV measurements, and the sensitivities of modified and unmodified electrodes are given in Table 2. Duration in the table means the time required for the modification of the electrode using the electroanalytical technique (chronoamperometry). After comparing the results, bare (unmodified) electrodes showed better sensitivity and signal than the modified electrodes. The sensitivity of the bare electrodes is almost 75% higher than that of the modified electrodes. The standard deviation (SD) in this case is referred to as a measure of repeatability of the modified or not modified sensor. The bare electrode is much more sensitive by far, and the slight increase in SD with respect to Au-deposited electrodes is not a demerit in the future electroanalytical performance of this bare electrode, when compared with the deposited ones. Here, the deposition with gold does not offer improvement in Trp detection. Thus, the determination of Trp was carried out using an unmodified SNGCE.

Table 2. Sensitivity of modified and unmodified SNGCE.

SNGCE	Duration (s)	Sensitivity ($n = 2$) (μA·L/mg)	SD
Gold solution electrodeposited ($HAuCl_4$)	200	0.18	0.009
	600	0.11	0.006
	1000	0.14	0.012
Gold nanoparticles drop-casted (AuNPs)		0.13	0.005
Bare (unmodified)		0.43	0.042

SNGCE: Sonogel–carbon electrode; SD: standard deviation.

It was also observed that the reproducibility of the electrode was improved when the electrode was repolished and repolarized after each measurement and having a comparable voltammogram (Figure 1) from the CV scanning.

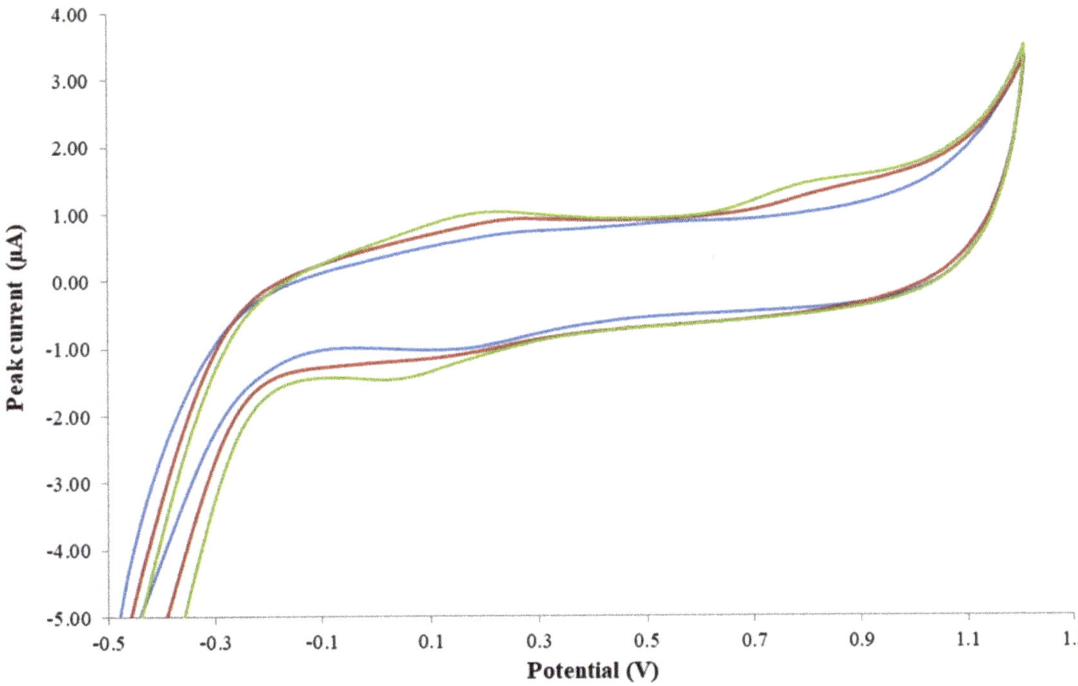

Figure 1. Voltammograms of different SNGCEs obtained from CV scanning: 'blue: after polishing and polarizing 1′, 'red: after polishing and polarizing 2′, 'green: after polishing and polarizing 3′.

In order to avoid the early oxidation of other species in the medium at potential values lower than 0.3, which could affect the Trp signal, and to obtain a good shape for the peak signal, the initial potential was set at 0.3 V and the end potential at 1.2 V for all the electroanalytical measurements. After that, a peak current corresponding to Trp was observed at a potential of about 0.749 V, and a typical voltammogram obtained during the calibration analysis is shown in Figure 2.

3.1.2. Optimization of DPV Conditions Using the Box–Behnken Design

The differential pulse voltammetric conditions were optimized using the BBD. The optimization process was carried out according to a previous work [32] in order to maximize the current intensity (used as the response variable) (see Section 2.9 for more details). Hence, after performing 15 experimental runs (Table 3), the data obtained were used to build a model (adjustment of 90%), where the optimal values extracted for the variables were as follows: modulation amplitude = 0.1 V, step potential = 0.016 V, and interval time = 0.6 s. Under these conditions, improved signals (with optimized current intensities or peak currents) were obtained. The surface plot obtained from the experiments is shown in Figure 3. The response surface indicates that the anodic current intensity is maximized at the highest modulation amplitude and step potential values using intermediate values of the interval time variable.

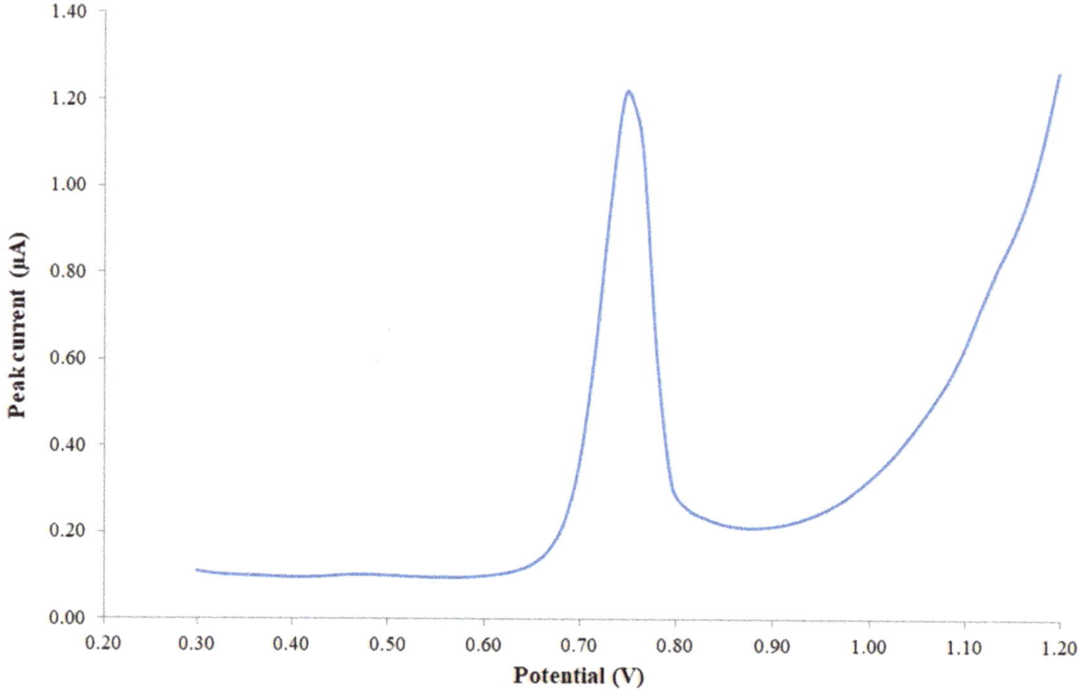

Figure 2. Voltammogram with the final peak signal of tryptophan (2 mg/L) after adjusting the measurement conditions.

Table 3. Box–Behnken design for optimizing electroanalytical experimental conditions.

Experiments	Modulation Amplitude (MA, mV)	Step Potential (ST, mV)	Interval Time (IT, s)	Coding Variables (MA, ST, and IT)			Current Intensity (µA)
1	100	16	0.4	+1	+1	0	3.50×10^{-2}
2	10	10	0.2	−1	0	−1	6.03×10^{-3}
3	100	10	0.6	+1	0	+1	4.66×10^{-2}
4	55	16	0.2	0	+1	−1	1.42×10^{-2}
5	55	4	0.2	0	−1	−1	1.12×10^{-2}
6	55	10	0.4	0	0	0	2.84×10^{-2}
7	10	4	0.4	−1	−1	0	4.53×10^{-3}
8	10	16	0.4	−1	+1	0	5.98×10^{-3}
9	55	10	0.4	0	0	0	3.13×10^{-2}
10	55	4	0.6	0	−1	+1	3.17×10^{-2}
11	55	16	0.6	0	+1	+1	3.84×10^{-2}
12	100	4	0.4	+1	−1	0	3.52×10^{-2}
13	55	10	0.4	0	0	0	2.76×10^{-2}
14	100	10	0.2	+1	0	−1	1.37×10^{-2}
15	10	10	0.6	−1	0	+1	4.81×10^{-3}

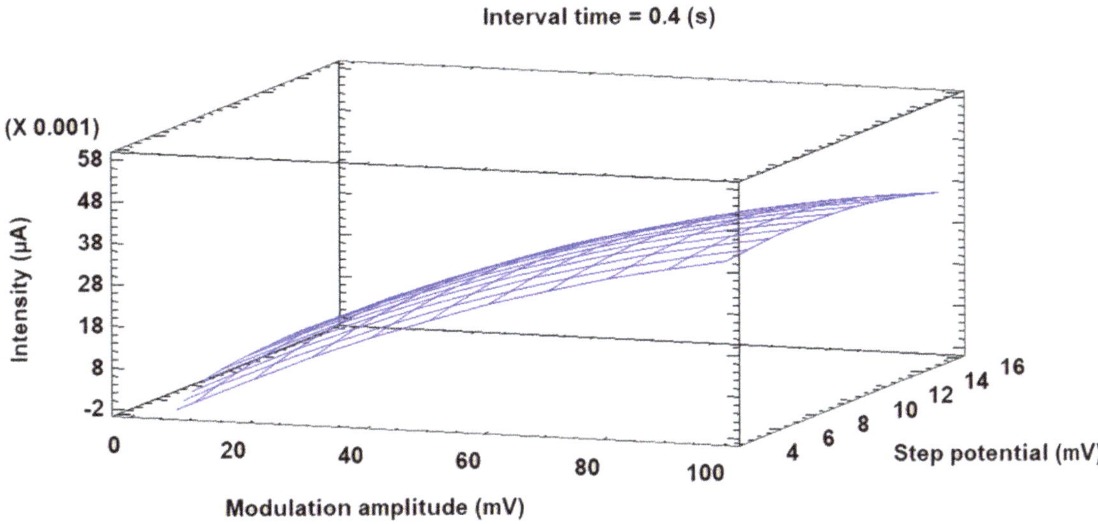

Figure 3. Surface plot of current response.

3.2. Validation of the Electrochemical Method

Linearity, LOD, LOQ, and intraday and interday precision were used to validate the new method. A Trp standard solution was analyzed at 11 concentration levels in the range of 0.1 to 5 mg/L, and a good linear regression was obtained between peak current and concentration. The correlation coefficient, y-intercept, and slope of the regression line were then determined. It was observed that using the optimized conditions, the slope (i.e., the sensitivity) was increased from the starting measures (Table 2). The analytical characteristics for the different concentration ranges are given in Table 4.

Table 4. Analytical characteristics for the determination of tryptophan.

Linearity Range (mg/L)	0.1–5
Regression equation	Y = 0.6273 × −0.2523
Determination coefficient (r^2)	0.9880
Correlation coefficient (r)	0.9940
Intercept (b) (µA)	0.2523
LOD (mg/L)	0.33
LOQ (mg/L)	1.09

LOD and LOQ were determined from the calibration data using regression analysis and were found to be 0.33 and 1.09 mg/L, respectively. The sensitivity of the electrode expressed as the slope of the calibration curve (0.63 µA·L/mg) was also improved as compared with the sensitivity (0.43 µA·L/mg) shown in Table 2.

The values of RSD were calculated to determine intraday and interday precision. The results showed that the RSD values for intraday and interday precision of the Trp standard solution were found to be less than 3.15% and 2.75%, respectively. The RSD value for the intraday precision of a real sample was 4.06% (Table 5). Therefore, the repeatability of the detection of the Trp concentration on the electrode was excellent.

Table 5. Intra- and interday precision for tryptophan (SD: standard deviation; RSD: relative standard deviation).

Intraday Precision (n = 10)			Interday Precision (n = 3)	
Concentration (mg/L)	SD	RSD (%)	SD	RSD (%)
2			1.19×10^{-8}	1.21
2.5	3.94×10^{-8}	3.12	3.19×10^{-8}	2.51
3			4.54×10^{-8}	2.73
Real sample (n = 7)	3.06×10^{-8}	4.06		

The accuracy for Trp was determined by standard addition method as percent recovery. Recovery values (%) were calculated from the peak current, and the average recovery (n = 3) was found to be 99.01%, which indicates that excellent recovery was obtained. Thus, the developed method may be successfully used for the determination of Trp in fruit juices with adequate accuracy.

It is also very important to remark that after consecutive measurements, the signal of Trp decreases due to fouling of the analyte on the electrode surface. Previous studies [33] reported that Trp, like most other organic molecules, can be easily adsorbed onto electrodes and can foul the electrode during successive scans. The fouling effect of bare glassy carbon electrodes (GCEs) was higher than that of modified electrodes. There was a decrease of almost 30% of the initial response after the 10th consecutive cycle in a tryptophan standard solution using the modified electrode, whereas the decrease of the same degree was observed after only the 3rd cycle using bare GCEs. This demonstrates that the fouling effect was much higher in bare GCEs than in modified electrode. In our case, it seems that this situation can be also corroborated (see Section 3.4). The redox mechanism involving Trp oxidation at a SNGCE, as concluded after CV results varying the scan rate, indicates that Trp might need not only to diffuse from a solution to the electrode surface, but also to interact with the surface before suffering the oxidation process, and that this interaction could control the rate of the redox process. This fact makes too much probable the existence of a fouling effect during Trp determination at a sonogel–carbon electrode. Therefore, adequate methodology, as the one used here, based on polishing, polarization, and CV conditioning of SNGCEs, was necessary before determining the analyte with our sensing system.

3.3. *Study of the Electrochemical Reaction Mechanism*

The results obtained in this section support the new method proposed to determine Trp. To understand the electrochemical reaction mechanism of Trp, the effect of scan rate on the oxidation peak currents of Trp at an unmodified SNGCE was investigated. The electrode was well polished and polarized in 25 mL of 0.1 M H_2SO_4 using amperometric conditions. Then, the CV measurement for a Trp standard solution of 2 mg/L in a BR buffer solution (pH 3.6) was performed at different scan rates of 50, 75, 100, 125, 150, and 200 mV s^{-1}. Subsequently, the signals were recorded and the oxidation peak current measured (see Figure 4a). It was observed that the oxidation peak potential of Trp shifts positively with the increase in scan rate, as reported in the literature [34,35]. These shifts may be attributed to the accumulation of the oxidation or reduction products on the electrode surface [35]. As shown in Figure 4b, a linear relationship between the oxidation peak current (I_p) and the scan rate (v) was obtained: $I_p = 7.3248v + 0.5501$ (I_p: μA, v: V s^{-1}), with a correlation coefficient of R = 0.9938, in the range of 50–200 mV s^{-1} and in a BR buffer solution (pH 3.6). This fact indicates that the electrochemical oxidation of Trp at the SNGCE was an adsorption-controlled process [1,34,36,37]. Similarly, a linear relationship was obtained between the oxidation peak current and the square root of the scan rate: $I_p = 4.8953v - 0.2285$ (I_p: μA, v: V s^{-1}), with a correlation coefficient of R = 0.9788 (Figure 4c), suggesting that the electrochemical oxidation of Trp was a diffusion-controlled process. Besides, the plot of logarithm of the peak current versus the logarithm of the scan rate gave a straight line with a slope close to a theoretical value of 0.6 (log I_p = 0.5738 log v − 5.3088; R^2 = 0.9816), which indicates a reaction with a mechanism controlled by a mixed adsorption/diffusion electrode process [35]. According to the results obtained and due to the fouling suffered by the electrode surface, it can be concluded that the electrochemical process of Trp consists of a mixture of both adsorption and diffusion-controlled processes. These results serve to corroborate the responses obtained and the electroanalytical methodology proposed for Trp determination.

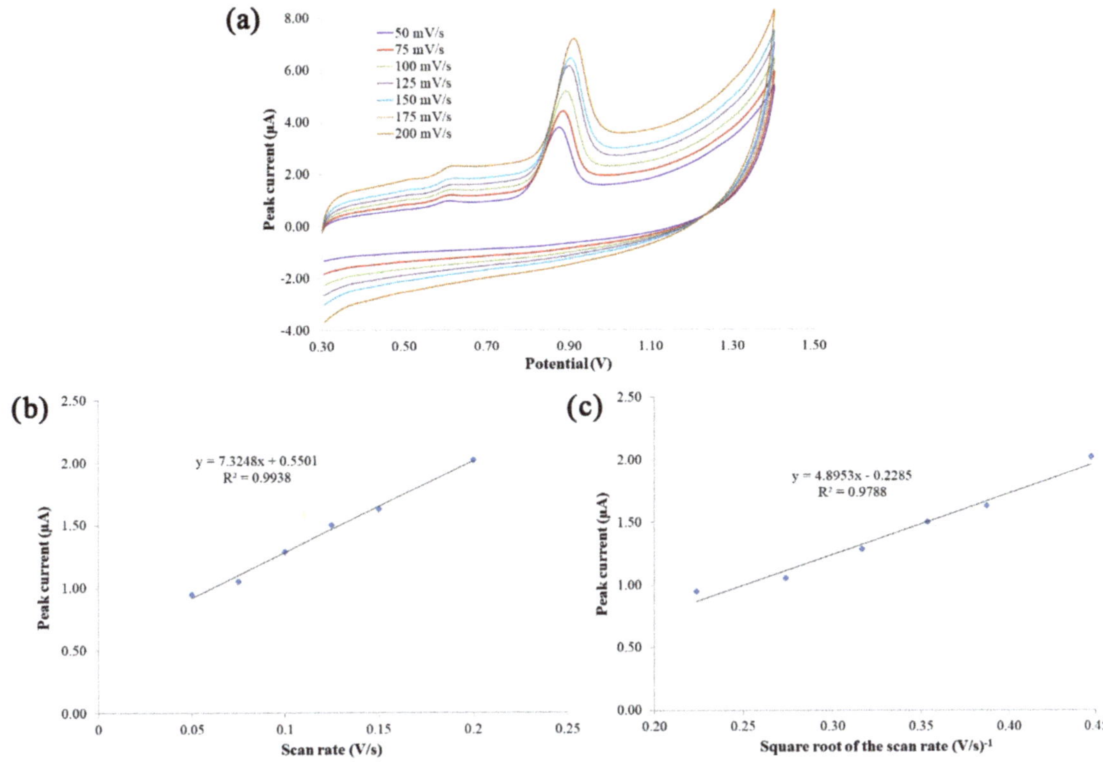

Figure 4. Cyclic voltammograms of tryptophan at different scan rates (**a**) (50 (blue), 75 (red), 100 (light green), 125 (purple), 150 (light blue), and 200 (orange) mV s^{-1}), plot of peak current vs. scan rate (**b**), and plot of peak current vs. square root of scan rate (**c**).

3.4. UPLC Determination of Tryptophan in Fruit Juice Samples

Based on the previously developed and validated UPLC method [38] with a fluorescence detector, Trp was quantified in different fruit juice samples. The retention time of Trp was found to be about 2.64 min (see Figures S1 and S2 in the Supplementary Materials section).

The UPLC method was validated for linearity, precision, accuracy, LOD, and LOQ according to the ICH and FDA guidelines.

The linearity of the method was determined by analyzing a series of standard solutions of tryptophan following the same conditions stated in Section 2.9. The signals obtained were proportional to their concentrations within these ranges. The equation for the calibration curve was y = 411,296 × −258,786 with a linear correlation coefficient (r) of 0.9996.

LOD and LOQ for Trp were calculated from the equation of regression analysis obtained from the calibration curve based on the standard deviation of the response and the slope of the calibration curve. The limit of detection and limit of quantitation were found to be 0.09 and 0.29 mg/L respectively.

The precision of the UPLC method was determined as repeatability (intraday precision) from 10 independent analyses (*n* = 10) of the fruit juice samples on the same day. Repeatability was also determined from 5 independent analyses (*n* = 5) by running the blank between each sample. The percentage relative standard deviations were 1.77% and 1.01%, respectively.

Moreover, the effect of the filtration of fruit juices on the recovery was assessed by spiking a standard solution of 1 mL (5 mg/L) with 5 mL of a pineapple juice sample before filtration and after filtration in triplicate (*n* = 3). It was observed that good recoveries were obtained in both cases. The spiking of the standard solution before the filtration of the real sample and then the filtering through 0.45 and 0.22 μL filters gave a recovery of 117%. Meanwhile, the spiking of the standard solution after the filtration of the real sample with 0.45 μL and then the filtering of the spiked sample through 0.22 μL gave a recovery of 115%. This ensured that the filtration of the fruit juice samples before spiking or after spiking has no significant effect on the analysis results. Thus, the method can be successfully used for the determination of Trp in fruit juice with adequate accuracy.

3.5. Comparison of UPLC and Electrochemical Methods

The developed electrochemical method was compared with the reference UPLC method with respect to method validation parameters (Table 6), and the results showed that both methods are reproducible and can be successfully used in a food quality control laboratory. As shown in the table below, UPLC has lower LOD and LOQ and, similarly, better recovery and intraday precision as compared with the electrochemical method. However, the electrochemical method is faster and requires less expensive systems.

Table 6. Method validation parameters for the determination of tryptophan in fruit juices.

Parameter	UPLC	Electrochemical
Linear range concentration (mg/L)	0.09–5	1.09–5
Coefficient of determination (r^2)	0.999	0.988
Intraday precision (*n* = 9), RSD (%), real sample	1.77	4.06
Intraday precision (*n* = 10), RSD (%), standard		3.12
Recovery (%)	116.18	99.01
LOD (mg/L)	0.09	0.33
LOQ (mg/L)	0.29	1.09

The developed electrochemical method was used for quantifying tryptophan in fruit juice samples, as an additional step in the validation process. The extent of our knowledge, this is the first time that tryptophan has been determined electrochemically in fruit juices. Results from the UPLC method were used as a reference method. Tryptophan was quantified in fruit juices that were packed in a container containing 200 mL, and the results are shown in Table 7. A fruit juice containing a mixture of pineapple and grape showed the maximum concentration, and an apple juice sample showed the minimum concentration in both the reference method (UPLC) and the developed electrochemical method. It must be noted that for those samples containing apple juice, either pure or mixed with other juices, the electrochemical method produced higher results than the chromatographic method. It can be concluded that the samples elaborated with apple juice contained some kind of interference that affected the electrochemical signal. Those interferences were separated using the chromatographic method.

Table 7. Concentration of tryptophan in some analyzed fruit juices.

Real Samples	Electrochemical	UPLC
	Conc. (mg/L)	Conc. (mg/L)
Pineapple	2.56	1.84
Peach + grape	1.31	1.15
Apple	1.34	0.45
Peach + apple + grape	1.22	0.65
Peach	1.13	1.21

4. Conclusions

An electrochemical method using a sonogel–carbon electrode was developed, optimized, and validated for the determination of Trp in fruit juices using ultra performance liquid chromatography as a reference method.

Sonogel–carbon electrodes were prepared as an electrochemical sensor for the determination of Trp. The developed electrochemical method was sensitive, rapid, cheap, and easy to use, and this is the first time that Trp determination was accomplished in fruit juices using the proposed methodology. Moreover, this methodology avoids fouling effects due to the Trp nature with an adequate combination of mechanical and electrochemical cleaning procedures of the electrode surface; that is why electrode surface modification is not recommended. The sample preparation needs only a previous filtration step as in UPLC; no other pretreatment is performed on the juice sample. The proposed method has shown many desirable properties for the determination of Trp, including good limit of detection and linear range, good reproducibility and repeatability, good recovery, and good sensitivity. The LOD and LOQ were 0.33 and 1.09 mg/L, respectively, for the developed electrochemical method. Additionally, the recovery was 99.01%, and the RSDs for intraday and interday precisions were 3.12% and lower than 2.75%, respectively. Comparison with the reference method demonstrated good agreement for the results of Trp for those fruit juices without apple juice; therefore, it can be successfully used in food quality control for most fruit juices.

Supplementary Materials: The following supporting information can be downloaded at: https://www.mdpi.com/article/10.3390/foods11142149/s1: Figure S1: Chromatogram of tryptophan obtained during the calibration curve; Figure S2: Chromatogram of tryptophan obtained during the analysis of pineapple juice; Table S1: Some of the previously reported methods and samples analyzed for the determination of Trp; References of Table S1. References [39–46] are cited in the supplementary materials.

Author Contributions: Conceptualization and methodology, J.M.P.-S. and M.P.; validation, A.T., J.M.P.-S. and M.P.; investigation, A.T.; original draft writing, A.T.; review and editing, J.M.P.-S. and M.P.; supervision, J.M.P.-S. and M.P.; funding acquisition, J.M.P.-S. and M.P. All authors have read and agreed to the published version of the manuscript.

Funding: Assefa Takele thanks the European Commission (EACEA) and the EMQAL consortium for supporting his study and research activity inside the Erasmus Mundus Master in Quality in Analytical Laboratories (EMQAL) through an Erasmus Mundus studentship. J.M. Palacios-Santander acknowledges the Junta de Andalucía (PAIDI2020) and Institute of Research on Electron Microscopy and Materials (IMEYMAT–APPLIEDSENS, BIOSENSEP, POLYBIOSENS, and NANO4(BIO)SENS projects) for their financial support. This research was cofinanced by the Proyecto Singular 'Ceia3 Instrumento Estratégico Hacia Un Tejido Productivo Agroalimentario, Moderno, Innovador Y Sostenible: Motor Del Territorio Rural Andaluz' supported by the Government of Andalusia (Project reference: PAI-TAN-AT2019-AGROMIS-EC) and by the 'Programa de Fomento e Impulso de la Investigación y de la Transferencia de la Universidad de Cadiz 2020–2021' with the funds given through the project PR2020-013 (Proyectos de Investigación-Puente 2020).

Institutional Review Board Statement: Not applicable.

Informed Consent Statement: Not applicable.

Data Availability Statement: Data is contained within the article or supplementary material.

Conflicts of Interest: The authors declare no conflict of interest.

References

1. Dehdashtian, S.; Shamsipur, M.; Gholivand, M.B. Fabrication of a novel electrochemical sensor based on an electrosynthesized indolyldihydroxyquinone as a bio-based modifier for sensitive and selective direct electrochemical determination of tryptophan. *J. Electroanal. Chem.* **2016**, *780*, 119–125. [CrossRef]
2. Zhen, Q.; Xu, B.; Ma, L.; Tian, G.; Tang, X.; Ding, M. Simultaneous determination of tryptophan, kynurenine and 5-hydroxytryptamine by HPLC: Application in uremic patients undergoing hemodialysis. *Clin. Biochem.* **2011**, *44*, 226–230. [CrossRef] [PubMed]
3. Pinhati, R.; Polonini, H.; Brandão, M. Quantification of tryptophan in plasma by high performance liquid chromatography. *Quim. Nova* **2012**, *35*, 623–626. [CrossRef]
4. Ye, D.; Luo, L.; Ding, Y.; Liu, B.; Liu, X. Fabrication of Co_3O_4 nanoparticles-decorated graphene composite for determination of l-tryptophan. *Analyst* **2012**, *137*, 2840–2845. [CrossRef] [PubMed]
5. Albu, C.; Radu, G.L. Development and Application of a HPLC-PDA-FL Method for the Determination of Melatonin and its Precursors in Infant Formulas. *Food Anal. Methods* **2018**, *11*, 951–958. [CrossRef]
6. Boulet, L.; Faure, P.; Flore, P.; Montérémal, J.; Ducros, V. Simultaneous determination of tryptophan and 8 metabolites in human plasma by liquid chromatography/tandem mass spectrometry. *J. Chromatogr. B Anal. Technol. Biomed. Life Sci.* **2017**, *1054*, 36–43. [CrossRef] [PubMed]
7. Kia, M.; Islamnezhad, A.; Shariati, S.; Biparva, P. Preparation of voltammetric biosensor for tryptophan using multi-walled carbon nanotubes. *Korean J. Chem. Eng.* **2011**, *28*, 2064–2068. [CrossRef]
8. Fiorucci, A.R.; Cavalheiro, E.T.G. The use of carbon paste electrode in the direct voltammetric determination of tryptophan in pharmaceutical formulations. *J. Pharm. Biomed. Anal.* **2002**, *28*, 909–915. [CrossRef]
9. Liang, Y.D.; Song, J.F. Flow-injection chemiluminescence determination of tryptophan through its peroxidation and epoxidation by peroxynitrous acid. *J. Pharm. Biomed. Anal.* **2005**, *38*, 100–106. [CrossRef] [PubMed]
10. Silber, B.Y.; Schmitt, J.A.J. Effects of tryptophan loading on human cognition, mood, and sleep. *Neurosci. Biobehav. Rev.* **2010**, *34*, 387–407. [CrossRef]
11. Çevikkalp, S.A.; Löker, G.B.; Yaman, M.; Amoutzopoulos, B. A simplified HPLC method for determination of tryptophan in some cereals and legumes. *Food Chem.* **2016**, *193*, 26–29. [CrossRef] [PubMed]
12. Chen, H.; Li, L.; Zhou, M.; Ma, Y.J. Flow-injection chemiluminescence determination of tryptophan using galangin-potassium permanganate-polyphosphoric acid system. *Chin. Chem. Lett.* **2008**, *19*, 203–206. [CrossRef]
13. Wu, Y.; Wang, T.; Zhang, C.; Xing, X.H. A rapid and specific colorimetric method for free tryptophan quantification. *Talanta* **2018**, *176*, 604–609. [CrossRef] [PubMed]
14. Farghaly, O.A.; Hameed, R.A.; Abu-Nawwas, A.A.H. Analytical Application Using Modern Electrochemical Techniques. *Int. J. Electrochem. Sci.* **2014**, *9*, 3287–3318. [CrossRef]
15. Jandrić, Z.; Cannavan, A. Authentication of Fruit Juices by Metabolomics Using UPLC-QTOF MS. *Fruit Juices Extr. Compos. Qual. Anal.* **2017**, *148*, 779–804. [CrossRef]
16. Anastácio, M.; dos Santos, A.P.M.; Aschner, M.; Mateus, L. Determination of trace metals in fruit juices in the Portuguese market. *Toxicol. Rep.* **2018**, *5*, 434–439. [CrossRef]
17. Rajauria, G.; Tiwari, B.K. Fruit Juices: An Overview. In *Fruit Juices: Extraction, Composition, Quality and Analysis*, 1st ed.; Rajauria, G., Tiwari, B.K., Eds.; Academic Press: Dublin, Ireland, 2018; pp. 3–13. [CrossRef]
18. Cubillana-Aguilera, L.M.; Palacios-Santander, J.M.; Naranjo-Rodríguez, I.; Hidalgo-Hidalgo-de-Cisneros, J.L. Study of the influence of the graphite powder particle size on the structure of the Sonogel-Carbon materials. *J. Sol.-Gel. Sci. Techn.* **2006**, *40*, 55–64. [CrossRef]
19. García Guzmán, J.J.; Cubillana-Aguilera, L.M.; Bellido-Milla, D.; Naranjo-Rodríguez, I.; Lete, C.; Palacios-Santander, J.M.; Lupu, S. Development of Sonogel-Carbon based biosensors using sinusoidal voltages and currents methods. *Sens. Actuators B Chem.* **2018**, *255*, 1525–1535. [CrossRef]
20. Pundir, C.S.; Lata, S.; Narwal, V. Biosensors for determination of D and L- amino acids: A review. *Biosens. Bioelectron.* **2018**, *117*, 373–384. [CrossRef]
21. Hilali, N.; Ghanam, A.; Mohammadi, H.; Amine, A.; García-Guzmán, J.J.; Cubillana-Aguilera, L.; Palacios-Santander, J.M. Comparison between modified and unmodified carbon paste electrodes for hexavalent chromium determination. *Electroanalysis* **2018**, *30*, 2750–2759. [CrossRef]
22. Carrera, C.; Aliaño-González, M.J.; Valaityte, M.; Ferreiro-González, M.; Barbero, G.F.; Palma, M. A Novel Ultrasound-Assisted Extraction Method for the Analysis of Anthocyanins in Potatoes (*Solanum tuberosum* L.). *Antioxidants* **2021**, *10*, 1375. [CrossRef] [PubMed]
23. González-de-Peredo, A.V.; Vázquez-Espinosa, M.; Espada-Bellido, E.; Ferreiro-González, M.; Carrera, C.; Barbero, G.F.; Palma, M. Development of Optimized Ultrasound-Assisted Extraction Methods for the Recovery of Total Phenolic Compounds and Anthocyanins from Onion Bulbs. *Antioxidants* **2021**, *10*, 1755. [CrossRef] [PubMed]

24. Ferreira, S.L.C.; Bruns, R.E.; Ferreira, H.S.; Matos, G.D.; David, J.M.; Brandão, G.C.; da Silva, E.G.P.; Portugal, L.A.; dos Reis, P.S.; Souza, A.S.; et al. Box-Behnken design: An alternative for the optimization of analytical methods. *Anal. Chim. Acta* **2007**, *597*, 179–186. [CrossRef]
25. Qiu, P.; Cui, M.; Kang, K.; Park, B.; Son, Y.; Khim, E.; Jang, M.; Khim, J. Application of Box-Behnken design with response surface methodology for modeling and optimizing ultrasonic oxidation of arsenite with H_2O_2. *Cent. Eur. J. Chem.* **2014**, *12*, 164–172. [CrossRef]
26. Tekindal, M.A.; Bayrak, H.; Ozkaya, B.; Genc, Y. Box- behnken experimental design in factorial experiments: The importance of bread for nutrition and health. *Turk. J. F. Crop.* **2012**, *17*, 115–123.
27. U.S. Food and Drug Administration/Center for Biologics Evaluation and Research. Guideline for Industry Q2A Text on Validation of Analytical Procedures. *Food Drug Adm.* **1995**, 1–9. Available online: https://www.fda.gov/ucm/groups/fdagov-public/@fdagov-drugs-gen/documents/document/ucm073381.pdf (accessed on 19 July 2022).
28. ICH. Guidance for industry: Q2B validation of analytical procedures: Methodology. *Int. Conf. Harmon. Tech. Requir. Regist. Tripart. Guidel.* **1996**, *13*, 62, FR 27464.
29. ICH. ICH Topic Q2 (R1) Validation of Analytical Procedures: Text and Methodology. *Int. Conf. Harmon.* **1995**, *1994*, 17. Available online: https://www.ema.europa.eu/en/documents/scientific-guideline/ich-q-2-r1-validation-analytical-procedures-text-methodology-step-5_en.pdf (accessed on 19 July 2022).
30. Shrivastava, A.; Gupta, V.B. Methods for the determination of limit of detection and limit of quantitation of the analytical methods. *Chron. Young Sci.* **2011**, *2*, 21–25. [CrossRef]
31. Pop, A.; Fizeșan, I.; Vlase, L.; Rusu, M.E.; Cherfan, J.; Babota, M.; Gheldiu, A.-M.; Tomuta, I.; Popa, D.-S. Enhanced Recovery of Phenolic and Tocopherolic Compounds from Walnut (*Juglans regia* L.) Male Flowers Based on Process Optimization of Ultrasonic Assisted-Extraction: Phytochemical Profile and Biological Activities. *Antioxidants* **2021**, *10*, 607. [CrossRef]
32. Ajaero, C.; Yahia, M.; Abdelrahim, M.; Palacios-Santander, J.M.; Gil, M.L.A.; Naranjo-Rodríguez, I.; Hidalgo-Hidalgo de Cisneros, J.L.; Cubillana-Aguilera, L.M. Comparative study of the electrocatalytic activity of different types of gold nanoparticles using Sonogel-Carbon material as supporting electrode. *Sens. Actuators B. Chem.* **2012**, *171–172*, 1244–1256. [CrossRef]
33. Guo, Y.; Guo, S.; Fang, Y.; Dong, S. Gold nanoparticle/carbon nanotube hybrids as an enhanced material for sensitive amperometric determination of tryptophan. *Electrochim. Acta* **2010**, *55*, 3927–3931. [CrossRef]
34. Wang, C.; Li, T.; Liu, Z.; Guo, Y.; Li, C.; Dong, C.; Shuang, S. An ultra-sensitive sensor based on β-cyclodextrin modified magnetic graphene oxide for detection of tryptophan. *J. Electroanal. Chem.* **2016**, *781*, 363–370. [CrossRef]
35. Elqudaby, H.M.; Mohamed, G.G.; El Din, G.M.G. Electrochemical behaviour of trimebutine at activated glassy carbon electrode and its direct determination in urine and pharmaceutics by square wave and differential pulse voltammetry. *Int. J. Electrochem. Sci.* **2014**, *9*, 856–869.
36. Li, J.; Kuang, D.; Feng, Y.; Zhang, F.; Xu, Z.; Liu, M.; Wang, D. Green synthesis of silver nanoparticles-graphene oxide nanocomposite and its application in electrochemical sensing of tryptophan. *Biosens. Bioelectron.* **2013**, *42*, 198–206. [CrossRef] [PubMed]
37. Khaleghi, F.; Irai, A.E.; Gupta, V.K.; Agarwal, S.; Bijad, M.; Abbasghorbani, M. Highly sensitive nanostructure voltammetric sensor employing Pt/CNTs and 1-butyl-3-methylimidazolium hexafluoro phosphate for determination of tryptophan in food and pharmaceutical samples. *J. Mol. Liq.* **2016**, *223*, 431–435. [CrossRef]
38. Setyaningsih, W.; Saputro, I.E.; Carrera, C.A.; Palma, M.; Barroso, C.G. Multiresponse optimization of a UPLC method for the simultaneous determination of tryptophan and 15 tryptophan-derived compounds using a Box-Behnken design with a desirability function. *Food Chem.* **2017**, *225*, 1–9. [CrossRef]
39. Delgado-Andrade, C.; Rufián-Henares, J.A.; Jiménez-Pérez, S.; Morales, F.J. Tryptophan determination in milk-based ingredients and dried sport supplements by liquid chromatography with fluorescence detection. *Food Chem.* **2006**, *98*, 580–585. [CrossRef]
40. Alegría, A.; Barberfi, R.; Farr, R.; Lagarda, M.J.; Ldpez, J.C. Isocratic high-performance liquid chromatographic determination of tryptophan in infant formulas. *J. Chromatogr. A.* **1996**, *721*, 83–88. [CrossRef]
41. Biasiolo, M.; Bertazzo, A.; Costa, C.; Beghetto, A.; Allegri, G. Determination of nonprotein tryptophan in yoghurts by selective fluorescence and HPLC. *Food Chem.* **1995**, *52*, 87–92. [CrossRef]
42. Dario, M.F.; Freire, T.B.; Pinto, C.A.S.d.; Prado, M.S.A.; Baby, A.R.; Velasco, M.V.R. Tryptophan and kynurenine determination in human hair by liquid chromatography. J. Chromatogr. B Anal. Technol. Biomed. *Life Sci.* **2017**, *1065–1066*, 59–62. [CrossRef]
43. Yılmaz, C.; Gökmen, V. Determination of tryptophan derivatives in kynurenine pathway in fermented foods using liquid chromatography tandem mass spectrometry. *Food Chem.* **2018**, *243*, 420–427. [CrossRef] [PubMed]
44. Chen, G.y.; Zhong, W.; Zhou, Z.; Zhang, Q. Simultaneous determination of tryptophan and its 31 catabolites in mouse tissues by polarity switching UHPLC-SRM-MS. *Anal. Chim. Acta* **2018**, *1037*, 200–210. [CrossRef] [PubMed]
45. Beitollahi, H.; Gholami, A.; Ganjali, M.R. Preparation, characterization and electrochemical application of Ag-ZnO nanoplates for voltammetric determination of glutathione and tryptophan using modified carbon paste electrode. *Mater. Sci. Eng. C.* **2015**, *57*, 107–112. [CrossRef] [PubMed]
46. Majidi, M.R.; Omidi, Y.; Karami, P.; Johari-Ahar, M. Reusable potentiometric screen-printed sensor and label-free aptasensor with pseudo-reference electrode for determination of tryptophan in the presence of tyrosine. *Talanta* **2016**, *150*, 425–433. [CrossRef] [PubMed]

foods

Article

Simultaneous Determination of Multiple Contaminants in Chicken Liver Using Dispersive Liquid-Liquid Microextraction (DLLME) Detected by LC-HRMS/MS

Belete Eshetu Gebreyohannes [1], Simiso Dube [1] and Mathew Muzi Nindi [2,*]

[1] Department of Chemistry, The Science Campus, College of Science Engineering and Technology, University of South Africa, Corner of Christiaan de Wet Road & Pioneer Avenue, Florida 1709, South Africa; 48221783@mylife.unisa.ac.za (B.E.G.); dubes@unisa.ac.za (S.D.)

[2] Institute for Nanotechnology and Water Sustainability, The Science Campus, College of Science Engineering and Technology, University of South Africa, Corner of Christiaan de Wet Road & Pioneer Avenue, Florida 1709, South Africa

* Correspondence: nindimm@unisa.ac.za; Tel.: +27-11-670-9303

Abstract: Simultaneous determination of a mixture of food contaminants, including pesticides, sulphonamides, fluoroquinolones, anthelmintics, and aflatoxin B1, in solid biological samples (chicken liver) by dispersive liquid-liquid microextraction/liquid chromatography-high resolution mass spectrometry (DLLME/LC-HRMS) is presented. Previous work focused on the application of DLLME to single-class contaminants. In this work, the DLLME extraction method has been extended to complex multiresidues in the biological matrix. The first part of this study was the selection of an appropriate solvent that enabled the dissolution of analytes from the chicken livers. The matrix-matched calibration curves showed good linearity in the range 0.5–50.0 $\mu g\ kg^{-1}$ for aflatoxin B1 and 50–500 $\mu g\ kg^{-1}$ for pesticides, fluoroquinolones, sulphonamides, and anthelmintics, with a coefficient of determination (R^2) values of 0.9916–0.9967. The mean recoveries were in the range of 80.4–96.3%, and the relative standard deviation (RSD) values were in the range of 1.53–8.98%. The limit of detection (LOD) and the limit of quantification (LOQ) values were 0.03 $\mu g\ kg^{-1}$ and 0.09 $\mu g\ kg^{-1}$, respectively, for aflatoxin B1, and for pesticides, fluoroquinolones, sulphonamides, and anthelmintics, they were in the range of 0.011–1.197 $\mu g\ kg^{-1}$ and 0.150–2.579 $\mu g\ kg^{-1}$, respectively. The developed method was compared with the standard solid phase extraction (SPE) method, and there was no significant difference between the two methods.

Keywords: pesticides; sulphonamides; fluoroquinolones; anthelmintics; Aflatoxin B1; dispersive liquid-liquid microextraction; biological samples; solid phase extraction

Citation: Gebreyohannes, B.E.; Dube, S.; Nindi, M.M. Simultaneous Determination of Multiple Contaminants in Chicken Liver Using Dispersive Liquid-Liquid Microextraction (DLLME) Detected by LC-HRMS/MS. *Foods* **2023**, *12*, 2594. https://doi.org/10.3390/foods12132594

Academic Editor: Gianfranco Picone

Received: 19 May 2023
Revised: 15 June 2023
Accepted: 27 June 2023
Published: 4 July 2023

1. Introduction

The safety of food is important in our everyday lives because of its impact on our health. Globally, an increase in scientific knowledge on chemical contamination of food destined for human consumption has been observed. The concern for food safety is also of great importance to world trade due to public health. Billions of people in the world are at risk from unsafe food due to physical, chemical, and biological contamination, especially organic contaminants such as pesticides, veterinary drugs, persistent environmental chemicals, and naturally occurring toxicants [1,2]. Contamination of foods can occur during production [3], storage, transportation, food processing [4], and the use of veterinary drugs in food-producing animals [5]. In addition, the safety and quality of food products have become a growing concern for consumers, governments, and producers because the presence of unwanted and/or threatening contaminants in foods such as meat, animal products, etc. impacts both the local and export economies [6–8]. Several institutions worldwide have put systems in place to protect humans from exposure to many of the chemicals identified as unsuitable for human consumption. The European Union (EU), the United States Environmental Protection Agency (USEPA), the United States Food and Drug Administration (FDA), Codex Alimentarius (Codex Alimentarius Commission, FAO-WHO), as well as other public health agencies around the world, have set maximum residue limits for contaminants in animal food and products. As an example, the use of veterinary drugs in the EU is regulated through Council Regulation 2377/90/EC (European Commission (EC) 2002). In general, regulatory bodies have set maximum residue limits (MRLs) for mycotoxins, veterinary drugs, and pesticides in products of animal or vegetable origin that are intended for human or animal consumption in order to assure human food safety (Table 1).

Table 1. Maximum residue limits (MRLs) set by the European Commission and South Africa for selected pesticides, sulphonamides, fluoroquinolones, anthelmintics, and aflatoxin B1 [9–11].

Compounds	Species	Maximum Residue Limit (MRL) ($\mu g\ kg^{-1}$)	
		SA	EU
Aflatoxin B1	Chicken liver	-	2
Albendazole	All food-producing animals	5000	
Atrazine	All food-producing animals	-	-
Danofloxacin	Chicken liver	400	400
Enrofloxacin	Chicken liver	50	200
Fenbendazole	All food-producing animals	500	-
Mebendazole	All food-producing animals	-	-
Simazine	All food-producing animals	-	-
Sulphachloropyridazine	All food-producing animals	100	100
Sulphadiazine	All food-producing animals	100	100
Sulphamerazine	All food-producing animals	100	100
Sulphaquinoxaline	All food-producing animals	100	100
Sulphapyridine	All food-producing animals	100	100
Terbutryn	All food-producing animals	-	-
Thiabendazole	All food-producing animals	-	-

-: not stated.

Therefore, in order to control and monitor contaminants in food, analytical methodologies must accurately identify and quantify the occurrence of harmful chemical substances in food samples. Sample preparation steps are key to any analytical methodology, as reflected by the time and cost of this step. The primary goal of any extraction technique is the isolation and/or preconcentration of analytes of interest from the complex sample matrix. In most cases, it is difficult, if not impossible, to determine analytes of interest from the food sample directly without sample preparation methods [12]. Some methods for comprehensive analysis of harmful chemicals (mycotoxins, veterinary medications, and pesticides) have been developed; however, they are time-consuming, expensive, and harmful to the environment [12–17]. In recent decades, researchers have focused on the miniaturisation of extraction techniques and the use of solvents that are less harmful to the environment (green solvents) in an attempt to address environmental issues as well as challenges associated with biological matrices. These new analytical techniques are compliant with green analytical chemistry principles [18,19]. One such method that has drawn the interest of many researchers is dispersive liquid microextraction due to its advantages of being simple, cheap, rapid, green, and having high efficiencies. This miniaturized sample preparation method is attractive since it uses microlitre volumes of organic solvents yet is capable of achieving high enrichment factors and producing clean extracts. In the past, the approach to residue analysis has targeted single-class organic contaminants in food. For example, a number of DLLME applications in biological matrices are targeted at single-class contaminants using different analytical instrumentation. Deng et al. [20] extracted sulphonamides, while Moema et al. [21] extracted fluoroquinolones from chicken livers using DLLME. In another research work, Liu and co-workers [22] extracted clenbuterol from porcine tissues, and Vinas et al. [23] also reported the extraction of thiamines from foods using DLLME. However, the continuous detection of various types of contaminants in food necessitates the development of analytical methods that can handle multiclass residues. This current work extends the capabilities of DLLME to a mixture of various classes of contaminants in response to recent developments in FAO and Codex.

In this work, a rapid, cheap, simple, and green extraction method, DLLME, was developed and validated for the simultaneous determination of multiple contaminants in chicken liver samples using LC MS/MS. To the best of our knowledge, this is the first DLLME method reported for the simultaneous determination of multiple contaminants like aflatoxin B1, pesticides, fluoroquinolones, sulphonamides, and anthelmintics in biological chicken matrices.

2. Experimental

2.1. Chemicals, Reagents, and Materials

All reagents and solvents were analytical and LC-MS grade. The following individual standards (>97% purity) and internal standards (IS) used in this study were purchased from Sigma-Aldrich (Steinheim, Germany): aflatoxin B1, albendazole, atrazine, danofloxacin, enrofloxacin, fenbendazole, mebendazole, simazine, sulphachloropyridazine, sulphadiazine, sulphamerazine, sulphaquinoxaline, sulphapyridine, terbutryn, thiabendazole, Aflatoxin B1-13C17 (IS), Albendazole-d3 (IS), Atrazine-d5 (IS), Danofloxacin-(methyl-d3) (IS), Enrofloxacin-d5 (IS), Fenbendazole-d3 (IS), Mebendazole-d3 (IS), Simazine-d10 (IS), Sulfamethazine-d4 (IS), Sulfaquinoxaline-d4 (IS), Sulfapyridine-d4, Terbutryn-d5 (IS). Reagents used in the experiments were HPLC and LC-MS-grade solvents and were purchased from Sigma Aldrich (Steinheim, Germany). These include acetone, chlorobenzene (C_6H_5Cl), 1,1,2,2-tetrachloroethane (C_2Cl_4), HPLC grade, while methanol, acetonitrile (LC-MS grade), and formic acid were purchased from Romil Ltd. (Cambridge, UK). Ultrapure water (18.2 MΩ cm at 25 °C) was processed by the Milli-Q® Reference Water Purification System (Merck Millipore, Bedford, MA, USA). Nitrogen gas (N_2) of 99.9% purity was generated using a Genius 1022 Nitrogen Generator (Peak Scientific Inc., Billerica, MA, USA). The Pierce™ Calibration Solutions (Pierce™ ESI Negative Ion Calibration Solution and Pierce™ LTQ ESI Positive Ion Calibration Solution) (10 mL) were obtained from Thermo Fisher Scientific (Rockford, IL, USA). Table S1 lists the structures, CAS numbers, pKa values, and K_{ow} values (listed in the table as log P) of target analytes used in this study.

2.2. Preparation of Standard Solutions

Individual stock solutions of 1000 µg mL^{-1} of aflatoxin B1, SQ, SPD, SDZ, SCP, SMZ, DFX, ENR [24–26], SIMZ, TRB, ATR, TBZ, FBZ, ABZ, and MBZ were prepared by accurately weighing out standards using a Mettler Toledo XP6U Micro Comparator balance (Greifensee, Switzerland) and dissolving them in either ACN, MeOH, water, or dimethyl sulfoxide depending on the solubility of each compound. Isotopically labelled internal standards (IS) solutions of aflatoxin B1-13C17, Albendazole-d3, Atrazine-d5, Danofloxacin-(methyl-d3), Enrofloxacin-d5, Fenbendazole-d3, Mebendazole-d3, Simazine-d10, Sulfamethazine-d4, Sulfaquinoxaline-d4, Sulfapyridine-d4, Terbutryn-d5 were prepared separately but in the same way as the working standard solution mixture. The stock solutions of the individual standards were used to prepare working mixture solutions and calibration standard solutions.

Matrix-matched calibration standard solutions were prepared by spiking the blank liver samples with appropriate volumes of the target analyte working standards such that the concentration range was 5 to 500 µg kg^{-1} for pesticides, sulphonamides, fluoroquinolones, and anthelmintics and 0.5 to 50 µg kg^{-1} for aflatoxin. This was followed by adding isotope-labelled standards to each matrix matched calibration standard sample. The matrix-matched calibration standard samples were then treated with the DLLME procedure as described in Sections 2.4 and 2.5 below. The standard solutions were stored at a temperature of 2–8 °C until ready for analysis.

2.3. LC-HRMS Analysis

2.3.1. Mass Spectrometry

A Thermo Scientific™ Q Exactive™ Plus Hybrid Quadrupole-Orbitrap™ Mass Spectrometer coupled to a Thermo Scientific™ Dionex UltiMate™3000 UHPLC system (Thermo Fisher Scientific, Waltham, MA, USA) was used in this study. Detection was carried out using Exactive™ Plus LC-MS/MS equipped with a heated electrospray ionisation (HESI) probe in multiple reaction monitoring (MRM) mode. The optimum ion source conditions compatible with the HPLC flow rate were used (capillary temperature of 290 °C; sheath gas flow, 50 arbitrary units (AU); spray voltage, 3.5 kV; auxiliary temperature, 400 °C). Analysis was performed in full MS, single ion monitoring (SIM) mode, and all-ion fragmentation (AIF) in positive ion mode over a scan range from m/z 80 to 750 with a mass accuracy of <5 ppm. The mass spectrometer was operated at a mass resolution offset of 70,000 full width at half maximum (FWHM), with the automatic gain control (AGC) target set at 1.0×10^6 and a maximum injection time (IT) of 100 ms. The mass spectrometer was calibrated weekly for mass accuracy using Thermo Scientific™ Pierce™ Calibration Solutions (Pierce™ ESI Negative Ion Calibration Solution and Pierce™ LTQ ESI Positive Ion Calibration Solution).

2.3.2. HPLC Separation

A Waters® XBridge™ C18 (3.5 µm, 4.6 × 75 mm) column was used, with 0.1% (v/v) formic acid in water as mobile phase A and 0.1% (v/v) formic acid in acetonitrile as mobile phase B. Linear gradient elution was used for the separation, starting from 2% to 45% B in 6 min, then held for 0.5 min, and decreased again to 2% B in 1.5 min, followed by a re-equilibration time of 1 min for the next run; the total run time was 8 min, at a flow rate of 0.5 mL min^{-1}, a sample injection volume of 10 µL, and the column temperature was set at 35 °C. The data analysis and processing were carried out using the Qual/Quan Browser TraceFinder software package (TraceFinder, Thermo Fisher Scientific, Waltham, MA, USA).

2.4. Sampling and Sample Pre-Treatment

The chicken liver biological matrix was utilised for method optimisation and validation and was acquired from local supermarkets in Gauteng province, South Africa. In this study, chicken liver blank samples were obtained from non-commercial, informal organic farmers. Blank samples were screened and confirmed to be free of residues of the analytes of interest (fluoroquinolones, pesticides, sulphonamides, anthelmintics, and aflatoxin B1). The samples were wrapped in aluminium foil to protect them from photo-degradation and then stored at 4–8 °C. The procedure for the extraction of the analytes of interest from chicken livers consisted of sample pre-treatment and DLLME. The pre-treatment procedure was performed as reported by Moema et al. [21], with some modifications. Chicken livers were chopped up into smaller pieces and homogenised, using a food processor to produce a puree. Homogenised liver samples, i.e., 5 g of the homogenate were weighed into 50 mL Falcon tubes. The blank chicken liver samples were spiked with standard solutions and internal standard solutions of analytes (fluoroquinolones, pesticides, sulphonamides, anthelmintics, and aflatoxin B1). The samples were then treated with 5 mL of 0.08% HF:MeCN (15:85) and mixed on a vortex mixer for 30 s, then centrifuged for 10 min at 4000 rpm. Optimisation studies were carried out using various concentrations of formic acid. The supernatant (acetonitrile extract) was transferred into a vial, and a 1 mL aliquot of the acetonitrile was used for the DLLME procedure.

2.5. Dispersive Liquid-Liquid Microextraction

For the DLLME extraction procedure, 5 mL of UHP water was first placed into a 15 mL centrifuge tube (Figure S1). Thereafter, a mixture of 1 mL of the acetonitrile sample extract (used as a disperser solvent) obtained from the earlier procedure and 400 µL of tetrachloroethane (the extraction solvent) were rapidly injected into the tube using a 5 mL syringe with a sharp needle to induce the formation of a cloudy solution that results from the dispersion of fine droplets of the extraction solvent in the water sample. The content was then centrifuged at 4000 rpm for 10 min to separate the organic phase (sedimented bottom layer) from the aqueous phase (upper layer). The lower/organic phase was withdrawn with a microsyringe and transferred into a 1.5 mL vial for evaporation of the solvent under nitrogen. The residue was reconstituted in 500 μL of the mobile phase and injected into the LC-MS/MS for analysis.

2.6. Solid Phase Extraction

A solid-phase extraction technique was adopted by Zhang et al. (2018) with some modifications [24]. A chicken liver sample (5.0 $\pm$ 0.01 g) was weighed into a 50-mL conical centrifuge tube. An amount of 5 mL of 0.08% formic acid: MeCN (15:85) was added and the tube was vortexed for 10 s. The sample was sonicated for about 30 min, then centrifuged at 4000 rpm for 20 min. The resulting supernatant solution was slowly transferred into a 15-mL centrifuge tube. A Waters Oasis HLB cartridge (12 cc, 500 mg) was set up for pass-through filtration. Before extraction, each Waters HLB cartridge was pre-conditioned with 3 mL of methanol and then rinsed with 3 mL of deionized water on an SPE manifold. An extract sample was then passed through the HLB cartridge. After extraction, the cartridge was washed with 1 mL of 5% methanol in water and subsequently air-dried under a vacuum for at least 20 min. The residues were then eluted from the cartridge with 2 portions of 5 mL of MeCN (LC grade). All the extracts were completely evaporated to dryness by a gentle stream of nitrogen. The dried sample under a gentle stream of nitrogen was followed by reconstitution in 500 μL of acetonitrile, and 10-μL aliquots were injected into HR-LC MS/MS system.

3. Results and Discussion

3.1. Mass Spectrometry Optimisation

The mass spectrometry (MS) method development was conducted through the introduction of target analytes into the mass spectrometry system via direct infusion of the standard solutions. All the target analytes and internal standards were found to be more sensitive in the positive ion mode. Previous studies confirmed the presence of the precursor ion $[M+H]^+$ of aflatoxin B1 [25], sulphonamides (SQ, SPD, SDZ, SCP, and SMZ) [26–30], fluoroquinolones (DFX and ENR) [31–33], pesticides (SIMZ, TRB, ATR, and TBZ) [34,35], and anthelmintics (FBZ, ABZ, and MBZ) [34,35]. Solutions of individual analytes in acetonitrile containing 0.1% (v/v) formic acid were infused at a flow rate of 10 $\mu L\ min^{-1}$ to determine the MS conditions for each analyte. Multiple reaction monitoring (MRM) was used, and a summary of precursor and product ions and collision energies for each compound is presented in Table 2. The developed separation method (Figure S4) in combination with MRM conditions generated typical ion chromatograms of the mixture of mixed contaminants (Figure S5).

3.2. Optimisation of DLLME Conditions

The DLLME extraction method was optimised for aflatoxin B1, pesticides, fluoroquinolones, sulphonamides, and anthelmintics in the chicken livers. The biological matrix extraction method was adopted from Moema et al. [21]. Various parameters that affect extraction efficiency, such as the type and volume of extraction solvents and the type and volume of disperser solvents, and the sample pH, were optimised. In DLLME extraction recovery (ER) was used to evaluate the extraction efficiency to obtain optimized extraction conditions. The extraction recovery was calculated: $ER = \{C_{sed} \times V_{sed}\}/\{V_o \times V_{eq,}\} \times 100$ where: (C_{sed}) and (C_0) initial concentrations of analytes within the sample and concentrations in the sediment phase, and V_{sed} and V_{aq} are the volumes of the sediment phase and sample solution, respectively. The blank chicken liver samples were spiked with the analytes at a concentration of 100 $\mu g\ kg^{-1}$ for pesticides, fluoroquinolones, sulphonamides, and anthelmintics, and at a concentration of 10 μg kg^{-1} for aflatoxin B1, and treated with 0.08% formic acid in water and acetonitrile (MeCN). The effect of the percent composition of formic acid in water/acetonitrile on the extraction efficiencies for mixed contaminants from blank chicken liver samples was investigated. The best ratio of formic acid in water (HF) and acetonitrile (MeCN) was found to be 15:85 (Figure 1). This solution was further used as the disperser solvent. Five mL of UHP water were placed into the Falcon tube, and the disperser and extraction solvents were rapidly added to the tube. For all the optimisation studies, peak areas were used to evaluate the extraction efficiencies.

3.2.1. Selection of Disperser Solvent

The disperser solvent is one of the ternary solvents that plays a key role in the DLLME extraction process, and it is a very important parameter to be optimise. It is a prerequisite that it should be miscible with both the aqueous and organic phases. Additionally, it is necessary that the disperser solvents disperse the extraction solvent into very fine droplets in the aqueous sample to increase contact area; the increased surface area of the droplets assists in the instantaneous partitioning of the analytes from the aqueous phase into the organic phase. The disperser solvent is critical for the formation of cloudiness due to the presence of fine droplets of extraction solvent dispersed throughout the aqueous phase. In this, acetone, acetonitrile, and methanol were evaluated as possible disperser solvents. The results showed the best extraction efficiency from the tested solvents in acetonitrile (Figure 2). Aflatoxin B1 was found to be most amenable to extraction, with recovery efficiencies ranging from 26–41%. This observation was very interesting considering the diversity of the analytes in this work and can be attributed to the high compatibility of acetonitrile with the aqueous solution in comparison with acetone and methanol. Therefore, acetonitrile was selected as the disperser solvent for all further experiments.

Table 2. Multiple reaction monitoring (MRM) conditions.

Analyte	Precursor Ion $[M + H]^+$, *m/z*	Quantifier Ion, *m/z*	Qualifier Ion, *m/z*	Collision Energies
Aflatoxin B_1	313.0707	285.0755	269.9604	28
Aflatoxin B1-13C17 (IS)	330.1836	234.0700	191.0152	28
Albendazole	266.0955	234.0702	264.0775	15
Albendazole-d3 (IS)	269.1146	234.0691	191.0152	13
Atrazine	216.1011	216.1017	174.0547	10
Atrazine-d5 (IS)	221.1324	216.1015	221.1328	10
Danofloxacin	358.1562	258.1568	81.01760	25
Danofloxacin-(methyl-d3) (IS)	361.1749	360.1725	316.1826	30
Enrofloxacin	360.1718	245.1725	202.0438	37
Enrofloxacin-d5 (IS)	365.2032	360.1725	202.0438	20
Fenbendazole	300.0801	186.0814	242.1442	10
Fenbendazole-d3 (IS)	303.0989	186.0812	242.1438	11
Mebendazole-d3 (IS)	299.1211	264.0773	265.0799	13
Mebendazole	296.1025	264.0764	296.1035	10
Simazine	202.0854	202.0858	84.0015	10
Simazine-d10 (IS)	212.1482	212.1486	313.0712	25
Sulphachloropyridazine	285.0208	156.0116	108.0448	10
Sulphadiazine	251.0597	251.0603	108.0449	10
Sulphamerazine	265.0754	156.0759	108.0449	13
Sulfamethazine-d4 (IS)	283.1161	283.1167	112.0700	13
Sulphaquinoxaline	301.0754	156.0100	301.0760	10
Sulfaquinoxaline-d4 (IS)	305.1005	305.2044	301.0760	13
Sulphapyridine	250.0645	250.0650	108.0449	10
Sulfapyridine-d4 (IS)	254.0896	254.0894	250.0644	10
Terbutryn	242.1429	186.0804	242.1438	10
Terbutryn-D5 (IS)	247.1748	189.0995	245.1621	10

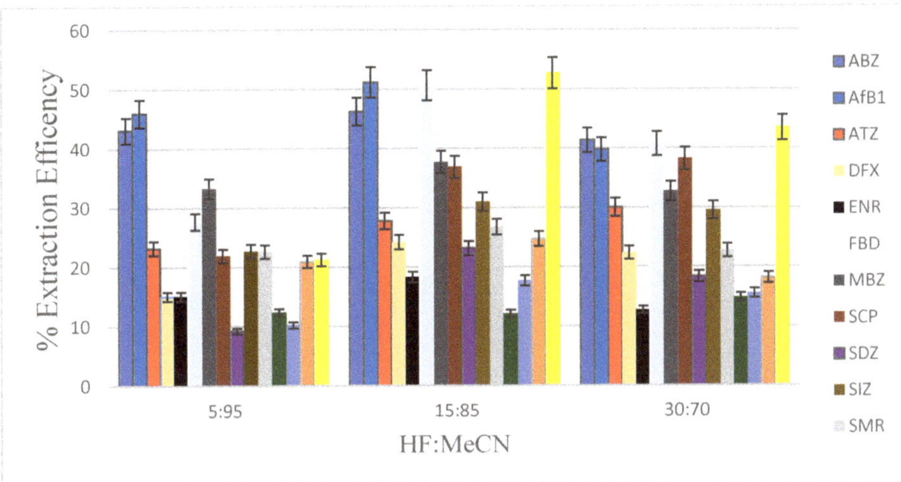

Figure 1. The extraction efficiencies of mixed contaminants (concentration of 100 μg kg^{-1} was used for all sulphonamides, fluoroquinolones, pesticides, anthelmintics; concentration of 10 μg kg^{-1} for aflatoxin) from spiked blank chicken liver biological matrix.

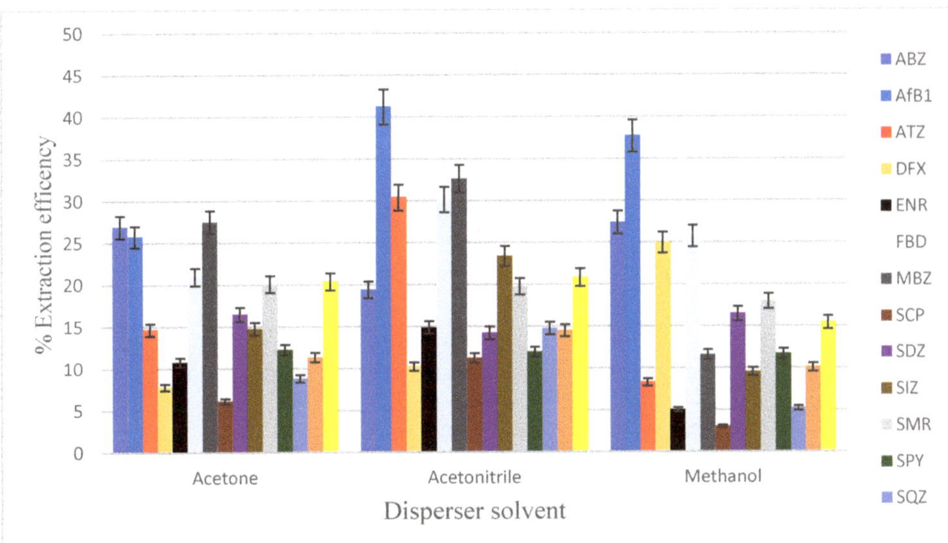

Figure 2. Effect of types of disperser solvent on extraction efficiency in the extraction of mixed contaminants (concentration of 100 μg kg^{-1} for sulphonamides, fluoroquinolones, pesticides, anthelmintics; concentration of 10 μg kg^{-1} for aflatoxin B1) in DLLME. Extraction conditions: 5 mL of UHP water; 400 μL of extraction solvent (tetrachloroethane); 1000 μL of disperser solvent (acetone, acetonitrile, methanol).

3.2.2. Extraction Solvent Selection

The selection of an appropriate extraction solvent is also important for the successful and efficient use of the DLLME. In the conversional DLLME, the extraction solvent has a density that is higher than that of water [36]. This facilitates the separation via centrifugation of the extractant analyte from the aqueous environment. In addition, the solvent must have good extraction capability for the compounds of interest, good chromatographic behaviour, and miscibility with water [37]. In this study, three organic solvents, with their densities given in parentheses, including tetrachloroethylene (1.62 g cm^{-3}), chloroform (1.49 g cm^{-3}), and tetrachloroethane (1.59 g cm^{-3}), were investigated as potential extraction solvents. Spiked chicken liver biological matrix samples were exposed to these different extraction solvents according to the procedure given in the experimental section (Section 2.5). Figure 3 clearly shows that tetrachloroethane was the most efficient extraction solvent in comparison to the other two solvents. Therefore, tetrachloroethane was used for further extraction work.

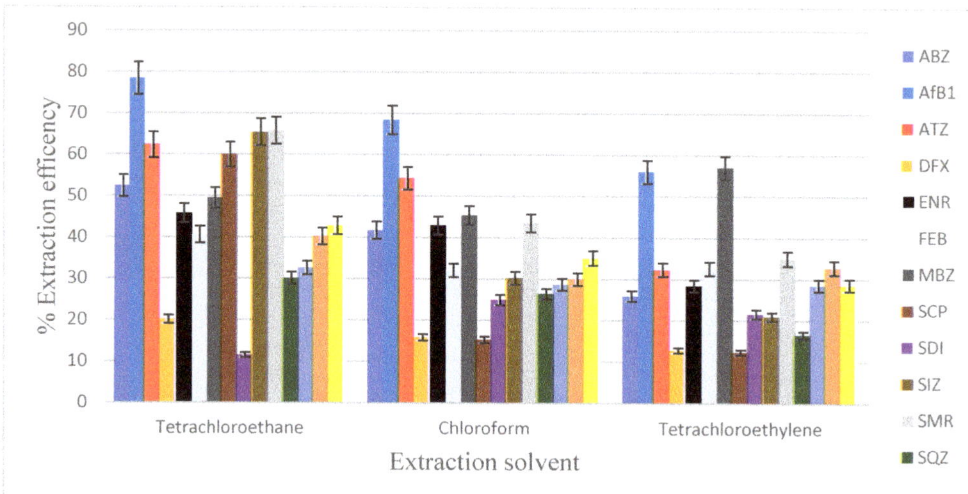

Figure 3. Effect of types of extraction solvent on extraction efficiency in extracting mixed contaminants (concentration of 100 μg kg^{-1} for sulphonamides, fluoroquinolones, pesticides, anthelmintics; concentration of 10 μg kg^{-1} for aflatoxin B1) in DLLME. Extraction conditions: 5 mL of UHP water; 400 μL of extraction solvent (tetrachloroethane); 1000 μL of disperser solvent (acetone, acetonitrile, methanol).

3.2.3. Optimisation of pH of the Sample

The pH is important in microextraction techniques, and in this study, pH optimisation was carried out. The distribution ratio of target analytes between the aqueous and organic phases is pH-dependant. Therefore, optimisation of pH is an attempt to get as many of the analytes in complex biological matrices into an extractable form [38–40], especially considering the diversity of the analytes of interest in this study. For example, amphoteric sulphonamides, with their pKa values in the pH ranges of 1.97–2.14 and 4.3–6.99, tend to exist as anionic forms in alkaline solutions [39,40]. The quinolone class of antibiotics is also amphoteric, with pKa values ranging from 5.63–6.73 and 5.69–6.68). In addition, multiple forms (cationic, anionic, zwitterionic, and neutral) could be expected in basic donor solutions [40,41]. The target pesticides are also ionisable.

In this study, the effect of sample pH on the extraction efficiencies for the target analytes was evaluated by varying the pH values from 5 to 10 using NaOH and HF (Figure 4). It was observed that all target analytes were extracted above 11% within the pH range of extraction. At pH ranges between 6 and 7, it was observed that most compounds were extracted with efficiencies > 50%. Extraction efficiencies of more than 50% for eight compounds were achieved at pH 6, while 11 compounds were extracted at efficiencies > 60% at pH 7 (Figure 4). The optimum pH for the extraction of analytes of interest was pH 7, where five analytes (ABZ, ATZ, FBD, SIZ, and AFB1) were extracted with efficiencies > 80%. This observation was an indication that the neutral forms of the analytes were more efficiently extractable than their ionic forms. In addition, it appears the stability of the target analytes was maintained in weakly acidic and alkaline environments. The degradation of target analytes in strongly acidic and alkaline conditions that has been suggested [38–40] did not impact this study. For the extraction of analytes of interest, the sample pH was therefore maintained at pH 7.

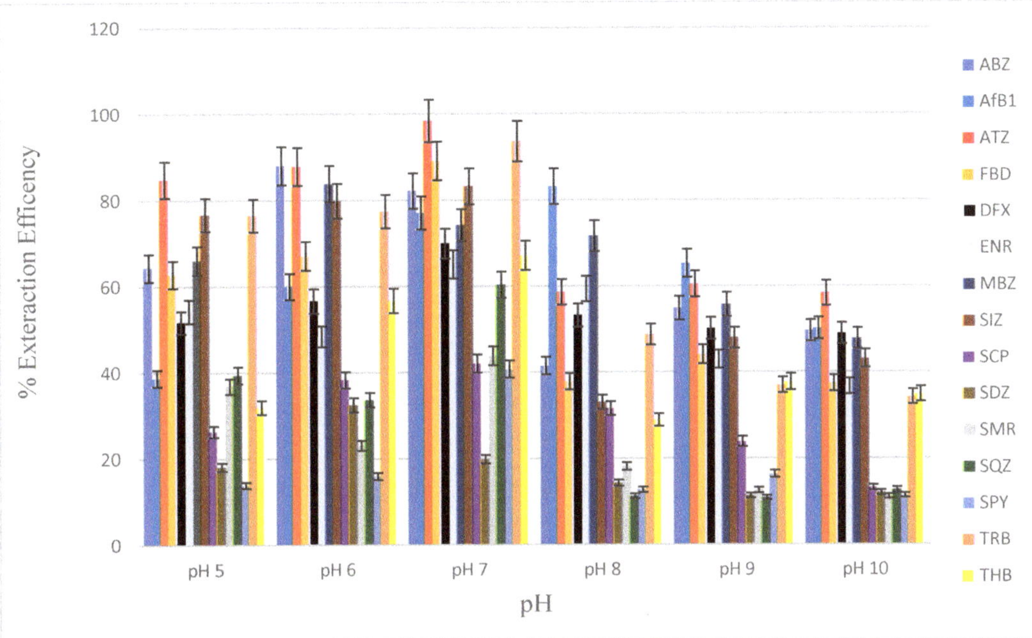

Figure 4. Effect of sample pH on extraction efficiency in extracting mixed contaminants (concentration of 100 µg kg^{-1} for sulphonamides, fluoroquinolones, pesticides, anthelmintics; concentration of 10 µg kg^{-1} for aflatoxin B1) in DLLME. Extraction conditions: 5 mL of UHP water; 400 µL of extraction solvent (tetrachloroethane), 1000 µL of disperser solvent (acetonitrile).

3.2.4. The Effect of Extraction Solvent Volume

The volume of extraction solvent is an important parameter and has a major effect on the extraction efficiency [42,43]. As the extraction solvent volume increased, dilution of the analyte was observed. The extraction efficiency remains constant, resulting in a decrease in the sensitivity of the determination for the target compounds [44,45]. In this work, the effect of extraction solvent volume was investigated by varying the volume from 100 to 500 µL, while maintaining all other parameters constant. The results showed that with an increase in the solvent volume, an increase in the extraction efficiencies was observed, up to a maximum at 400 µL (Figure S2). The results also showed that at volumes >400 µL there was a slight decrease in the extraction efficiencies. The observed decrease was due to the dilution of the analytes caused by the increase in the volume of the organic (sedimented) phase. The extraction efficiencies ranged from 22–95% for all the analytes, with an extraction efficiency of >60% observed for 10 compounds and >78% for eight compounds. Hence, 400 µL of extraction solvent volume was used for all subsequent experiments.

3.2.5. The Effect of Disperser Solvent Volume

The cloudiness of the solution, the degree of dispersion of the extraction solvent in the aqueous phase, and the extraction efficiency are dependent on the disperser solvent volume [44]. The effect of the volume of the disperser solvent, i.e., acetonitrile, on the extraction efficiency was investigated by varying the volume from 250 to 1500 μL, while maintaining all other parameters constant. The results showed an increase in extraction efficiencies was observed with an increase from 500 to 1000 μL, followed by a decrease in extraction efficiencies at volumes >1000 μL (Figure S3). The results showed that extraction efficiencies of 38 to 96% for all analytes were achieved at a MeCN volume of 1000 μL. It was observed that 11 compounds had extraction efficiency >59%, whereas seven compounds had an extraction efficiency >79% [45]. This is because at significantly high volumes, the volume of the sedimented phase is increased, thus lowering the partitioning of the analytes into tetrachloroethane [46]. Similarly, when the disperser solvent volume was too low, the cloudiness was low, and therefore the recovery of analytes was also low. However, when the disperser solvent volume was too high, the solubility of the analytes in the aqueous phase increased, and therefore the extraction efficiency decreased; the decrease in distribution coefficient (D) or partition coefficient (P) plays a role here. The optimum volume of the disperser solvent was taken to be 1000 μL.

3.3. Method Validation

As defined by the ISO/IEC 17025 guideline, validation is "the confirmation by examination and the provision of objective evidence that the particular requirements for a specific intended use are fulfilled" [47]. In-house method validation was performed to investigate the applicability of the proposed method according to the main reference documents, namely Commission Decision (EC) No. 2002/657/EC, SANTE/12682/2019 Guidance Document on analytical quality control and method validation procedures for pesticide residues and analysis in food and feed, and the Eurachem Guide: The Fitness for Purpose of Analytical Methods [48–50].

Using the optimum extraction conditions of the developed method, the limits of detection, quantification, linearity, recovery, and precision were investigated using spiked liver samples. The linearity of the method was assessed using nine-point matrix-matched calibration curves. It should be noted that at each point, a maximum of three replicates were used. An internal standard was added to each sample. The blank liver samples were spiked with the target analytes in the range from 5 to 500 μg kg^{-1} for pesticides, sulphonamides, fluoroquinolones, and anthelmintics, and in the range of 0.5 to 50 μg kg^{-1} for aflatoxin. Calibration curves were constructed using an analyte/internal standard peak area ratio vs. concentration of analyte. The least-squares regression equations from the calibration curves were used to calculate the limit of detection (LOD) and limit of quantification (LOQ). LODs and LOQs were calculated from the calibration data and regression statistics using the formulas $X_{LOD} = 3\ S_{y/x}/b$ and $X_{LOQ} = 10\ S_{y/x}/b$, respectively, where Sy/x is the standard error and b is the slope of the regression line [51]. The LODs ranged from 0.03 to 1.197 μg kg^{-1}, whereas the LOQs ranged from 0.036 to 2.99 μg kg^{-1}. Table 3 shows the calibration curve data, together with the coefficient of determination (R^2), LODs, and LOQs. The calibration curves gave good linearity, at various ranges, with the coefficient of determination (R^2) ≥ 0.9916 for all target analytes.

In any extraction method, the recovery is an important parameter to assess its suitability. In this work, three concentration levels ranging from 20, 50, and 200 μg kg^{-1} for fluoroquinolones, pesticides, sulphonamides, and anthelmintics and from 2, 5, and 20 μg kg^{-1} for aflatoxin B1 were used for recovery studies. The precision expressed as a relative standard deviation (RSD) of the method was also calculated at these concentration levels. Five replicates of each sample were analysed within a day (intra-day precision), and the procedure was repeated over three consecutive days (inter-day precision) (Table 4).

Table 3. Analytical performance parameters for the determination of mixed contaminants in chicken liver samples using DLLME method.

Compound	Linear Range (μg kg^{-1})	R^2	LOD (μg kg^{-1})	LOQ (μg kg^{-1})
Aflatoxin B1	0.5–50	0.9938	0.026	0.086
Albendazole	5–500	0.9939	0.263	0.876
Atrazine	5–500	0.9967	0.292	0.975
Danofloxacin	5–500	0.9925	0.665	2.172
Enrofloxacin	5–500	0.9916	0.774	2.579
Mebendazole	5–500	0.9926	0.109	0.364
Fenbendazole	5–500	0.9949	0.026	0.085
Simazine	5–500	0.9916	0.011	0.036
Sulphaquinoxaline	5–500	0.9927	0.234	0.779
Sulphadiazine	5–500	0.9916	0.465	1.548
Sulphamerazine	5–500	0.9976	0.166	0.597
Sulphachloropyridazine	5–500	0.9917	1.197	2.99
Sulphapyridine	5–500	0.9925	0.278	0.927
Terbutryn	5–500	0.9916	0.040	0.150
Thiabendazole	5–500	0.9947	0.274	0.913

3.4. Comparison of DLLME with SPE

In this work, a paired *t*-test was used to compare the developed and validated DLLME method with the standard SPE method (Waters Oasis HLB cartridge) to assess if the methods were significantly different. For this purpose, the comparison was done using mean recoveries at three concentration levels of mixed contaminant analytes ranging from 20, 50, and 200 μg kg^{-1} for fluoroquinolones, pesticides, sulphonamides, and anthelmintic, and 2, 5, and 20 μg kg^{-1} for aflatoxin B1. Table 5 shows the results of comparison of recoveries using the two extraction methods.

The calculated t-value for all analytes is less than the t-critical value of 2.45, indicating that the results obtained by DLLME and SPE methods do not differ significantly in terms of accuracy. Furthermore, of the two methods, DLLME has the advantages of very short extraction times and the fact that the extraction equilibrium is attained very quickly (a few seconds) compared to the SPE method. Overall, DLLME has the advantages of being very simple, rapid, inexpensive, easy to use, benign to the environment, and not involving any labour-intensive steps compared to solid phase extraction.

3.5. Quantification of Mixed Contaminants

The optimised and validated DLLME method was applied to real samples obtained from several local supermarkets in Gauteng province, South Africa. Processed by DLLME extraction and analytes quantified by high resolution LC-MS/MS as described above (Sections 2.3 and 2.4).

Table 6 shows a summary of determination of the mixed contaminants in 12 different chicken liver samples. Aflatoxin B1, SCP, and ENR were detected in five of the 12 chicken liver samples assayed. The other mixed contaminants were not detected in any of the samples, which might be due to the fact that the concentrations of target analytes in the samples were below the limit of detection (LOD) and limit of quantification (LOQ). All the mixed contaminants detected were found to be below the stipulated South African MRL range and EU MRL range, which might be an indication that proper withdrawal times were observed by farmers. However, since this was a small sample population, more samples still need to be analysed for confirmation.

Table 4. Recovery and precision of method to determine mixed contaminants in chicken liver samples spiked at three different concentration levels.

Target Compounds	Added (μg kg^{-1})	Detection	% Recovery	Precision	
				Intra-Day % RSD (n = 18)	Inter-Day % RSD (n = 54)
Aflatoxin B1	2	1.80	90.00	6.57	5.78
	5	4.70	940.0	7.21	6.23
	20	19.00	95.00	2.38	4.52
Albendazole	20	17.77	88.84	4.75	3.87
	50	44.94	89.88	6.22	7.22
	200	158.80	79.40	2.40	4.89
Atrazine	20	17.40	87.00	2.21	3.87
	50	47.80	95.60	1.77	2.77
	200	174.20	87.10	7.45	8.98
Danofloxacin	20	19.70	98.50	6.66	6.75
	50	43.70	87.40	7.14	4.23
	200	168.60	84.30	2.50	5.52
Enrofloxacin	20	19.45	97.25	4.34	8.23
	50	41.75	83.50	8.82	8.66
	200	176.85	88.43	3.01	6.20
Fenbendazole	20	19.54	97.71	6.23	4.44
	50	19.54	39.09	6.23	6.23
	200	182.59	91.30	3.57	4.66
Mebendazole	20	18.43	92.17	6.34	2.56
	50	47.57	95.13	6.05	7.52
	200	166.70	83.35	5.25	6.47
Simazine	20	17.17	85.83	5.54	6.57
	50	48.12	96.24	5.87	7.21
	200	190.02	95.01	7.25	2.38
Sulphachloropyridazine	20	17.81	89.05	5.42	4.75
	50	44.18	88.37	6.49	6.22
	200	164.39	82.20	4.42	2.40
Sulphadiazine	20	18.36	91.79	2.57	2.21
	50	40.41	80.83	2.45	1.77
	200	182.76	91.38	4.34	9.45
Sulphamerazine	20	17.25	86.25	5.53	6.66
	50	43.45	86.90	8.00	7.14
	200	176.85	88.43	8.61	2.50
Sulphaquinoxaline	20	16.17	80.87	2.81	4.34
	50	40.52	81.04	3.09	8.82
	200	173.57	86.78	2.86	3.01
Sulphapyridine	20	16.87	84.33	4.33	6.23
	50	46.18	92.37	5.93	5.23
	200	162.85	81.43	3.07	3.57
Terbutryn	20	16.16	80.82	1.53	6.34
	50	40.39	80.78	2.30	6.05
	200	167.33	83.66	2.27	6.57
Thiabendazole	20	19.07	95.35	1.85	7.21
	50	44.72	89.44	8.53	5.38
	200	186.82	93.41	4.83	6.52

Table 5. Comparison of DLLME with SPE method using paired *t*-test, mean recoveries of each of mixed contaminants were compared at three different concentration levels.

Target Compounds	Standard Deviation (SD)	Standard Error of the Mean (SE)	t-Value	t-Critical (0.05)
Aflatoxin B1	4.44	2.56	1.55	2.45
Albendazole	1.06	0.61	−3.17	2.45
Atrazine	1.31	0.76	−0.64	2.45
Danofloxacin	2.99	1.73	−014	2.45
Enrofloxacin	2.99	1.73	−2.54	2.45
Fenbendazole	2.95	1.71	−0.29	2.45
Mebendazole	2.78	1.61	0.78	2.45
Simazine	6.81	3.93	−0.59	2.45
Sulfachloropyridazine	2.66	1.53	−1.49	2.45
Sulfadiazine	1.67	0.96	2.14	2.45
Sulfamerazine	1.98	1.14	−2.81	2.45
Sulfaquinoxaline	4.02	2.32	2.08	2.45
Sulphapyridine	1.87	1.08	−2.27	2.45
Terbutryn	1.28	0.74	0.73	2.45

Table 6. Detection of fluoroquinolones, pesticides, sulphonamides, anthelmintics and aflatoxin B1 in chicken liver samples.

Sample	Analytes														
	AFB1	ABZ	ATZ	DFX	ENR	MEB	FEB	SIZ	SCP	SDZ	SMR	SPD	SQ	TER	TBZ
A	<LOD	<LOD	<LOD	<LOD	<LOD	<LOD	<LOD	<LOD	<LOD	<LOD	<LOD	<LOD	<LOD	<LOD	<LOD
B	<LOD	<LOD	<LOD	<LOD	<LOD	<LOD	<LOD	<LOD	<LOD	<LOD	<LOD	<LOD	<LOD	<LOD	<LOD
C	<LOD	<LOD	<LOD	<LOD	41.02	<LOD	<LOD	<LOD	<LOD	<LOD	<LOD	<LOD	<LOD	<LOD	<LOD
D	0.23	<LOD	<LOD	<LOD	<LOD	<LOD	<LOD	<LOD	92.11	<LOD	<LOD	<LOD	<LOD	<LOD	<LOD
E	<LOD	<LOD	<LOD	<LOD	<LOD	<LOD	<LOD	<LOD	<LOD	<LOD	<LOD	<LOD	<LOD	<LOD	<LOD
F	<LOD	<LOD	<LOD	<LOD	<LOD	<LOD	<LOD	<LOD	26.50	<LOD	<LOD	<LOD	<LOD	<LOD	<LOD
G	0.94	<LOD	<LOD	<LOD	31.85	<LOD	<LOD	<LOD	<LOD	<LOD	<LOD	<LOD	<LOD	<LOD	<LOD
H	<LOD	<LOD	<LOD	<LOD	<LOD	<LOD	<LOD	<LOD	<LOD	<LOD	<LOD	<LOD	<LOD	<LOD	<LOD
I	<LOD	<LOD	<LOD	<LOD	<LOD	<LOD	<LOD	<LOD	<LOD	<LOD	<LOD	<LOD	<LOD	<LOD	<LOD
J	<LOD	<LOD	<LOD	<LOD	<LOD	<LOD	<LOD	<LOD	<LOD	<LOD	<LOD	<LOD	<LOD	<LOD	<LOD
K	<LOD	<LOD	<LOD	<LOD	<LOD	<LOD	<LOD	<LOD	<LOD	<LOD	<LOD	<LOD	<LOD	<LOD	<LOD
L	<LOD	<LOD	<LOD	<LOD	<LOD	<LOD	<LOD	<LOD	<LOD	<LOD	<LOD	<LOD	<LOD	<LOD	<LOD

<LOD (Limit of detection), Concentration in μg kg^{-1}.

4. Conclusions

The presented DLLME sample preparation method was successful in efficiently extracting a rather complex mixture of food contaminants, which are usually challenging due to their diverse chemical properties. The method was successfully developed, validated, and applied to a biological food matrix such as chicken livers. The developed method was validated with a coefficient of determination (R^2) range of 0.9916–0.9967, LOD and LOQ of 0.03 μg kg^{-1} and 0.09 μg kg^{-1}, respectively, for aflatoxin B1, and LOD and LOQ for pesticides, fluoroquinolones, sulphonamides, and anthelmintics that ranged from 0.011–1.197 μg kg^{-1} and 0.150–2.579 μg kg^{-1}, respectively. The mean recoveries were in the range of 80.4–96.3%, and the relative standard deviations (RSDs) were in the range of 1.53–8.98%. Comparison with standard SPE methods shows that DLLME provides acceptable accuracy and thus could be considered as an alternative fast, simpler, and green method for the extraction of multiclass contaminants in food matrices. Therefore, the newly developed method could be used as a routine method for the determination of mixed contaminants in chicken liver samples due to its advantages over other methods.

Supplementary Materials: The following supporting information can be downloaded at: https://www.mdpi.com/article/10.3390/foods12132594/s1, Figure S1: Effect of volume of extraction solvent on extraction efficiency in DLLME. Figure S2: Effect of volumes of disperser solvent on extraction efficiencies in DLLME; Figure S3: Effect of volumes of disperser solvent on extraction efficiencies in DLLME; Figure S4: Chromatogram of mixed contaminants (sulphonamides, fluoroquinolones, pesticides, anthelmintic and aflatoxin B1); Figure S5: Selected ion Chromatograms for individual compounds; Table S1: Recoveries and RSDs of target compounds at different spiked levels in chicken liver using solid phase extraction; Table S2: Physicochemical properties of fluoroquinolones, pesticides, sulphonamides, anthelmintics and aflatoxin B1 [52].

Author Contributions: Conceptualization of research, M.M.N. and S.D.; methodology, B.E.G.; validation, B.E.G.; formal analysis, B.E.G.; investigation, B.E.G.; resources, M.M.N. and S.D.; writing—original draft preparation, B.E.G.; writing—review and editing, M.M.N. and S.D. supervision, M.M.N. and S.D.; project administration, M.M.N.; funding acquisition, M.M.N. and S.D. All authors have read and agreed to the published version of the manuscript.

Funding: BEG was funded by the UNISA Postgraduate Bursary.

Data Availability Statement: The datasets generated for this study are available on request to the corresponding author.

Acknowledgments: The authors would like to thank the Chemistry Department and the Unisa Postgraduate Bursary for their support of BEG. We would like to thank our Botswana National Veterinary Laboratory colleagues for their assistance with the initial standards of this project.

Conflicts of Interest: The authors declare no conflict of interest.

References

1. Fung, F.; Wang, H.S.; Menon, S. Food safety in the 21st century. *Biomed. J.* **2018**, *41*, 88–95. [CrossRef] [PubMed]
2. Bogialli, S.; Di Corcia, A. Matrix solid-phase dispersion as a valuable tool for extracting contaminants from foodstuffs. *J. Biochem. Biophys. Methods* **2007**, *70*, 163–179. [CrossRef] [PubMed]
3. De Brabander, H.F.; Noppe, H.; Verheyden, K.; Bussche, J.V.; Wille, K.; Okerman, L.; Vanhaecke, L.; Reybroeck, W.; Ooghe, S.; Croubels, S. Residue analysis: Future trends from a historical perspective. *J. Chromatogr. A* **2009**, *1216*, 7964–7976. [CrossRef]
4. Villa, P.; Markaki, P. Aflatoxin B1 and ochratoxin A in breakfast cereals from athens market: Occurrence and risk assessment. *Food Control* **2009**, *20*, 455–461. [CrossRef]
5. Shankar, B.P.; Prabhu, B.H.M.; Chandan, S.; Ranjith, D.; Shivakumar, V. Rapid methods for detection of veterinary drug residues in meat. *Vet. World* **2010**, *3*, 241–246. [CrossRef]
6. Antignac, J.P.; Courant, F.; Pinel, G.; Bichon, E.; Monteau, F.; Elliott, C.; Le Bizec, B. Mass spectrometry-based metabolomics applied to the chemical safety of food. *TrAC Trends Anal. Chem.* **2011**, *30*, 292–301. [CrossRef]
7. Castro-Puyana, M.; Herrero, M. Metabolomics approaches based on mass spectrometry for food safety, quality and traceability. *TrAC Trends Anal. Chem.* **2013**, *52*, 74–87. [CrossRef]
8. Gallart-Ayala, H.; Moyano, E.; Galceran, M.T. Liquid chromatography/tandem mass spectrometry (highly selective selected reaction monitoring) for the analysis of isopropylthioxanthone in packaged food. *J. Chromatogr. A* **2008**, *1208*, 182–188. [CrossRef]
9. European Commission (EC). Corrigendum to Regulation (EC) No 1181/2002 of 1 July 2002, amending Annex 1 of Council Regulation (EEC) No 2377/90 laying down a Community procedure for the establishment of maximum residue limits of veterinary medicinal products in. *J. Eur. Comm.* **2002**, *L172*, 13–20.
10. South Africa Goverment Gazette. Foodstuffs, Cosmetics and Disinfectants Act 1972, Regulations Governing the Maximum Limits for Veterinary Medicine and Stock Remedy Residues that May Be Present in Foodstuffs: Amendment. Regulations No. R1543 of 2002. 2002. Available online: https://www.gov.za/ (accessed on 14 March 2022).
11. EC. Regulation No 396 2005, European Union Regulation (EC) 396/2005 of the European parliament and of the council of 23 February 2005 on maximum residue levels of pesticides in or on food and feed of plant and animal origin and amending Council Directive 91/414/EEC. *Off. J. Eur. Communities* **2005**, *L269*, 1–15.
12. Mol, H.G.J.; Plaza-Bolaños, P.; Zomer, P.; De Rijk, T.C.; Stolker, A.A.M.; Mulder, P.P.J. Toward a generic extraction method for simultaneous determination of pesticides, mycotoxins, plant toxins, and veterinary drugs in feed and food matrixes. *Anal. Chem.* **2008**, *80*, 9450–9459. [CrossRef] [PubMed]
13. Frenich, A.G.; Romero-González, R.; del Mar Aguilera-Luiz, M. Comprehensive analysis of toxics (pesticides, veterinary drugs and mycotoxins) in food by UHPLC-MS. *TrAC Trends Anal. Chem.* **2014**, *63*, 158–169. [CrossRef]
14. MGómez-Pérez, L.; Romero-González, R.; Martínez, V.J.L.; Frenich, A.G. Analysis of pesticide and veterinary drug residues in baby food by liquid chromatography coupled to Orbitrap high resolution mass spectrometry. *Talanta* **2015**, *131*, 1–7. [CrossRef] [PubMed]
15. Amate, C.F.; Unterluggauer, H.; Fischer, R.J.; Fernández-Alba, A.R.; Masselter, S. Development and validation of a LC-MS/MS method for the simultaneous determination of aflatoxins, dyes and pesticides in spices. *Anal. Bioanal. Chem.* **2010**, *397*, 93–107. [CrossRef]

16. PPérez-Ortega; Gilbert-López, B.; García-Reyes, J.F.; Ramos-Martos, N.; Molina-Díaz, A. Generic sample treatment method for simultaneous determination of multiclass pesticides and mycotoxins in wines by liquid chromatography-mass spectrometry. *J. Chromatogr. A* **2012**, *1249*, 32–40. [CrossRef]
17. Zhan, J.; Yu, X.J.; Zhong, Y.Y.; Zhang, Z.T.; Cui, X.M.; Peng, J.F.; Feng, R.; Liu, X.T.; Zhu, Y. Generic and rapid determination of veterinary drug residues and other contaminants in raw milk by ultra performance liquid chromatography-tandem mass spectrometry. *J. Chromatogr. B Anal. Technol. Biomed. Life Sci.* **2012**, *906*, 48–57. [CrossRef]
18. Armenta, S.; Esteve-Turrillas, F.A.; Garrigues, S.; de la Guardia, M. Green Analytical Chemistry: The Role of Green Extraction Techniques. *Compr. Anal. Chem.* **2017**, *76*, 1–25. [CrossRef]
19. O'Mahony, J.; Clarke, L.; Whelan, M.; O'Kennedy, R.; Lehotay, S.J.; Danaher, M. The use of ultra-high pressure liquid chromatography with tandem mass spectrometric detection in the analysis of agrochemical residues and mycotoxins in food-Challenges and applications. *J. Chromatogr. A* **2013**, *1292*, 83–95. [CrossRef]
20. Deng, K.J.; Lan, X.H.; Sun, G.; Ji, L.Y.; Zheng, X.L. Determination of Sulfonamide Residues in Chicken Liver Using High-Performance Liquid Chromatography. *Food Anal. Methods* **2016**, *9*, 3337–3344. [CrossRef]
21. Moema, D.; Nindi, M.M.; Dube, S. Development of a dispersive liquid–liquid microextraction method for the determination of fluoroquinolones in chicken liver by high performance liquid chromatography. *Anal. Chim. Acta* **2012**, *730*, 80–86. [CrossRef]
22. Liu, B.; Yan, H.; Qiao, F.; Geng, Y. Determination of clenbuterol in porcine tissues using solid-phase extraction combined with ultrasound-assisted dispersive liquid-liquid microextraction and HPLC-UV detection. *J. Chromatogr. B Anal. Technol. Biomed. Life Sci.* **2011**, *879*, 90–94. [CrossRef] [PubMed]
23. Viñas, P.; López-García, I.; Bravo-Bravo, M.; Briceño, M.; Hernández-Córdoba, M. Dispersive liquid-liquid microextraction coupled to liquid chromatography for thiamine determination in foods. *Anal. Bioanal. Chem.* **2012**, *403*, 1059–1066. [CrossRef] [PubMed]
24. Zhang, Y.; Xue, X.; Su, S.; Guo, Z.; Wang, J.; Ding, L.; Liu, Y.; Zhu, J. A Multi-Class, Multi-Residue Method for Detection of Veterinary Drugs in Multiple Meat Using a Pass-Through Cleanup SPE Technique and UPLC-MS/MS Analysis. *Food Anal. Methods* **2018**, *11*, 2865–2884. [CrossRef]
25. Salim, S.A.; Sukor, R.; Ismail, M.N.; Selamat, J. Dispersive Liquid-Liquid Microextraction (DLLME) and LC-MS/MS Analysis for Multi-Mycotoxin in Rice Bran: Method development, optimization and validation. *Toxins* **2021**, *13*, 280. [CrossRef]
26. Rocha, D.G.; Santos, F.A.; da Silva, J.C.C.; Augusti, R.; Faria, A.F. Multiresidue determination of fluoroquinolones in poultry muscle and kidney according to the regulation 2002/657/EC. A systematic comparison of two different approaches: Liquid chromatography coupled to high-resolution mass spectrometry or tandem mass spe. *J. Chromatogr. A* **2015**, *1379*, 83–91. [CrossRef]
27. Yassine, M.; Rifai, A.; Doumyati, S.; Trivella, A.; Mazellier, P.; Budzinski, H.; Al Iskandarani, M. Oxidation of danofloxacin by free chlorine—Kinetic study, structural identification of by-products by LC–MS/MS and potential toxicity of by-products using in silico test. *Environ. Sci. Pollut. Res.* **2017**, *24*, 7982–7993. [CrossRef] [PubMed]
28. Barbieri, M.V.; Postigo, C.; Guillem-Argiles, N.; Monllor-Alcaraz, L.S.; Simionato, J.I.; Stella, E.; Barceló, D.; de Alda, M.L. Analysis of 52 pesticides in fresh fish muscle by QuEChERS extraction followed by LC-MS/MS determination. *Sci. Total Environ.* **2019**, *653*, 958–967. [CrossRef] [PubMed]
29. Blanchoud, H.; Alliot, F.; Chen, N.; Valdes, D. Rapid SPE–LC MS/MS analysis for atrazine, its by-products, simazine and S metolachlor in groundwater samples. *MethodsX* **2020**, *7*, 100824. [CrossRef]
30. Chen, D.; Tao, Y.; Zhang, H.; Pan, Y.; Liu, Z.; Huang, L.; Wang, Y.; Peng, D.; Wang, X.; Dai, M.; et al. Development of a liquid chromatography-tandem mass spectrometry with pressurized liquid extraction method for the determination of benzimidazole residues in edible tissues. *J. Chromatogr. B Anal. Technol. Biomed. Life Sci.* **2011**, *879*, 1659–1667. [CrossRef]
31. Pokrant, E.; Medina, F.; Maddaleno, A.; Martín, B.S.; Cornejo, J. Determination of sulfachloropyridazine residue levels in feathers from broiler chickens after oral administration using liquid chromatography coupled to tandem mass spectrometry. *PLoS ONE* **2018**, *13*, e0200206. [CrossRef]
32. Hu, F.Y.; He, L.M.; Yang, J.W.; Bian, K.; Wang, Z.N.; Yang, H.C.; Liu, Y.H. Determination of 26 veterinary antibiotics residues in water matrices by lyophilization in combination with LC-MS/MS. *J. Chromatogr. B Anal. Technol. Biomed. Life Sci.* **2014**, *949–950*, 79–86. [CrossRef]
33. Patyra, E.; Nebot, C.; Gavilán, R.E.; Cepeda, A.; Kwiatek, K. Development and validation of an LC-MS/MS method for the quantification of tiamulin, trimethoprim, tylosin, sulfadiazine and sulfamethazine in medicated feed. *Food Addit. Contam. Part A Chem. Anal. Control Expo. Risk Assess.* **2018**, *35*, 882–891. [CrossRef]
34. Busatto, Z.; da Silva, A.F.B.; de Freitas, O.; Paschoal, J.A.R. LC-MS/MS methods for albendazole analysis in feed and its metabolite residues in fish fillet and a leaching study in feed after an alternative procedure for drug incorporation. *Food Addit. Contam.-Part A Chem. Anal. Control Expo. Risk Assess.* **2017**, *34*, 509–519. [CrossRef] [PubMed]
35. Fattahi, N.; Assadi, Y.; Hosseini, M.R.M.; Jahromi, E.Z. Determination of chlorophenols in water samples using simultaneous dispersive liquid-liquid microextraction and derivatization followed by gas chromatography-electron-capture detection. *J. Chromatogr. A* **2007**, *1157*, 23–29. [CrossRef] [PubMed]
36. Kozani, R.R.; Assadi, Y.; Shemirani, F.; Hosseini, M.R.M.; Jamali, M.R. Part-per-trillion determination of chlorobenzenes in water using dispersive liquid-liquid microextraction combined gas chromatography-electron capture detection. *Talanta* **2007**, *72*, 387–393. [CrossRef] [PubMed]

37. Rezaee, M.; Assadi, Y.; Hosseini, M.R.M.; Aghaee, E.; Ahmadi, F.; Berijani, S. Determination of organic compounds in water using dispersive liquid-liquid microextraction. *J. Chromatogr. A* **2006**, *1116*, 1–9. [CrossRef] [PubMed]
38. Hendriks, G.; Uges, D.R.A.; Franke, J.P. Reconsideration of sample pH adjustment in bioanalytical liquid-liquid extraction of ionisable compounds. *J. Chromatogr. B Anal. Technol. Biomed. Life Sci.* **2007**, *853*, 234–241. [CrossRef] [PubMed]
39. Herrera-Herrera, A.V.; Hernández-Borges, J.; Borges-Miquel, T.M.; Rodríguez-Delgado, M.A. Dispersive liquid-liquid microextraction combined with nonaqueous capillary electrophoresis for the determination of fluoroquinolone antibiotics in waters. *Electrophoresis* **2010**, *31*, 3457–3465. [CrossRef]
40. Campone, L.; Piccinelli, A.L.; Celano, R.; Rastrelli, L. Analytica Chimica Acta pH-controlled dispersive liquid–liquid microextraction for the analysis of ionisable compounds in complex matrices: Case study of ochratoxin A in cereals. *Anal. Chim. Acta* **2012**, *754*, 61–66. [CrossRef]
41. Bedassa, T.; Gure, A.; Megersa, N. The QuEChERS analytical method combined with low density solvent based dispersive liquid–liquid microextraction for quantitative extraction of multiclass pesticide residues in cereals. *Bull. Chem. Soc. Ethiop.* **2017**, *31*, 1–15. [CrossRef]
42. Farajzadeh, M.A.; Seyedi, S.E.; Shalamzari, M.S.; Bamorowat, M. Dispersive liquid-liquid microextraction using extraction solvent lighter than water. *J. Sep. Sci.* **2009**, *32*, 3191–3200. [CrossRef]
43. Beltran, J.; López, F.J.; Hernández, F. Solid-phase microextraction in pesticide residue analysis. *J. Chromatogr. A* **2000**, *885*, 389–404. [CrossRef] [PubMed]
44. Rajabi, M.; Ghanbari, H.; Barfi, B.; Asghari, A.; Haji-Esfandiari, S. Ionic liquid-based ultrasound-assisted surfactant-emulsified microextraction for simultaneous determination of three important flavoring compounds in plant extracts and urine samples. *Food Res. Int.* **2014**, *62*, 761–770. [CrossRef]
45. Poliwoda, A.; Krzyzak, M.; Wieczorek, P.P. Supported liquid membrane extraction with single hollow fiber for the analysis of fluoroquinolones from environmental surface water samples. *J. Chromatogr. A* **2010**, *1217*, 3590–3597. [CrossRef]
46. Nerín, C.; Salafranca, J.; Aznar, M.; Batlle, R. Critical review on recent developments in solventless techniques for extraction of analytes. *Anal. Bioanal. Chem.* **2009**, *393*, 809–833. [CrossRef] [PubMed]
47. *ISO/IEC 17025:2017*; General Requirements for the Competence of Testing and Calibration Laboratories. Institute for Standardization of Serbia: Belgrade, Serbia, 2017; pp. 1–27.
48. Council Directive 96/23/EC. Commission Decision of 12 August 2002 implementing Council Directive 96/23/EC concerning the performance of analytical methods and the interpretation of results. *Off. J. Eur. Communities* **2002**, *L221*, 8–36.
49. SANTE. Analytical Quality Control and Method Validation Procedures for Pestice Residues Analysis in Food and Feed, SANTE/11813/2017 2019. Available online: https://www.eurl-pesticides.eu/userfiles/file/EurlALL/SANTE_11813_2017-fin.pdf.a (accessed on 22 March 2022).
50. Eurachem Guide. *The Fitness for Purpose of Analytical Methods: A Laboratory Guide to Method Validation and Related Topics*; Eurachem: Gembloux, Belgium, 1998; 61p.
51. Wenzl, T.; Haedrich, J.; Schaechtele, A.; Robouch, P.; Stroka, J. Guidance Document on the Estimation of LOD and LOQ for Measurements in the Field of Contaminants in Feed and Food. EUR 28099 EN. 2016. Available online: https://publications.jrc.ec.europa.eu/repository/handle/JRC102946 (accessed on 14 March 2022).
52. Zgoła-Grześkowiak, A.; Grześkowiak, T. Dispersive liquid-liquid microextraction. *TrAC Trends Anal. Chem.* **2011**, *30*, 1382–1399. [CrossRef]

foods

MDPI

Article

Development of Extraction Method for Determination of Saponins in Soybean-Based Yoghurt Alternatives: Effect of Sample pH

Anastassia Bljahhina [1,2], Maria Kuhtinskaja [2] and Tiina Kriščiunaite [1,*]

[1] Center of Food and Fermentation Technologies (TFTAK), Mäealuse 2/4, 12618 Tallinn, Estonia; anastassia.bljahhina@tftak.eu

[2] Department of Chemistry and Biotechnology, Tallinn University of Technology, Akadeemia tee 15, 12618 Tallinn, Estonia

* Correspondence: tiina@tftak.eu

Abstract: The number of plant-based dairy alternative products on the market is growing rapidly. In the case of soybean-based yoghurt alternatives, it is important to trace the content of saponins, the phytomicronutrients with a disputable health effect, which are likely to be responsible for the bitter off-taste of the products. We present a new sample extraction method followed by hydrophilic interaction liquid chromatography with mass spectrometric detection (HILIC-MS) for identifying and quantifying soyasaponins in soybean-based yoghurt alternatives. Soyasaponin Bb, soyasaponin Ba, soyasaponin Aa, and soyasaponin Ab were quantified using commercially available standard compounds and with asperosaponin VI as the internal standard. As the recoveries of soyasaponins were unacceptable in yoghurt alternatives at their natural acidic pH, the adjustment of pH was performed as one of the first steps in the sample extraction procedure to achieve the optimum solubility of soyasaponins. The validation of the method included the assessment of linearity, precision, limit of detection and limit of quantification (LOQ), recovery, and matrix effect. The average concentrations of soyasaponin Bb, soyasaponin Ba, soyasaponin Ab, and soyasaponin Aa in several measured soybean-based yoghurt alternatives utilising the developed method were 12.6 ± 1.2, 3.2 ± 0.7, 6.0 ± 2.4 mg/100 g, and below the LOQ, respectively. This method provides an efficient and relatively simple procedure for extracting soyasaponins from yoghurt alternatives followed by rapid quantification using HILIC-MS and could find a rightful application in the development of healthier and better-tasting dairy alternatives.

Keywords: bitterness; *Glycine max*; plant-based foods; plant proteins; LC-MS

Citation: Bljahhina, A.; Kuhtinskaja, M.; Kriščiunaite, T. Development of Extraction Method for Determination of Saponins in Soybean-Based Yoghurt Alternatives: Effect of Sample pH. *Foods* **2023**, *12*, 2164. https://doi.org/10.3390/foods12112164

Academic Editors: Gianfranco Picone and Antonello Santini

Received: 30 March 2023
Revised: 23 May 2023
Accepted: 25 May 2023
Published: 27 May 2023

1. Introduction

In recent decades, the market for plant-based dairy alternatives has vastly expanded. In addition to cereals, pseudocereals, and nuts, legumes are typically used to produce plant-based dairy alternatives. Due to their high protein content and quality, legumes such as soybeans (*Glycine max* L.) are widely used to manufacture dairy alternatives [1]. However, there is still a lack of quantitative data on the migration of phytonutrients during food processing from plant-based protein sources to the final consumable products. Along with macronutrients, soybeans contain several classes of biologically active compounds, including naturally occurring complex oleanane triterpenoid glycoside saponins [2]. Chitisankul et al. studied saponin content in nine soybean varieties and fourteen different soybean-based milk alternatives. The average total soyasaponin content reported was 246 ± 92 and 269 ± 140 μmol per 100 g dry weight basis (dwb), respectively [3,4], suggesting a transfer of saponins from the dry matter throughout the production chain of plant-based milk alternatives.

The dietary preferences of many consumers are shifting towards plant-based products due to environmental, health, and ethical reasons. Thus, from a nutritional point of view, it is important to quantify phytochemicals from emerging plant-based alternatives. Although human cells are not able to degrade saponins [5], some bacteria from gut microbiota convert saponins into sapogenols [6] and enter the bloodstream [7]. Until now, the data on the effects of saponins on human health are controversial. Negative consequences of high saponin consumption have been proven in livestock; e.g., health issues in the digestive tract of ruminants as well as decreases in wool, milk, and egg production were observed [8]. In addition, high concentrations of saponins may lead to the inefficient absorption of fat-soluble vitamins and damage the membrane of the intestinal inner epithelial wall [9]. On the contrary, several in vitro and in vivo studies have shown the positive immunological and antiviral effects of soyasaponins [10]. In addition, anti-cancerogenic [6], hepato-protective [11], anti-inflammatory [12], and anti-obesity effects [13,14] have been reported. The beneficial and deleterious nutritional properties of saponins are likely to be dose- and diet-dependent.

Soyasaponins are amphiphilic compounds composed of polar sugar moieties attached to a nonpolar pentacyclic ring [15,16]. Soyasaponins are generally distributed between group A and B depending on the glycosylation positions of soyasapogenol A and soyasapogenol B [17,18]. Soyasaponin Aa and soyasaponin Ab are glycosylated at the C-3 and C-22 position of soyasapogenol A (group A), while soyasaponin Ba and soyasaponin Bb are glycosylated at the C-3 position of soyasapogenol B (group B). The structures of the studied soyasaponins are shown in Figure 1.

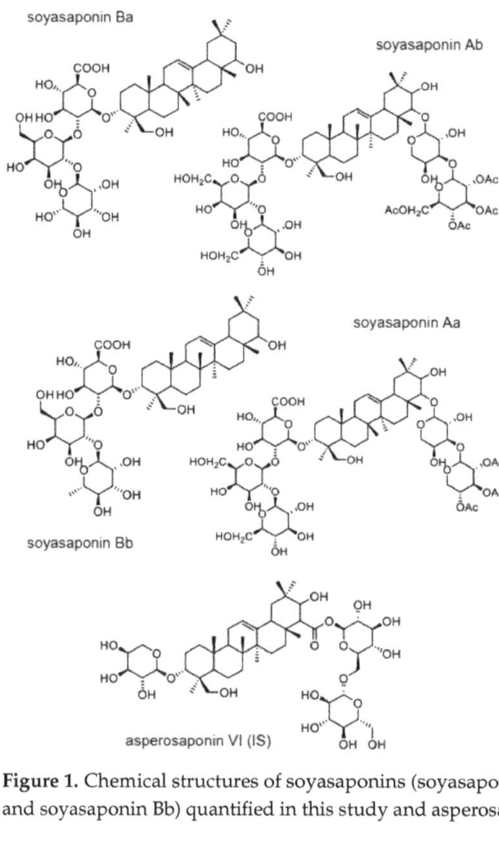

Figure 1. Chemical structures of soyasaponins (soyasaponin Aa, soyasaponin Ab, soyasaponin Ba, and soyasaponin Bb) quantified in this study and asperosaponin VI (used as internal standard (IS)).

In addition to structural differences, group A soyasaponins contribute more to a bitter sensation than group B saponins [19], causing a major unpleasant taste of soybean-based dairy alternatives [20,21]. The group A soyasaponins are located both in soybean seed hypocotyls and cotyledons [3]. The removal of hypocotyls is usually performed during the production of soybean-based milk alternatives but is not enough to fully discard group A soyasaponins from the end products [3,4]. Hence, the residual soyasaponin concentration might still influence the bitterness of a product and thus limit the consumer acceptance.

Researchers have widely characterised the molecular structures of several forms of soyasaponins and have reported different methods for their quantification [4,22–29]. Indeed, saponin quantification is considered challenging due to the lack of chromophores in their molecular structure, leaving out the possibility of using UV light at a specific wavelength for quantification. Liquid chromatography (LC) coupled to electrospray ionisation mass spectrometry (ESI/MS) is an alternative approach providing significant selectivity and specificity without a need for the derivatization of analytes [16,25]. Despite extensive research carried out on different soybean foods, only a few studies have identified or relatively quantified the levels of soyasaponins in soybean-based dairy alternatives [4,29]. In the case of previously published methods, the liquid samples were initially pre-processed before extraction by being either freeze-dried [4] or dried by rotary evaporation [29]. The time-consuming application of these techniques may be considered the major drawback of previously reported quantification methods impeding the direct analysis of liquid samples. New extraction procedures are required to mitigate the issues with traditional analysis methods, allowing to save on equipment resources, increase the analysis throughput, and overall, facilitate the implementation of quality control throughout the development of new soybean-based dairy alternative products.

This study aimed to develop a selective extraction and quantification method for the determination of soyasaponins (soyasaponin Aa, soyasaponin Ab, soyasaponin Ba, and soyasaponin Bb) from a soybean-based yoghurt alternative matrix using hydrophilic interaction liquid chromatography with mass spectrometric detection (HILIC-MS). To our knowledge, there are no studies presenting the soyasaponin quantification method in which sample extraction has been performed directly from liquid soybean-based dairy alternative samples.

2. Materials and Methods

2.1. Food Samples

Soybean-based drink (SBD) and five soybean-based yoghurt alternatives (YA1, YA2, YA3, YA4, and YA5) from different producers were purchased from the local supermarket. Supplementary Table S1 provides nutritional and compositional information available on the label of the products. Samples were aliquoted and stored at −20 °C.

2.2. Chemicals and Materials

All solvents were HPLC grade and were purchased from Honeywell (Charlotte, NC, USA). Formic acid (FA) (98% for MS) and the ammonia solution (25% for LC-MS) were from Honeywell (Charlotte, NC, USA) and Merck KGaA (Darmstadt, Germany), respectively. The standard compounds soyasaponin Aa, soyasaponin Bb, soyasaponin Ba, and asperosaponin VI Phyproof® Reference substances were from PhytoLab GmbH & Co. KG (Dutendorfer, Germany), and soyasaponin Ab was from MedChemExpress (Monmouth Junction, NJ, USA). Biotage Isolute® PLD+ (100 mg/mL) cartridges were obtained from Biotage Sweden AB (Uppsala, Sweden). Ultrapure water (18.2 mΩ·cm) was prepared using MilliQ® HX 7040SD equipped with MilliQ LC-Pak (Merck KGaA, Darmstadt, Germany).

2.3. Extraction Method for Samples

Soyasaponins were extracted according to the previously published method developed for pea and oat saponins [30] with some modifications. Briefly, the thawed homogeneous liquid sample was weighed (0.35–0.40 g) into a 5 mL volumetric flask (n = 3). Ultrapure water was added to the line and mixed thoroughly. The sample solution (native pH of yoghurt alternative was ~4.6) was alkalised to reach the sample pH 8 ± 0.25 using aqueous ammonia solution (5%, *v/v*) or aqueous FA (25%, *v/v*). Samples were incubated on a tube rotator Stuart SB3 (Bibby Scientific Ltd, Staffordshire, UK) at room temperature for 30 min. After incubation, samples were centrifuged at 17,000 *g* at 10 °C for 10 min. After transferring the supernatant to a new Eppendorf tube, an equal volume of pure acetonitrile was added (MeCN, 1:1, *v:v*). The solution was mixed thoroughly and centrifuged at 14,800× *g* at 10 °C for 10 min to remove precipitated proteins. The supernatant (1000 µL) was passed through a PLD+ column using a vacuum manifold (VacMaster 10, Biotage Sweden AB, Uppsala, Sweden) at –0.5 bar. The filtrate (100 µL) was combined with an IS working solution (asperosaponin VI; 100 µL) and injected into the LC-MS.

2.4. Liquid Chromatography Mass Spectrometry

Analysis was performed as described previously [30] with adaptations to the analysis of soyasaponins. Briefly, a Waters UPLC® system (Waters Corporation, Milford, MA, USA) attached to a Waters Quattro Premier XE Mass Spectrometer equipped with ZSpray™ Source was used to analyse the samples. The equipment was controlled by Waters MassLynx™ 4.1 (V4.1 SCN805, Waters Corporation, Milford, MA, USA). Mobile phase A consisted of ultrapure water containing 0.1% FA, and mobile phase B consisted of MeCN containing 0.1% FA. The gradient was changed as follows: 0–0.17 min at 10% A; 0.17–1.5 min linear gradient 10–70% A; 1.5–4.17 min at 70% A; 4.18 min switch to 10% A; 4.18–6.0 min at 10% A. The mobile phases were pumped at 200 µL/min flow rate. A BEH Amide column (1.0 × 50 mm, 1.7 µm) coupled with BEH Amide VanGuard Pre-column (2.1 × 5 mm) from Waters Corporation (Milford, MA, USA) were used to retain saponins. The autosampler and column heater were set at 8 °C and 50 °C, respectively.

The MS part of the method proposed for determination of oat and pea saponins [30] was adapted to target the quantification of soyasaponins. Based on a scan-type experiment of external standards, the deprotonated molecules $[M-H]^-$ were chosen. The capillary voltage was set to −2.5 kV; cone voltages were optimised separately for every compound. The analysis was performed using negative electrospray ionisation (ESI^-) mode using single-ion-recording (SIR) mass-to-charge ratios shown in Table 1. High-purity nitrogen was set as a cone and as desolvation gas at a rate of 25 L/h and 600 L/h, respectively. The temperature of the desolvation gas was set to 350 °C. Data acquisition and target analyte quantification were performed in Waters MassLynx™ and QuanLynx™ V4.1 (SCN805, Waters Corporation, Milford, MA, USA) and Microsoft Excel® (Microsoft 365 Apps for enterprise, Microsoft Corporation, Richmond, WA, USA). Other parameters for MS were employed according to the description provided previously [30].

Table 1. The used m/z values and cone voltages for analytes.

Analyte	$[M-H]^-$ $m/z \pm 0.5$ Da	Cone Voltage (V)
Soyasaponin Bb	941.5	100
Soyasaponin Ba	957.3	120
Soyasaponin Aa	1364.3	120
Soyasaponin Ab	1435.6	120
Asperosaponin VI	927.5	120

2.5. Calibration and Quantification

The stock solutions of soyasaponin Aa, soyasaponin Ab, and soyasaponin Bb (1000 mg/L) were prepared in ethanol (EtOH; 99.9% purity). The stock solution of soyasaponin Ba (1000 mg/L) was dissolved in ethanol:methanol solution (EtOH:MeOH; 1:1, *v/v*). The stock solution of asperosaponin VI used as IS (1000 mg/L) was prepared in ultrapure water. All solutions were aliquoted and stored at −80 °C. The working solution of asperosaponin VI (30 mg/L) was made freshly before the analysis using the aqueous MeCN (50%, *v/v*). The calibration curve standard solutions were diluted in MeCN:H_2O:EtOH solution (50:36:14, *v/v*).

The IS (asperosaponin VI) working solution was added to the calibration curve standard solutions and the sample solutions before the injection, keeping the concentration of the IS constant. Calibration curve solutions were built for all soyasaponins and were run in triplicate (0.01–2.5 mg/L). Seven-point calibration curves of soyasaponins were prepared by plotting peak area ratios of soyasaponins/IS against the concentration of the external standard compound. The linear regression approach led to linear responses showing correlation coefficients of >0.99 for all analytes.

2.6. Validation of the Method

The European Medicines Agency (EMA) validation guideline was used to evaluate the following parameters during method validation: selectivity, specificity, calibration curve and range, limit of detection (LOD), limit of quantification (LOQ), precision, sample extraction recoveries, and matrix effect (ME) [31].

The calibration curve and range were evaluated via repeated measurements of standard solutions of soyasaponins consisting of eight individual points obtained from serial dilution of stock solutions. The LODs and LOQs were calculated using a previously published tutorial [32].

To determine the intra- and interday precision of the instrumental method, the standard solution and the IS were both injected six times and across three independent days to affirm the stability of the retention times (RTs) and peak areas. In addition, the repeatability (intraday) and intermediate precision (interday) of the whole method was investigated using YA2. Repeatability analysis was performed by six replicate analyses of samples on the same day. The intermediate precision of the method was by analysis of six replicates on three different days over four weeks under the same experimental conditions.

The total recoveries of analytes were assessed by spiking YA2 with a known amount of soyasaponins at four different concentration levels (unspiked, lower LOQ, middle LOQ, and upper LOQ) and performing the extraction methods as described above [33].

ME was evaluated by post-extraction spiking of sample extracts with calibration curve standard solutions and comparing the solvent-matched calibration curve slopes with matrix-matched slopes [32].

2.7. Statistical Analysis

Data analysis for sample extraction method development was performed in R 4.2.2 (The R Foundation for Statistical Computing, Vienna, Austria). ANOVA followed by Tukey–Kramer post hoc test was performed with R package 'agricolae' 1.3–5. The significance level was set to 0.05. The results are presented as mean with standard deviation (SD) or relative standard deviation (RSD). All analyses were repeated in triplicate if not marked otherwise.

3. Results and Discussion

3.1. Development of Liquid Chromatography Mass Spectrometry Method

Previously reported LC-MS methods for the quantification of saponins varied from 6 to 80 min [2,30,34,35]. The shortest method with some modifications in the gradient and a total runtime of six minutes was used as a basis in our study. The SIR chromatograms shown in Figure 2 were obtained following an analysis of the soyasaponin standards, the sample of soybean-based yoghurt alternative, and the IS using the optimised analytical method described in Section 2.4. During the method development, we tested the multiple reaction monitoring (MRM) experimental conditions on our instrumentation. However, it did not enhance selectivity; instead, it notably decreased sensitivity by failing to generate consistent fragments. Nevertheless, the reasonably rapid retention of soyasaponins on the column and high-resolution peaks were achieved using the SIR mode. The proposed chromatographic method is more environmentally friendly and sustainable than previous approaches as i has a shorter duration, high-throughput nature, and reduced solvent usage.

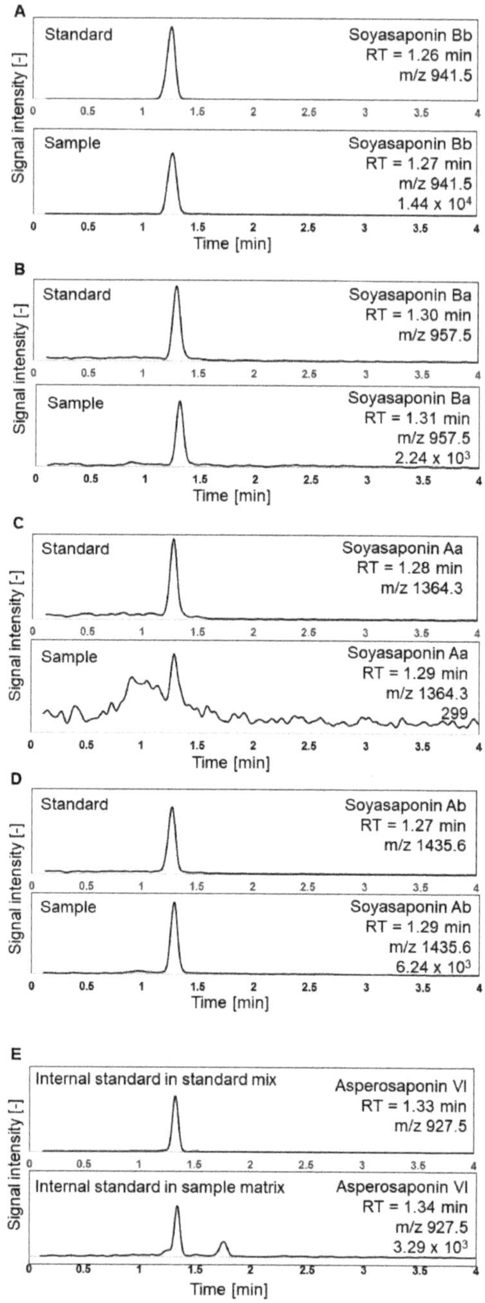

Figure 2. LC-MS chromatograms of external standard compounds and YA1 sample (SIR; ESI-): (**A**) soyasaponin Bb, (**B**) soyasaponin Ba, (**C**) soyasaponin Aa, (**D**) soyasaponin Ab, and (**E**) internal standard asperosaponin VI.

Although soyasaponins include over one hundred different compounds [4], only the forms relevant to soybean-based yoghurt alternatives, including soyasaponin Bb, soyasaponin Ba, soyasaponin Aa, and soyasaponin Ab, were selected for total quantification. During method development, we also screened soybean-based yoghurt alternatives for the possible semi-quantification of the DDMP-conjugated form, but these compounds were not identified in the matrix. Plant-based yoghurt alternatives are typically pasteurised at 95 °C or undergo an ultra-high-temperature treatment above 100 °C during the production process, which helps to manage microbiological concerns and prolong the shelf life [1]. Under heat treatment, the thermo-sensitive DDMP conjugates of group B soyasaponins may degrade into non-DDMP saponin species. Hu et al. showed that the DDMP-conjugated B group saponins started to decrease already when heated at 65 °C [2]. Indeed, the range of possible analytes to quantitate could be potentially expanded by total synthesis or fractioning other soyasaponin compounds from raw materials, but in both cases, it is time-consuming, not cost-effective, and impractical for routine analysis in laboratories.

Asperosaponin VI was chosen as the IS for soyasaponins in this method based on its structural similarity to the triterpenoid core [36] (Figure 1) and LC-MS retention similar to the targeted analytes. Ideally, each soyasaponin target compound should be quantified using its corresponding isotopically labelled internal standard when these become more readily available, without the need for the custom total synthesis of standards or the cultivation of isotopically labelled soybeans.

3.2. The Influence of Sample pH on Saponin Extraction

A previously published method for the measurement of saponins in oat- and pea-based drinks [30] was used as a starting point for the development of an extraction method for soyasaponins from soybean-based yoghurt alternatives. Traditionally, saponin extraction is performed using ethanol or methanol from a solid fat-free sample before subsequent LC-MS analysis, the whole procedure starting from a Soxhlet-assisted fat-removing step, followed by the solvent extraction. The simplified procedure in this recently proposed method allowed the extraction of saponins from liquid samples with a minimal number of extraction steps and a small volume of solvents. The comprehensive comparison of the performance of this extraction method with selected traditional ones has been provided elsewhere [30].

In the present study, we focused on the exploration of the effect of the pH of the soybean-based yoghurt alternatives on the quantification of soyasaponins as the pH of these products is considerably lower than that of the SBD. The native pH of the SBD and the soybean-based yoghurt alternatives (YA1 and YA2) were 8.8 and 4.6–4.7, respectively. The effect of the native pH of the products and the effect of the pH adjustment before extraction on the yield of the extracted saponins are reported in Table 2. The SBD and the soybean-based yoghurt alternatives were analysed as described in Section 2.3: unspiked and spiked with all four soyasaponins and with or without a pH adjustment included in the sample extraction protocol. Indeed, in the samples at their native pH, the recoveries of soyasaponins in the SBD ranged from 80 to 109%, while for both yoghurt alternatives, the concentrations and recoveries were significantly lower than those observed in the SBD. Moreover, the recoveries of soyasaponins at native pH were similar among yoghurt alternatives. These results suggest that soyasaponin recoveries could be pH-dependent.

Table 2. The relationship between sample pH and recovery (R) rates of soyasaponins during different extraction procedures. Samples: soybean-based drink (SBD); soybean-based yoghurt alternatives (YA1 and YA2). Each result represents mean ± standard deviation (n = 3). ANOVA statistical significance test was performed within sample matrices; means with different letters are significantly different at $p < 0.05$.

Sample	pH	Soyasaponin Bb		Soyasaponin Ba		Soyasaponin Aa		Soyasaponin Ab	
		mg/100 g	R [1] (%)	mg/100 g	R [2] (%)	mg/100 g	R [3] (%)	mg/100 g	R [4] (%)
SBD	8.8 (native)	12.6 ± 0.71 ^{a}	98 ± 4 ^{a}	2.43 ± 0.26 ^{a}	109 ± 8 ^{a}	<LOQ	85 ± 5 ^{a}	<LOQ	80 ± 5 ^{a}
	4.2 ± 0.2	1.2 ± 0.14 ^{b}	23 ± 5 ^{b}	0.82 ± 0.04 ^{b}	26 ± 5 ^{b}	<LOQ	54 ± 6 ^{b}	<LOQ	51 ± 5 ^{b}
YA1	4.7 (native)	0.84 ± 0.03 ^{b}	20 ± 2 ^{b}	0.56 ± 0.02 ^{b}	25 ± 2 ^{b}	<LOQ	43 ± 3 ^{b}	0.58 ± 0.06 ^{b}	41 ± 3 ^{b}
	7.0 ± 0.2	5.39 ± 0.53 ^{a}	77 ± 2 ^{a}	1.03 ± 0.15 ^{a}	109 ± 3 ^{a}	<LOQ	104 ± 1 ^{a}	1.82 ± 0.1 ^{a}	115 ± 1 ^{a}
YA2	4.6 (native)	1.3 ± 0.08 ^{b}	27 ± 2 ^{c}	0.54 ± 0.04 ^{b}	25 ± 1 ^{c}	<LOQ	48 ± 3 ^{c}	3.07 ± 0.07 ^{b}	63 ± 4 ^{b}
	7.0 ± 0.2	10.94 ± 0.63 ^{b}	85 ± 10 ^{b}	2.36 ± 0.03 ^{ab}	83 ± 8 ^{b}	<LOQ	91 ± 7 ^{ab}	8.09 ± 0.38b ^{a}	98 ± 6 ^{a}
	7.5 ± 0.2	13.63 ± 1.63 ^{a}	89 ± 4 ^{ab}	2.68 ± 0.06 ^{a}	102 ± 5 ^{ab}	<LOQ	86 ± 4 ^{b}	9.93 ± 1.38 ^{a}	103 ± 1 ^{a}
	8.0 ± 0.2	14.43 ± 0.61 ^{a}	100 ± 14 ^{ab}	2.54 ± 0.34 ^{a}	114 ± 16 ^{a}	<LOQ	107 ± 15 ^{a}	9.18 ± 0.36 ^{a}	110 ± 15 ^{a}
	8.5 ± 0.2	13.51 ± 1.97 ^{a}	111 ± 1 ^{a}	2.79 ± 0.4 ^{a}	99 ± 8 ^{ab}	<LOQ	74 ± 7 ^{b}	9.01 ± 1.15 ^{a}	91 ± 8 ^{a}

[1] soyasaponin Bb spike concentration: 2.10 mg/L. [2] soyasaponin Ba spike concentration: 1.86 mg/L. [3] soyasaponin Aa spike concentration: 1.91 mg/L. [4] soyasaponin Ab spike concentration: 2.03 mg/L.

As the composition and nutritional information of the SBD, YA1, and YA2 were very similar (see Table S1), the following experiments were conducted with adjusted pHs of the samples to test the hypotheses of pH effects on soyasaponin recoveries (the results are shown in Table 2). The pH of the SBD was acidified to mimic the pH of the yoghurt alternatives, while the yoghurt alternatives were alkalised to mimic the pH of SBD. The experiment indicated that the SBD acidified to pH 4.2 had unacceptable recoveries of soyasaponins, ranging from 23 to 54%. On the contrary, yoghurt alternatives that were alkalised (pH 7.0 ± 0.2) resulted in higher soyasaponin recoveries: from 77 to 115% and from 83 to 98% in YA1 and YA2, respectively.

Based on these findings, additional experiments were conducted to assess the pH value at which soyasaponins would result in the highest and most meaningful recovery. YA2 was analysed at three additional pH values: 7.5 ± 0.2, 8.0 ± 0.2, and 8.5 ± 0.2. The results indicated that pH values of 7–8.5 had a beneficial influence on the recovery of soyasaponins, but there was no strict pH optimum value. ANOVA showed a statistical difference in the recoveries of soyasaponins at analysed pH values in most cases. The most acceptable recoveries were achieved at pH 7.5 ± 0.2 and pH 8.0 ± 0.2. Therefore, for further analyses, the method's optimum pH value was chosen to be 8 ± 0.25.

Even though saponins are known as amphiphilic molecules, having a non-water soluble triterpene core and attached water-soluble sugar moieties, and are preferably soluble in organic solvents, soyasaponin Bb solubility is very low in the acidic region and increases drastically in the 6.5–7.3 pH region in aqueous buffers, having a solubility maximum in the range of 7 to 8 pH [37]. This fact elucidates the influence of different pH values of the samples on the soyasaponins recovery experiments. By adjusting the pH in soybean-based yogurt alternatives, the solubility issues of soyasaponins in acidic environments are overcome, enabling the direct analysis of liquid samples using a recently published method with modifications relevant to soyasaponins [30].

3.3. Validation of the Method

Validation was executed to assess the linear ranges, LODs and LOQs, precision, recoveries, and matrix effect of the proposed method for the determination of soyasaponins in yoghurt alternatives (Table 3). The calibration curves were constructed using a linear model with a weighing of 1/x. All four soyasaponins standards had high linearity ($R^2 > 0.99$) in the 0.01–2.52 mg/L concentration range. The estimated LOQs for soyasaponins were ≤33.4 µg/L. The results of the LOQs were either lower or in accordance with previous research [34,35].

Table 3. The linear range, calibration curve, limits of detection (LODs), and limits of quantification (LOQs) of soyasaponins.

Analyte	Linear Range (mg/L)	Calibration Curve	R^2	LOD (µg/L)	LOQ (µg/L)
Soyasaponin Bb	0.01–2.52	y = 0.7699x + 0.0048	0.9930	0.2	12.6
Soyasaponin Ba	0.02–2.26	y = 0.2949x + 0.0025	0.9975	8.0	33.4
Soyasaponin Aa	0.02–2.33	y = 0.3994x + 0.0021	0.9965	7.0	27.0
Soyasaponin Ab	0.01– 2.48	y = 0.3259x + 0.0033	0.9943	1.4	25.1

The repeatability of the method was investigated after the linearity of the soyasaponins was defined as acceptable. The results of the experiments are shown in the Supplementary Materials (Table S2). The RSDs of the peak RTs and the peak areas did not exceed 2% and 4%, respectively. It was observed that the intra- and interday RSDs for the whole method were lower than 12% and suitable for the routine analysis of soybean-based products. The precision observed using this method agreed with results reported by other LC-MS methods [34,35].

The recoveries of the soyasaponins were determined by spiking the YA2 with the analytes. Table 4 shows the results of the recovery of the YA2 at three spiking levels. The recoveries ranged from 81 to 101%. The obtained recoveries were acceptable according to the guidelines [31] and comparable with the previously published methods [2,30,34,35].

Table 4. The recoveries of soyasaponins in soybean-based yoghurt alternative matrix (mean ± standard deviation (n = 3)).

Spiking Level	Soyasaponin Bb [1]		Soyasaponin Ba [2]		Soyasaponin Aa [3]		Soyasaponin Ab [4]	
	mg/L	R, %	mg/L	R, %	mg/L	R, %	mg/L	R, %
L1	0.03	95 ± 4	0.02	97 ± 5	0.02	101 ± 9	0.03	96 ± 3
L2	0.77	87 ± 4	0.68	87 ± 3	0.62	81 ± 1	0.8	90 ± 2
L3	1.54	82 ± 2	1.36	86 ± 2	1.23	81 ± 0	1.61	88 ± 0

[1] Unspiked matrix soyasaponin Bb concentration: 0.99 mg/L. [2] Unspiked matrix soyasaponin Ba concentration: 0.27 mg/L. [3] Unspiked matrix soyasaponin Aa concentration: 0.03 mg/L. [4] Unspiked matrix soyasaponin Ab concentration: 0.72 mg/L.

The experiment demonstrated that the soyasaponin Bb, soyasaponin Ba, soyasaponin Aa, and soyasaponin Ab MEs were 91%, 94%, 99%, and 94%, respectively. According to the guidelines, the achieved MEs were at an acceptable range [38], indicating sufficient sample clean-up.

According to validation guidelines, the method has confirmed sufficient validation performance regarding precision, recovery, sensitivity, and specificity. In addition, its efficiency and robustness for all the different yoghurt alternatives make the method valuable for screening and quality assurance.

3.4. Determined Concentrations of Soyasaponins in Soybean-Based Yoghurt Alternatives

We applied the developed and validated sample extraction method to quantify the soyasaponins in five soybean-based yoghurt alternatives (Table 5). All analysed samples had similar soyasaponin concentrations. The soyasaponin Bb, soyasaponin Ba, and soyasaponin Ab concentrations ranged from 11.7 to 14.5 mg/100 g, 2.6 to 4.2 mg/100 g, and 2.7 to 8.5 mg/100 g, respectively. The soyasaponin Aa concentration in all measured samples was below the LOQ.

Table 5. Soyasaponins content (mg/100 g) in soybean-based yoghurt alternatives (mean ± standard deviation (n = 3)). ANOVA statistical significance test was performed across all analysed samples; means with different letters are significantly different at $p < 0.05$.

Sample Code	mg/100 g Soyasaponin Bb	Soyasaponin Ba	Soyasaponin Aa	Soyasaponin Ab
YA1	14.5 ± 0.4 [a]	2.8 ± 0.2 [b]	<LOQ	3.5 ± 0.4 [bc]
YA2	11.9 ± 0.8 [b]	2.9 ± 0.2 [b]	<LOQ	7.7 ± 0.3 [a]
YA3	12.6 ± 0.5 [b]	4.0 ± 0.4 [a]	<LOQ	8.5 ± 0.4 [a]
YA4	11.7 ± 0.4 [b]	2.6 ± 0.1 [b]	<LOQ	2.7 ± 0.3 [c]
YA5	13.3 ± 1.0 [ab]	4.2 ± 0.6 [a]	<LOQ	4.5 ± 0.5 [b]

Until now, limited data on the contents of soyasaponins in soybean-based dairy alternatives, including yoghurt alternatives, were available. A recently published study focused on characterizing the soyasaponin composition of 39 food products, including an analysis of 14 soybean-based milk alternative product [4]. The study showed that soyasaponin Aa was below the LOQ, soyasaponin Bb ranged from 27 to 308 mg/100 g dwb, soyasaponin Ba was quantified to be up to 14 mg/100 g dwb, and soyasaponin Ab ranged from 1 to 44 mg/100 g dwb in these products. Considering the average dry weight (~11%) of the samples in our study, our results showed that the average values of soyasaponin Bb, soyasaponin Ba, and soyasaponin Ab in the samples were 114 mg/100 g dwb, 29 mg/100 g dwb, and 54 mg/100 g dwb, which are in agreement with previously reported concentrations [4]. In another study, soyasaponin content was investigated [17], and the average sum of soyasaponin content in soybean-based milk alternatives was 39 µmol/100 g. The estimated average sum of quantified soyasaponins in the present study was 21 μmol/100 g in soybean-based yoghurt alternatives. Considering different methods used for quantification and possible different soy varieties, our results supported both previously published soyasaponin studies [4,17].

Previously, nine varieties of soybean were studied [3]. They found that only four varieties contained soyasaponin Aa (from 22.3 to 97.5 mg/100 g dwb of seed), two varieties contained soyasaponin Ab (from 75.8 to 95.5 mg/100 g dwb of seed), seven varieties contained soyasaponin Ba (up to 6.4 mg/100 g dwb of seed), and all nine studied varieties contained soyasaponin Bb (from 8.7 to 21.3 mg/100 g dwb of seed) among other soyasaponins. Generally, hypocotyls contained larger quantities of soyasaponins than cotyledons. Since hypocotyls might be removed during the production of soybean-based dairy alternatives, there may be smaller amounts of soyasaponins than in the original soybean seed [3]. The seeds contained DDMP-conjugated soyasaponin Bb and soyasaponin Ba, which might degrade into respective non-conjugated forms during the production of soybean-based yoghurt alternatives [1,2], resulting in higher soyasaponin contents reported in our study.

In another study, the soyasaponin content was analysed in tofu, one of the popular consumed soybean-based foods [4]. The authors showed that tofu had a quite diverse soyasaponin composition. Among others, tofu contained soyasaponin Aa at 80 mg/100 g dwb, soyasaponin Ab from 23 to 136 mg/100 g dwb, soyasaponin Ba from 5 to 11 mg/100 g dwb, and soyasaponin Bb ranging from 112 to 312 mg/100 g dwb of the product. The quantities of soyasaponins in tofu were also similar to the results obtained in the current study in the soybean-based yoghurt alternatives.

Although the number of analysed samples in the present study was small, due to the limited number of soybean-based yoghurt alternatives available on the local market, it was possible to demonstrate the applicability of the developed method on real samples. In the case of commercial end products, it is not possible to make assumptions about the content of saponins in the soybean varieties used for the production or the effectiveness of starter culture bacteria involved in the technological process to degrade saponins. Analysing the entire production chain, from soybean seeds to the final dairy alternative products, would provide a more comprehensive understanding of the mitigation of phytonutrients and allow for a thorough investigation of the entire technological process.

4. Conclusions

In this study, a new sample extraction method for the direct analysis of liquid samples was developed for the determination of soyasaponins in soybean-based yoghurt alternatives using HILIC-MS. The rapid LC-MS method was able to quantify soyasaponin Bb, soyasaponin Ba, soyasaponin Aa, and soyasaponin Ab using asperosaponin VI as an internal standard. The results show that the acidic pH of the soybean-based yoghurt alternatives significantly affected the quantification of soyasaponins, leading to unsatisfactory soyasaponin recoveries. To address this issue, the effect of alkalisation on the extraction yield of saponins was evaluated, and the highest yield (from 100 to 114%) was achieved at pH 8.0 ± 0.25. By adjusting the pH at the beginning of the sample extraction process, it became possible to achieve satisfactory recoveries of soyasaponins in soybean-based yoghurt alternatives. The developed method was validated using a soybean-based yoghurt alternative as a test matrix. Overall, the inter-day precision of the method was below 12%. This validated method could be applied in the analysis of commercially available soybean-based yoghurt alternatives and used in technology and product development, e.g., for the high-throughput screening of fermentation processes to unveil the saponins-degrading ability of starter cultures. The application of the presented method has the potential to enhance the acceptance of emerging and developed plant-based dairy alternatives by consumers by improving the quality of the final product and the taste by controlling the taste-active compounds. This method could also be extended for analyses of soyasaponins in dairy alternative products produced from other legume species.

Supplementary Materials: The following supporting information can be downloaded at https://www.mdpi.com/article/10.3390/foods12112164/s1. Table S1: Nutritional information of analysed products. Table S2: Repeatability of retention times (RTs), peak areas of soyasaponins, and precision of the whole method.

Author Contributions: Conceptualization, A.B.; methodology, A.B.; software, A.B.; validation, A.B.; formal analysis, T.K.; investigation, T.K.; resources, T.K.; data curation, A.B.; writing—original draft preparation, A.B.; writing—review and editing, T.K. and M.K.; visualization, A.B.; supervision, T.K. and M.K.; project administration, T.K.; funding acquisition, T.K. All authors have read and agreed to the published version of the manuscript.

Funding: The financial support for this research was provided by the European Union European Regional Development Fund (ERDF) and the Estonian Research Council via project RESTA16.

Data Availability Statement: The data are contained within the article or the Supplementary Materials.

Acknowledgments: The authors would like to thank Dmitri Pismennõi and Georg Arju for their assistance with mass spectrometric equipment, Aleksei Kaleda (TFTAK) for his assistance in statistical analysis, and Kristina Joarand for performing the sample analysis and validation.

Conflicts of Interest: The authors declare no conflict of interest. The financial supporters had no role in the design of the study; in the collection, analyses, or interpretation of data; in the writing of the manuscript; or in the decision to publish the results.

References

1. Montemurro, M.; Pontonio, E.; Coda, R.; Rizzello, C.G. Plant-Based Alternatives to Yogurt: State-of-the-Art and Perspectives of New Biotechnological Challenges. *Foods* **2021**, *10*, 316. [CrossRef] [PubMed]
2. Hu, J.; Lee, S.-O.; Hendrich, S.; Murphy, P.A. Quantification of the Group B Soyasaponins by High-Performance Liquid Chromatography. *J. Agric. Food Chem.* **2002**, *50*, 2587–2594. [CrossRef]
3. Chitisankul, W.T.; Takada, Y.; Takahashi, Y.; Ito, A.; Itabashi, M.; Varanyanond, W.; Kikuchi, A.; Ishimoto, M.; Tsukamoto, C. Saponin Composition Complexities in Hypocotyls and Cotyledons of Nine Soybean Varieties. *LWT* **2018**, *89*, 93–103. [CrossRef]
4. Chitisankul, W.T.; Itabashi, M.; Son, H.; Takahashi, Y.; Ito, A.; Varanyanond, W.; Tsukamoto, C. Soyasaponin Composition Complexities in Soyfoods Relating Nutraceutical Properties and Undesirable Taste Characteristics. *LWT* **2021**, *146*, 111337. [CrossRef]
5. Amin, H.A.S.; Hanna, A.G.; Mohamed, S.S. Comparative Studies of Acidic and Enzymatic Hydrolysis for Production of Soyasapogenols from Soybean Saponin. *Biocatal. Biotransf.* **2011**, *29*, 311–319. [CrossRef]
6. Hu, J.; Reddy, M.B.; Hendrich, S.; Murphy, P.A. Soyasaponin I and Sapongenol B Have Limited Absorption by Caco-2 Intestinal Cells and Limited Bioavailability in Women. *J. Nutr.* **2004**, *134*, 1867–1873. [CrossRef]
7. Neacsu, M.; Raikos, V.; Benavides-Paz, Y.; Duncan, S.H.; Duncan, G.J.; Christie, J.S.; Johnstone, A.M.; Russell, W.R. Sapogenol Is a Major Microbial Metabolite in Human Plasma Associated with High Protein Soy-Based Diets: The Relevance for Functional Food Formulations. *Foods* **2020**, *9*, 422. [CrossRef]
8. Addisu, S.; Assefa, A. Role of Plant Containing Saponin on Livestock Production; A Review. *Adv. Biol. Res.* **2016**, *10*, 309–314.
9. Samtiya, M.; Aluko, R.E.; Dhewa, T. Plant Food Anti-Nutritional Factors and Their Reduction Strategies: An Overview. *Food Prod. Process. Nutr.* **2020**, *2*, 6. [CrossRef]
10. Yates, P.S.; Roberson, J.; Ramsue, L.K.; Song, B.-H. Bridging the Gaps between Plant and Human Health: A Systematic Review of Soyasaponins. *J. Agric. Food Chem.* **2021**, *69*, 14387–14401. [CrossRef]
11. Ishii, Y.; Tanizawa, H. Effects of Soyasaponins on Lipid Peroxidation through the Secretion of Thyroid Hormones. *Biol. Pharm. Bull.* **2006**, *29*, 1759–1763. [CrossRef]
12. Kuzuhara, H.; Nishiyama, S.; Minowa, N.; Sasaki, K.; Omoto, S. Protective Effects of Soyasapogenol A on Liver Injury Mediated by Immune Response in a Concanavalin A-Induced Hepatitis Model. *Eur. J. Pharmacol.* **2000**, *391*, 175–181. [CrossRef]
13. Yang, S.H.; Ahn, E.-K.; Lee, J.A.; Shin, T.-S.; Tsukamoto, C.; Suh, J.; Mei, I.; Chung, G. Soyasaponins Aa and Ab Exert an Anti-Obesity Effect in 3T3-L1 Adipocytes Through Downregulation of PPARγ. *Phytother. Res.* **2015**, *29*, 281–287. [CrossRef]
14. Kim, H.-J.; Hwang, J.-T.; Kim, M.J.; Yang, H.-J.; Sung, M.J.; Kim, S.-H.; Park, S.; Gu, E.-J.; Park, Y.; Kwon, D.Y. The Inhibitory Effect of Saponin Derived from Cheonggukjang on Adipocyte Differentiation In Vitro. *Food Sci. Biotechnol.* **2014**, *4*, 1273–1278. [CrossRef]
15. Guang, C.; Chen, J.; Sang, S.; Cheng, S. Biological Functionality of Soyasaponins and Soyasapogenols. *J. Agric. Food Chem.* **2014**, *62*, 8247–8255. [CrossRef] [PubMed]
16. Savarino, P.; Demeyer, M.; Decroo, C.; Colson, E.; Gerbaux, P. Mass Spectrometry Analysis of Saponins. *Mass Spectrom. Rev.* **2021**, *42*, 954–983. [CrossRef]
17. Kamo, S.; Suzuki, S.; Sato, T. The Content of Soyasaponin and Soyasapogenol in Soy Foods and Their Estimated Intake in the Japanese. *Food Sci. Nutr.* **2014**, *2*, 289–297. [CrossRef] [PubMed]

18. Shiraiwa, M.; Harada, K.; Okubo, K. Composition and Structure of "Group B Saponin" in Soybean Seed. *Agric. Biol. Chem.* **1991**, *55*, 911–917. [CrossRef]
19. Okubo, K.; Iijima, M.; Kobayashi, Y.; Yoshikoshi, M.; Uchida, T.; Kudou, S. Components Responsible for the Undesirable Taste of Soybean Seeds. *Biosci. Biotechnol. Biochem.* **1992**, *56*, 99–103. [CrossRef]
20. Guo, S.; Chen, C. Analysis on the Bitterness and Astringency of Soymilk Processed from Different Soybean Cultivars. *J. Food Nutr. Popul. Health* **2018**, *2*, 101. [CrossRef]
21. Takada, Y.; Sasama, H.; Sayama, T.; Kikuchi, A.; Kato, S.; Ishimoto, M.; Tsukamoto, C. Genetic and Chemical Analysis of a Key Biosynthetic Step for Soyasapogenol A, an Aglycone of Group A Saponins That Influence Soymilk Flavor. *Theor. Appl. Genet.* **2013**, *126*, 721–731. [CrossRef] [PubMed]
22. Dini, I.; Schettino, O.; Simioli, T.; Dini, A. Studies on the Constituents of Chenopodium Quinoa Seeds: Isolation and Characterization of New Triterpene Saponins. *J. Agric. Food Chem.* **2001**, *49*, 741–746. [CrossRef] [PubMed]
23. Decroos, K.; Vincken, J.-P.; Heng, L.; Bakker, R.; Gruppen, H.; Verstraete, W. Simultaneous Quantification of Differently Glycosylated, Acetylated, and 2,3-Dihydro-2,5-Dihydroxy-6-Methyl-4H-Pyran-4-One-Conjugated Soyasaponins Using Reversed-Phase High-Performance Liquid Chromatography with Evaporative Light Scattering Detection. *J. Chromatogr. A* **2005**, *1072*, 185–193. [CrossRef] [PubMed]
24. Oleszek, W.; Bialy, Z. Chromatographic Determination of Plant Saponins—An Update (2002–2005). *J. Chromatogr. A* **2006**, *1112*, 78–91. [CrossRef]
25. Decroo, C.; Colson, E.; Lemaur, V.; Caulier, G.; De Winter, J.; Cabrera-Barjas, G.; Cornil, J.; Flammang, P.; Gerbaux, P. Ion mobility mass spectrometry of saponin ions. *Rapid Commun. Mass Spectrom.* **2019**, *33*, 22–33. [CrossRef]
26. Berhow, M.A.; Singh, M.; Bowman, M.J.; Price, N.P.J.; Vaughn, S.F.; Liu, S.X. Quantitative NIR Determination of Isoflavone and Saponin Content of Ground Soybeans. *Food Chem.* **2020**, *317*, 126373. [CrossRef] [PubMed]
27. Chien, H.-J.; Wang, C.-S.; Chen, Y.-H.; Toh, J.-T.; Zheng, Y.-F.; Hong, X.-G.; Lin, H.-Y.; Lai, C.-C. Rapid Determination of Isoflavones and Other Bioactive Compounds in Soybean Using SWATH-MS. *Anal. Chim. Acta* **2020**, *1103*, 122–133. [CrossRef] [PubMed]
28. Abdelghani, D.Y.; Gad, A.I.; Orabi, M.M.; Abou-Taleb, K.A.; Mahmoud, E.A.; Al Amoudi, S.A.; Zari, A.; Althubaiti, E.H.; Edris, S.; Amin, S.A. Bioactivity of Organic Fermented Soymilk as Next-Generation Prebiotic/Probiotics Mixture. *Fermentation* **2022**, *8*, 513. [CrossRef]
29. Chang, S.Y.; Kim, D.-H.; Han, M.J. Physicochemical and Sensory Characteristics of Soy Yogurt Fermented with Bifidobacterium Breve K-110, Streptococcus Thermophilus 3781, or Lactobacillus Acidophilus Q509011. *Food Sci. Biotechnol.* **2010**, *19*, 107–113. [CrossRef]
30. Bljahhina, A.; Pismennõi, D.; Kriščiunaite, T.; Kuhtinskaja, M.; Kobrin, E.-G. Quantitative Analysis of Oat (*Avena sativa* L.) and Pea (*Pisum sativum* L.) Saponins in Plant-Based Food Products by Hydrophilic Interaction Liquid Chromatography Coupled with Mass Spectrometry. *Foods* **2023**, *12*, 991. [CrossRef]
31. EMA Bioanalytical Method Validation—Scientific Guideline. Available online: https://www.ema.europa.eu/en/bioanalytical-method-validation-scientific-guideline (accessed on 27 November 2022).
32. Kruve, A.; Rebane, R.; Kipper, K.; Oldekop, M.-L.; Evard, H.; Herodes, K.; Ravio, P.; Leito, I. Tutorial Review on Validation of Liquid Chromatography–Mass Spectrometry Methods: Part II. *Anal. Chim. Acta* **2015**, *870*, 8–28. [CrossRef] [PubMed]
33. AOAC. Guidelines for Standard Method Performance Requirements. Appendix F. 2016, p. 18. Available online: https://www.aoac.org/wp-content/uploads/2019/08/app_f.pdf (accessed on 10 January 2023).
34. Gu, L.; Tao, G.; Gu, W.; Prior, R.L. Determination of Soyasaponins in Soy with LC-MS Following Structural Unification by Partial Alkaline Degradation. *J. Agric. Food Chem.* **2002**, *50*, 6951–6959. [CrossRef] [PubMed]
35. Donat, P.V.; Caprioli, G.; Conti, P.; Maggi, F.; Ricciutelli, M.; Torregiani, E.; Vittori, S.; Sagratini, G. Rapid Quantification of Soyasaponins I and Bg in Italian Lentils by High-Performance Liquid Chromatography (HPLC)–Tandem Mass Spectrometry (MS/MS). *Food Anal. Methods* **2014**, *7*, 1024–1031. [CrossRef]
36. PubChem Asperosaponin VI. Available online: https://pubchem.ncbi.nlm.nih.gov/compound/14284436 (accessed on 26 February 2023).
37. Shimoyamada, M.; Osugi, Y.; Shiraiwa, M.; Okubo, K.; Watanabe, K. Solubilities of Soybean Saponins and Their Solubilization with a Bisdesmoside Saponin. *J. Jpn. Soc. Food Sci. Technol.* **1993**, *40*, 210–213. [CrossRef]
38. Grant, R.P.; Rappold, B.A. Development and Validation of Small Molecule Analytes by Liquid Chromatography-Tandem Mass Spectrometry. In *Principles and Applications of Clinical Mass Spectrometry*, 1st ed.; Rifai Horvath, A.R., Wittwer, C.T., Eds.; Elsevier: London, UK, 2018; pp. 115–179.

foods

Article

Low-Field Benchtop NMR Spectroscopy for Quantification of Aldehydic Lipid Oxidation Products in Culinary Oils during Shallow Frying Episodes

Miles Gibson, Benita Claire Percival, Mark Edgar and Martin Grootveld *

Leicester School of Pharmacy, De Montfort University, The Gateway, Leicester LE1 9BH, UK
* Correspondence: mgrootveld@dmu.ac.uk; Tel.: +44-(0)116-250-6443

Abstract: Introduction: Toxic aldehydic lipid oxidation products (LOPs) arise from the thermo-oxidative deterioration of unsaturated fatty acids present in heated culinary oils when exposed to high-temperature frying episodes, and currently these effects represent a major public health concern. Objectives: In this study, we investigated the applications of low-field (LF), benchtop NMR analysis to detect and quantify toxic aldehyde species in culinary oils following their exposure to laboratory-simulated shallow frying episodes (LSSFEs) at 180 °C. Four culinary oils of variable fatty acid (FA) composition were investigated to determine the analytical capabilities of the LF NMR instrument. Oil samples were also analysed using a medium-field (400 MHz) NMR facility for comparative purposes. Results: Aldehydes were quantified as total saturated and total α,β-unsaturated classes. The time-dependent production of α,β-unsaturated aldehydes decreased in the order chia > rapeseed ≈ soybean > olive oils, as might be expected from their polyunsaturated and monounsaturated FA (PUFA and MUFA, respectively) contents. A similar but inequivalent trend was found for saturated aldehyde concentrations. These data strongly correlated with medium-field ^{1}H NMR data obtained, although LF-determined levels were significantly lower in view of its inability to detect or quantify the more minor oxygenated aldehydic LOPs present. Lower limit of detection (LLOD) values for this spectrometer were 0.19 and 0.18 mmol/mol FA for *n*-hexanal and *trans*-2-octenal, respectively. Aldehydic lipid hydroperoxide precursors of aldehydic LOPs were also detectable in LF spectra. Conclusions: We therefore conclude that there is scope for application of these smaller, near-portable NMR facilities for commercial or 'on-site' quality control determination of toxic aldehydic LOPs in thermally stressed frying oils.

Keywords: benchtop NMR spectrometer; frying practices; cooking oils; PUFAs; MUFAs; lipid hydroperoxides; aldehyde toxins; quality control

Citation: Gibson, M.; Percival, B.C.; Edgar, M.; Grootveld, M. Low-Field Benchtop NMR Spectroscopy for Quantification of Aldehydic Lipid Oxidation Products in Culinary Oils during Shallow Frying Episodes. *Foods* **2023**, *12*, 1254. https://doi.org/10.3390/foods12061254

Academic Editor: Gianfranco Picone

Received: 15 February 2023
Revised: 7 March 2023
Accepted: 10 March 2023
Published: 15 March 2023

1. Introduction

High-field, high-resolution ^{1}H nuclear magnetic resonance (NMR) analysis is extensively applied in lipidomics, together with determinations of the full composition and molecular nature of many culinary oil products (including the saturation and unsaturation status of major fatty acids (FAs) present, for example), their longevities, and even geographic origins [1]. More recently, it has also been quite widely employed for the specific determination of toxic lipid oxidation products (LOPs) in these products, including a series of aldehydes which are formed in these commonly employed cooking oils during high-temperature frying practices [2–4]. Furthermore, additional studies have extensively documented the potential toxicological hazards that these aldehydic species may present, along with their overall public health implications [5–7]. Indeed, these aldehydes, especially the α,β-unsaturated classes, are highly chemically reactive, and they form relatively stable, latent source adducts with many critical biomolecules in vivo, for example selected proteins and DNA. Such DNA damage renders these toxins mutagenic, genotoxic and, at least in some cases, carcinogenic. Consistently, Weng et al. reported that aldehydes represent the dominant carcinogens present in tobacco cigarette smoke [8]. The causal links between reactive aldehyde species and non-communicable chronic disease (NCD) risks in humans are therefore of much pertinent importance, and to date there is much evidence available indicating associations between the consumption of fried foods and the development and progression of a range of serious NCDs, for example, coronary heart diseases [9] and prostate cancer [10]. Further studies, e.g., that reported in [11], have established associations between human exposure to Chinese-style cooking fumes, which contain high levels of toxic aldehydes such as acrolein, and the risk of developing lung cancer. Moreover, it has been demonstrated that aldehydic toxins are readily transferred to foods from the oils in which they are fried; these foods typically include potato chips, beef patties, and fried chicken, which are frequently consumed by humans [12]. However, these levels were found to be significantly lower in fried potato chips than they were in the frying oil itself (only ca. 5% or so in mol/kg units), and this is attributable to only a small amount of the food mass being accounted for by uptake of aldehyde-rich oil (typically 10–15% (*w/w*) [13]), and their known chemical reactions with different classes of food biomolecules. A range of other factors such as frying oil FA compositions and frying duration are also relevant.

To date, high-field (HF) NMR spectroscopy has been successfully utilised for the rapid multicomponent analysis of quite a wide variety of LOPs present in thermally stressed frying oils, which have included primary conjugated hydroperoxydiene isomers, their secondary fragmentation products (particularly saturated and unsaturated aldehydes), and epoxy-fatty acids, for example [4,14]. Moreover, the use of two-dimensional correlation spectroscopies, both homo- and heteronuclear, has been invaluable for confirming provisional LOP assignments made in 1D spectra [15]. Notably, our research group was the very first to report these applications and advantages as early as 1994 [16]. Such advances have recently led to the development of methods for the analysis of such LOPs in fried foods, approaches which feature a key lipid extraction stage [12].

Notwithstanding, the use of both medium-field (MF, with 300–400 MHz operating frequencies) and HF ^{1}H NMR spectroscopy for the simultaneous multicomponent analysis of aldehydic and further LOPs in such matrices is reliant upon a number of operational necessities such as the institutional accessibility of such expensive instruments, which are dependent on bulky cryogenically cooled superconducting magnets, the frequent use of high volumes of deuterated solvents such as deuterochloroform ($CDCl_3$) for sample preparation purposes, and requirements for the professional inputs of highly specialised operational technical staff, along with the availability of those with specialist spectral interpretational skills. Moreover, the high costs of such large HF facilities (e.g., with 500–700 MHz operating frequencies), together with their stringent demands for high volumes of cryogenic cooling gases [16], limits their applications in many academic institutions, or within the commercial sector. Comparatively, low-field (LF) 'benchtop' NMR instruments benefit from considerably lower power requirements, permanent magnets, and virtual portability. Thus, LF NMR analysis shows great promise for applications within industrial cooking oil production, or even restaurant settings, for the direct 'on-site' detection and determination of edible oil quality, together with screens for toxic LOPs generated within during the use of such frying media.

Currently, such LF NMR spectrometers are emerging, cost-effective and portable alternatives to their HF counterparts [17]. Indeed, recent studies have demonstrated the growing applications of LF NMR spectroscopy to numerous analytical sectors, including 'point-of-care' medical diagnostics [17], forensic science, synthetic chemical reaction monitoring [18], and versatile approaches towards chemical education [19]. Furthermore, selected hyperpolarisation techniques have been applied to specific investigations [20]. The research landscape of developing applications for LF NMR has also been shown to offer major advantages in the detection of vegetable oil adulteration [21,22], together with both one- (1D) and two-dimensional (2D) NMR approaches for probing and determining edible oil authenticities [23].

Previously, preliminary studies of the LF NMR detection and analysis of aldehyde species in used and reused cooking oils has only been documented by Grootveld et al. (2014) [24,25], wherein detectable levels of selected aldehyde classes were observed in heated sunflower oil samples, and olive oil collected from a 'real-life' restaurant site. In the current study, we report the further exploration and development of LF NMR techniques for evaluating frying oil qualities. In particular, we consider the analytical reliability and reproducibility of LF ^{1}H NMR determinations of aldehydic LOPs in oils heated according to shallow frying practices, and also limits for their detection and quantification. Comparisons of these analytical data to those acquired at MF strength are also made. We also consider the potential health risk status of these products through the reliable quantification of aldehydic LOP toxins. Overall, this LF NMR analysis technology highlights great promise for application at industrial food product manufacturing sites or restaurants for the detection and determination of these agents.

2. Materials and Methods

2.1. Reagents and ^{1}H NMR Solvents

All materials, including deuterochloroform ($CDCl_3$) and aldehyde calibration standards, were purchased from Sigma-Aldrich Chemical Co. (UK), unless otherwise stated; the $CDCl_3$ product contained the ^{1}H NMR reference standard tetramethylsilane (TMS) at an added level of 1.00% (*v*/*v*) (equivalent to 73.45 mmol/L on consideration of its density of 0.648 g/mL). NMR tubes were purchased from Norell (Morganton, NC, USA), and Eppendorf^{TM} microcentrifuge tubes were obtained from Fisher Scientific Ltd. (Loughborough, UK).

2.2. Culinary Oil Products and Their FA Compositions

Four culinary oils of differing triacylglycerol FA compositions were studied. These comprised a mixed-origin refined olive oil (MOO), soybean oil (SBO), rapeseed oil (RSO) and chia seed oil (CSO). With the exception of the SBO, which originated from a US retail source, all these products were purchased from reputable retail outlets based in the UK. The acylglycerol contents of FA classes present in the oils analysed were calculated using major acylglycerol resonance intensities from 400 MHz ^{1}H NMR spectra of the unheated control (0 min heating time-point) samples, according to the methodological equations reported in Ref. [4].

Samples were stored under dark conditions at ambient temperature to diminish photodegradative peroxidation during storage periods prior to analysis. The lipid content profiles of each oil analysed, which are represented as % (*w/w*) saturated (SAT), monounsaturated (MUFA) and polyunsaturated (PUFA) FA contents, were found to be MOO: 16% SAT, 76% MUFA and 8% (*w/w*) PUFA; SBO: 16% SAT, 24% MUFA and 60% (*w/w*) PUFA, the latter including 5% omega-3 FAs (predominantly linolenoylglycerols); RSO: 5% SAT, 67% MUFA and 28% (*w/w*) PUFA, the latter including 7% (*w/w*) omega-3 FAs (predominantly linolenoylglycerols); CSO: 9% SAT, 9% MUFA and 82% (*w/w*) PUFA, the latter including 65% (*w/w*) omega-3 FAs (again predominantly as linolenoylglycerols). For the SBO, RSO and CSO products, omega-3 FAs were analysed via electronic integration of their characteristic linolenoylglycerol-distinctive $\delta = 0.95$ ppm triplet resonance in 400 MHz spectra acquired on this oil.

Preliminary experiments conducted also featured a commercially available sample of a UK refined sunflower oil (SFO) product. This oil was found to have an FA content of 10% SAT, 31% MUFA and 59% PUFA.

2.3. Aldehydic LOP Calibration Standards

Aldehyde calibration standards were prepared using *n*-hexanal and *trans*-2-octenal in deuterochloroform ($CDCl_3$) solution, and increasing concentrations of these analyte solutions were added to unheated (control) olive oil (MOO product) samples. These preparations were conducted by adding a 0.20 mL aliquot of each aldehyde calibration solution (ranging from 0.00 to 54.98 mmol/L for *n*-hexanal, and 0.00 to 40.40 mmol/L for *trans*-2-octenal), 0.20 mL of unheated olive oil, 0.20 mL of additional $CDCl_3$ solvent, and 0.10 and 0.06 mL of solutions of the lipid-soluble antioxidant 2,5-di-*tert*-butylhydroquinone (2,5-DTBHQ) and the secondary ^{1}H NMR chemical shift reference and internal standard 1,3,5-trichlorobenzene (TCB) (67.0 and 19.8 mmol/L, respectively), both in $CDCl_3$. The densities of these aldehydes were accounted for when adding or diluting μL volumes of each. For experiments in which the signal-to-noise (STN) ratio was determined and evaluated, a corresponding series of aldehyde calibration standards were prepared in a medium without any added oil, and in these cases the 0.20 mL unheated olive oil constituent was replaced with an equivalent volume of additional $CDCl_3$.

2.4. H NMR Analysis

^{1}H NMR analysis of prepared aldehyde calibration standard solutions, and control and thermally stressed oil samples, was performed on a 60 MHz Magritek Spinsolve Benchtop system operating at a frequency of 61.67 MHz. Spectra were acquired using a 1D Proton+ sequence. Parameters employed for these analyses were 32K data points; 128 scans; acquisition time 6.4 s; repetition time 10 s; and a pulse angle of 90° (the total sample acquisition duration was ~21 min). These spectra were also acquired on a MF 400 MHz Bruker Avance AV400 NMR spectrometer (Leicester School of Pharmacy, De Montfort University, Leicester, UK) operating at a frequency of 399.93 MHz. For this facility, spectral acquisition parameters were 32K data points; 128 scans with 2 dummy scans; 3 μs pulses, spectral width 8278 Hz; and a receiver gain setting of 14.8.

2.5. Purity of Aldehyde Calibrations Standards Used

Reference ^{1}H NMR spectra of these aldehyde standards revealed that they contained 17 and 5 mol% of their corresponding carboxylic acid oxidation products, i.e., hexanoic acid and *trans*-2-octenoic acid, for *n*-hexanal and *trans*-2-octenal, respectively, and this was again considered when preparing their calibration standard solutions; purities reported by their manufacturer were 98 and 94% (*w/w*), respectively. The identities of these carboxylic acid oxidation products were confirmed by the observation of their characteristic ^{1}H NMR resonances in the aldehyde spectra acquired, i.e., that of the α-CH_2 protons of hexanoic acid (*t*, $\delta = 2.354$ ppm, $J = 7.67$ Hz) and those of the 2- and 3-positon olefinic protons in *trans*-2-octenoic acid (i.e., *dt*, $\delta = 5.831$ ppm, $J = 16.0, 1.3$ Hz, and *m*, $\delta = 6.660$ ppm, $J = 15.9, 7.0$ Hz, respectively) [26]. The concentrations and contaminating mol % values of these oxidation products in these standard solutions were determined via electronic integration of these resonances, together with the corresponding, more prominent signals arising from their parent aldehydes.

2.6. H NMR Determination of Different Classes of Aldehydes

Aldehyde concentrations were determined by electronic integration of their characteristic ^{1}H NMR -C$\underline{H}$O function resonances, and normalising these data to those of the total acylglycerol terminal-C$\underline{H}_3$ function resonance ($\delta = 0.82$–1.11 ppm) of total FAs from the added unheated olive oil co-calibrant at both 60 and 400 MHz operating frequencies, so that aldehyde concentrations are reported as mmol aldehyde per mol of total FA (mmol/mol FA units). Aldehydic LOP levels in control and thermally stressed culinary oil products were then determined via reference to the calibration plots shown in the Results Section below. However, it should be noted that for the purpose of comparative evaluations, aldehydes were simply classified as either total saturated or total α,β-unsaturated, since it was not possible to effectively resolve superimposing resonances of individual sub-classes of these analyte species within their overall saturated and α,β-unsaturated classes at LF strength (60 MHz).

2.7. Calibration and Bland–Altman Dominance Plots of Authentic Standard Aldehyde Solutions in Combined $CDCl_3$/Culinary Oil Media

Calibration and Bland–Altman style dominance plots of the ^1H NMR-determined concentrations of the saturated and α,β-unsaturated aldehydes *n*-hexanal and *trans*-2-octenal, respectively, involved matched analysis sample datasets, with determinations made on these two analytes at both 60 and 400 MHz operating frequencies. Units for these determinations, which were conducted in the combined $CDCl_3$/culinary oil medium, were mmol aldehyde/mol total acylglycerol FA. For these plots, Dixon's test was applied to detect any potential outlier samples, but none were found. Similarly, all determinations which were found to have none detectable (nd, specifically values below the specified lower limit of detection (i.e., <LLOD) at both operating frequencies utilised were removed from the datasets. As recommended [27], corresponding ^1H NMR profiles of blank samples, which were prepared as described above but with $CDCl_3$ in place of culinary oils, were acquired, and their 'noise' intensities at the appropriate δ values were included in these calibration plots. Spectra were acquired on replicate (n = 3) preparations of such blank samples for these purposes.

2.8. Estimation of Signal-to-Noise Ratios, and Lower Limits of Detection and Quantification Values, for Saturated and α,β-Unsaturated Aldehyde Analyte Solution Calibrants in Neat $CDCl_3$ and Combined $CDCl_3$/Culinary Oil Media

STN ratios for total saturated and α,β-unsaturated aldehydes were monitored by LF ^1H NMR analysis of the *n*-hexanal and *trans*-2-octenal calibrant standards for each of these two aldehyde classes, and which were fully resolvable and quantifiable at this field strength, were determined using built-in software scripts within the *MestreNova* software package (Feliciano Barrera 9B—Bajo, 15,706 Santiago de Compostela, Spain). The signal-free region used to calculate the noise level was δ = 10.50–11.50 ppm, and Equation (1) was employed to derive the standard deviation of this value (*Noise SD*), where N = number of datapoints within the signal-free region; Y_i = value of each digital point in the spectrum; and Y_m = mean value of the digital points within that region. Mean STN ratio values were then computed as the ratio of the aldehyde resonance peak heights to that of the *Noise SD* value.

$$Noise\ SD = \sqrt{\sum \frac{(Y_i - Y_m)^2}{(N-1)}} \tag{1}$$

Lower limits of detection and quantification (LLOD and LLOQ, respectively) were estimated as three-times and ten-times the mean STN ratio values (3(STN) and 10(STN), respectively).

2.9. Thermal Stressing of Culinary Oil Products According to Laboratory-Simulated Shallow Frying Episodes

Each culinary oil evaluated was exposed to laboratory-simulated shallow frying episodes (LSSFEs) for periods of 0–90 min as previously described in Ref. [12], although for the experiments outlined here, samples were collected for ^1H NMR analysis at the 0, 30, 60 and 90 min time-points. A total of n = 3 replicate samples for each culinary oil tested and each collection time-point were obtained. Volumes (6.00 mL) of each culinary oil were placed in air-dried 250 mL glass beakers within a thermostatted silicon oil bath, which was heated to 180 °C in the presence of atmospheric O_2 according to our LSSFEs in order to simulate shallow frying conditions. Aliquots (0.25 mL) of these oils were sampled for ^1H NMR analysis at each of the above time-points.

2.10. Sample Collection and Preparation for ^1H NMR Analysis

Aliquots (0.20 mL) of all oil samples collected were transferred to 1.5 mL micro-centrifuge tubes, and then 0.40 mL of $CDCl_3$ containing 1% (*v*/*v*) TMS as a chemical shift reference (δ = 0.00 ppm), and 0.06 and 0.10 mL volumes of $CDCl_3$ solutions of TCB (67.0 mmol/L) and the chain-breaking antioxidant 2,5-DTBHQ (19.8 mmol/L), respectively, were added. 2,5-DTBHQ was included in these preparations in order to prevent any artefactual peroxidation of culinary oil unsaturated FAs (UFAs) during periods of sample preparation and storage prior to analysis. Subsequently, samples were vortexed thoroughly and transferred to 5 mm diameter Norell NMR tubes for ^{1}H NMR analysis.

^{1}H NMR analysis of these oil samples was performed at both 60 and 400 MHz operating frequencies as described in Section 2.4

2.11. ANOVA Model for the Statistical Analysis of Experimental Datasets

Univariate statistical chemometrics analysis was conducted to detect any significant differences between the mean replicate values of both total saturated and total α,β-unsaturated aldehydic LOPs for each oil product evaluated, each LSSFE sampling time-point (60 versus 90 min) and for each instrumental operating frequency (60 versus 400 MHz) by an analysis-of-variance (ANOVA) model. The experimental design employed comprised a three-factor model with the fixed effects of culinary oil product (O_i), LSSFE sampling time-point (T_j) and spectrometer operating frequency (F_k). Also incorporated in the designs were first-order oil product × sampling time-point, oil product × spectrometer operating frequency, and sampling time-point × spectrometer operating frequency interaction effects (OT_{ij}, OF_{ik} and TF_{jk}, respectively). The mathematical model for this design is displayed in Equation (2), where y_{ijkl} represents each replicate aldehyde concentration, μ the overall sample mean aldehyde concentration in the absence of any possible explanatory sources of variation, and e_{ijkl} fundamental error. An ANOVA rather than an analysis-of-covariance (ANCOVA) model was selected for this analysis since the latter approach is dependent on linear relationships between response variables (aldehyde concentrations) and the quantitative sampling time-point covariable considered, and this was clearly not the case for the dataset acquired in this study, since it is well known that all such relationships are sigmoidal (S-shaped) and not linear [12]. Software module options utilised for the performance of these ANOVA models were those available from *XLSTAT2020* (Addinsoft, Paris, France). Post hoc evaluations of the different oil products tested involved comparisons of their mean values using the Bonferroni test.

$$y_{ijkl} = \mu + O_i + T_j + F_k + OT_{ij} + OF_{ik} + TF_{jk} + e_{ijkl} \quad (2)$$

Subsequent to application of this model, if all first-order interaction terms were found not to be statistically significant (as indeed they were for the total unsaturated but not the saturated aldehyde outcome variables), they were removed from the model, and their variance contributions were then transferred to that of the error term for the re-testing of the main factor effects only (Equation (3)).

$$y_{ijkl} = \mu + O_i + T_j + F_k + e_{ijkl} \quad (3)$$

Any missing data from the above ANOVA models were estimated by replacement with group mean or mode values, followed by reduction of error mean square degrees-of-freedom values by the number of such replacements made accordingly.

2.12. Computational Simulations of the ^{1}H NMR Spectra of Solution-Phase Triacylglycerols

Simulated ^{1}H NMR spectra of triacylglycerols, and that of their glycerol backbone, were obtained using *Bruker TopSpin NMR-SIM* software. Chemicals shift and coupling constant values, and example ^{1}H NMR spectra, were first obtained from the *Human Metabolome Database* (HMBD) [26]. A 400 MHz reference spectrum of glycerol in hexadeuterated dimethylsulphoxide (d6-DMSO), was first simulated, and these data were then applied and optimised to simulate the experimental spectra of a typical triacylglycerol species at both 60 and 400 MHz operating frequencies.

3. Results

3.1. H NMR Analysis of Major Acylglycerol FA Classes in Control (Unheated) Culinary Oils at 60 and 400 MHz Operating Frequencies

The ^{1}H NMR profiles of major acylglycerols present in control (unheated) culinary oil samples were obtained for all products tested, and the 60 MHz spectra obtained were more than adequate to allow determinations of different types of FAs present, specifically SATs, MUFAs and PUFAs. Although the terminal-CH_3 signal of ω-3 FAs (δ = 0.96 ppm) was at least partially distinguishable from that of other PUFAs (predominantly ω-6 FAs), MUFAs and SATs (δ = 0.91 ppm) at an operating frequency of 60 MHz, its electronic integration and quantification was precluded in view of resonance overlap with the latter more intense corresponding bulk lipid signal. Figure 1 shows typical spectral profiles of soybean oil (0.00–10.00 ppm), and characteristic resonances assigned and labelled are listed in Table 1, including those arising from oleoylglycerols, linoleoylglycerols and linolenoylglycerols, together with saturated FAs.

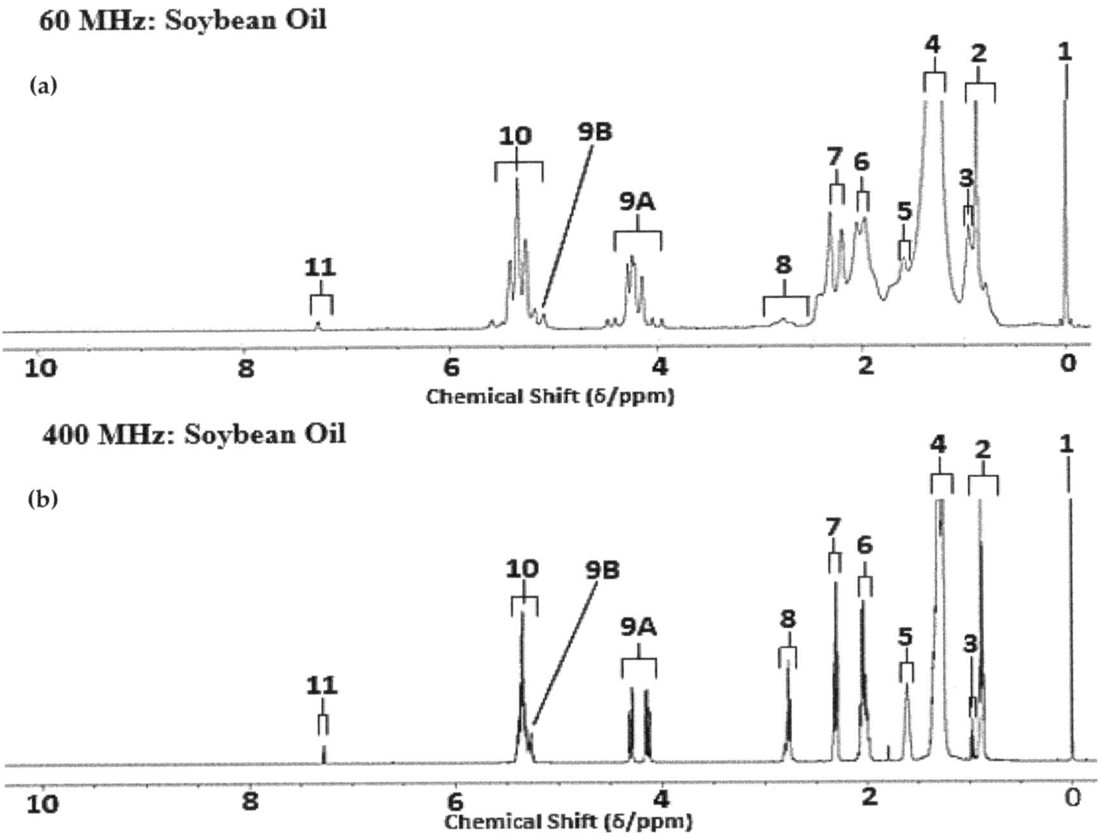

Figure 1. *Cont.*

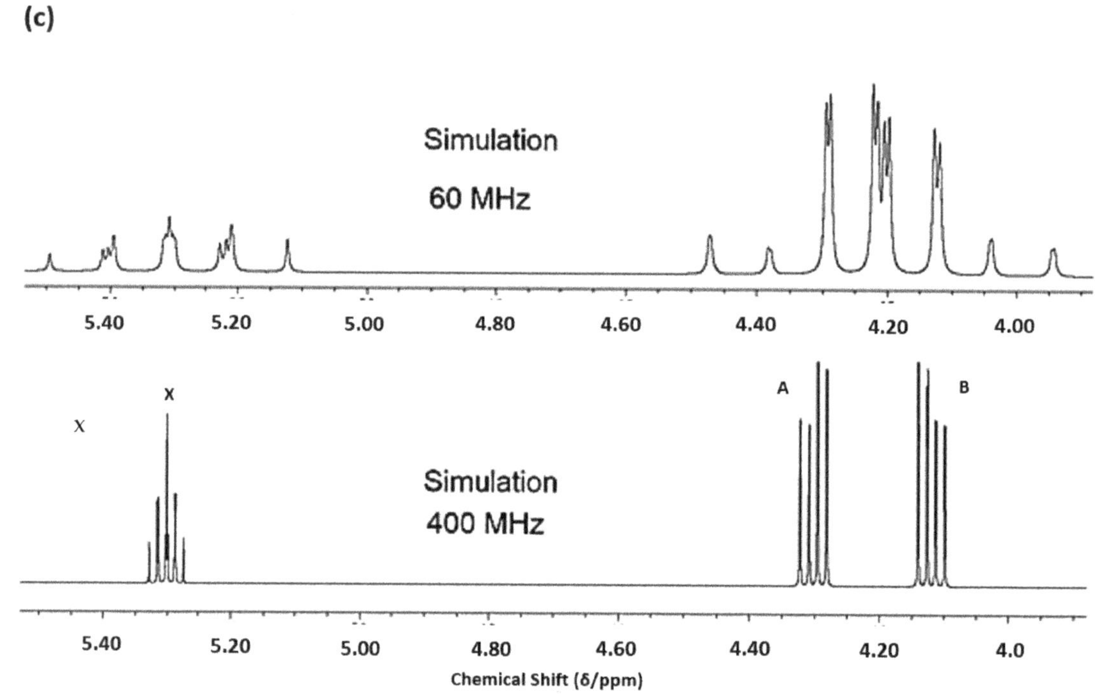

Figure 1. Comparative evaluation of the ^{1}H NMR spectral profiles of bulk major lipids in an unheated (control) sample of SBO acquired at (**a**) 60 and (**b**) 400 MHz operating frequencies. Assignment labels for resonances are available in Table 1. (**c**) Simulated partial ^{1}H NMR spectrum of the glycerol backbone resonances of a typical triacylglycerol species (expanded Resonance 9A and 9B regions (3.90–5.50 ppm)). Resonances A and B had centralised chemical shift values of 4.22 and 5.30 ppm for this simulation. The 2JAB coupling constant utilised was 10.8 Hz, with 3JAX = 6.0 Hz, and 3JBX = 5.2 Hz.

Table 1. Assignments for the major lipid resonances found in the ^{1}H NMR spectra of the culinary oils evaluated in this study at both 60 and 400 MHz operating frequencies. The spectral ranges and multiplicities of these signals are also provided in the second column.

Resonance Assignment Code	δ/ppm (*Coupling Pattern*)	Assignment
1	0.00 (*s*)	TMS-CH_3
2	0.85–0.92 (*t*)	Acyl chain Terminal-CH_3 (non-ω-3 FAs)
3	0.94–0.97 (*t*)	Acyl chain Terminal-CH_3 (ω-3 FAs)
4	1.20–1.38 (*m*)	Bulk acyl chain -$(CH_2)_n$-
5	1.56–1.66 (*m*)	Acyl chain -OCO-CH_2-$\underline{CH_2}$-
6	1.95–2.11 (*m*)	MUFA/PUFA Acyl chain -CH=CH-$\underline{CH_2}$-
7	2.26–2.36 (*dt*)	Acyl chain -OCO-$\underline{CH_2}$-
8	2.75–2.82 (*dd*)	PUFA Acyl chain -CH=CH-$\underline{CH_2}$-CH=CH-
9A	4.11–4.32 (<u>AB</u>X, *dd, dd*)	Glycerol backbone -$\underline{CH_2}$-OCO-
9B	5.20–5.23 (AB<u>X</u>, *dd*)	Glycerol backbone -C<u>H</u>-OCO-
10	5.23–5.28 (*m*)	MUFA/PUFA -C<u>H</u>=C<u>H</u>-
11	7.26 (s)/7.27 (s)	TCB Aromatic Protons/C<u>H</u>Cl_3

Although the 400 MHz ^{1}H NMR profile shows a clear AB portion of the ABX multiplet (<u>AB</u>X) of its triacylglycerol glycerol backbone -CH_2OCO- resonance pattern (δ = 4.11–4.32 ppm), at 60 MHz the appearance of this multiplet is significantly different, and this observation is attributable to the well-known 'roofing' effect that occurs when scalar *J*-coupled signals appear just a few Hz away from each other in spectra acquired. Whilst the chemical shift difference between the A and B signals of the glycerol backbone-C<u>H</u>$_2$OCO- group are the same at 60 and 400 MHz when measured in ppm, when determined in Hz, the difference in chemical shift between these markedly differs between these two operating frequencies (there are only 60 Hz per ppm at 60 MHz, but 400 Hz per ppm at 400 MHz). Therefore, the signals of A and B in this ABX system are spectrally 'closer' in Hz when measured at 60 MHz (15 Hz apart) compared to when spectra are acquired at 400 MHz (73 Hz apart). Given that the $2J$ AB scalar coupling for the two magnetically inequivalent -C<u>H</u>$_2$OCO- protons is 10.8 Hz, the chemical shift separation of these A and B protons in Hz is very similar to the size of the coupling constant between them at 60 MHz, and therefore this resonance appears as a distorted second-order multiplet. However, at 400 MHz, the separation between the chemical shift values of these protons is much larger than the coupling between them (i.e., a 73 Hz separation); therefore, at this MF strength, the multiplet appears as a 'classical' doublet of doublets for both the A and B portions of it. The low intensity signals located both slightly downfield and upfield of the multiplet at 60 MHz operating frequency (δ = 4.11–4.32 ppm) are therefore not unexpected, and have been be accurately calculated and simulated in this work, as shown in Figure 1c.

3.2. *Detection and Quantification of Distinctive Aldehyde Classes at Different ^{1}H NMR Operating Frequencies*

The original experimental design for this investigation primarily featured the heating of culinary oils according to LSSFEs at a temperature of 180 °C for durations of 0, 30, 60 and 90 min. However, although reliable aldehyde levels were provided at an operating frequency of 400 MHz, the majority of aldehyde levels in all cooking oils explored at the 30 min sampling time-point were below, or far below the LLOQ value determined for the LF 60 MHz spectrometer employed, and therefore we elected to completely remove the 30 min dataset from this investigation. Hence, the remaining sampling time-points for this investigation were the 0, 60 and 90 min heating exposure durations only. However, little or no aldehydic LOPs were detectable in the oil samples at the zero-control (unheated) time-point, even at an operating frequency of 400 MHz.

Aldehydic LOP species arising from the thermo-oxidation of UFAs in edible oils exposed to laboratory-simulated shallow frying episodes were readily detectable by LF ^{1}H NMR analysis, and the major signals identifiable by ^{1}H NMR analysis predominantly arose from *trans*-2-alkenals, *trans*-*trans*-2,4-alkadienals and *n*-alkanals (Figure 2). Although this 60 MHz technique was able to partially distinguish between these two major α,β-unsaturated aldehydes generated (doublets located at δ = 9.48 and 9.52 ppm, respectively), their individual quantification using this operating frequency was not possible. However, these two unsaturated aldehyde signals were indeed resolvable from that of *n*-alkanals (δ = 9.74 ppm, *t*) at this operating frequency [12]. Additionally, signals assignable to 4,5-epoxy-*trans*-2-alkenals, *cis,trans*-alka-2,4-dienals and low-molecular-mass *n*-alkanals in thermally stressed PUFA-rich oils investigated [12] were also observable in 60 MHz spectra, most especially at the 90 min heating time-point (Table 2), but again estimates of their individual concentrations were not feasible in view of the superimposition of their resonances with themselves and/or other aldehydic LOPs.

Notwithstanding, as expected, 400 MHz ^{1}H NMR analysis had the power to not only resolve the above 9.48 and 9.52 ppm α,β-unsaturated aldehyde resonances, but also both detect and resolve the lower intensity unsaturated aldehyde resonances present, specifically those of 4,5-epoxy-2-alkenals (δ = 9.55 ppm, *d*), *cis,trans*-alka-2,4-dienals (δ = 9.63 ppm, *d*), *cis*-2-alkenals (δ = 10.07 ppm, *d*), and a composite 4-hydroxy-*trans*-2-alkenal/4-hydroperoxy-*trans*-2-alkenal signal region (δ = 9.57 and 9.58 ppm, respectively, *d*), along with those assigned to 4-oxo-*n*-alkanals (δ = 9.78 ppm, *t*) and low-molecular-mass *n*-alkanals (δ = 9.79 ppm, *t*) [2–4,12,13,15,25]. Although selected resonances were visible and hence detectable, despite their decreased resolution and relatively low concentrations present in these thermally stressed oil media, these resonances were not quantitatively distinguishable from those of the more highly intense aldehydic signals in the 60 MHz spectra acquired, nor each other, and hence this series of aldehydes were not directly quantifiable at this field strength. As expected, the LF spectra acquired contained much spectrally broader overlapping resonances than those observed at the 400 MHz operating frequency. Comparative 60 and 400 MHz ^{1}H NMR profiles of the MOO product prior and subsequent to thermal stressing at 180 °C for 90 min are shown in Figure 3.

The minor oxygenated aldehydic LOPs detectable and quantifiable at MF strength are only generated from PUFA sources, along with much higher and higher concentrations of the di-unsaturated aldehydes *trans,trans*- and *cis,trans*-alka-2,4-dienals, respectively. Although MUFA-rich culinary oils such as the MOO product evaluated here, which contain only low levels of PUFAs, generate markedly lower levels of aldehydes in general (a consequence of the relative resistivity of MUFAs to peroxidation), higher or much higher proportionate concentrations of *n*-alkanals and *trans*-2-alkenals than these isomeric alka-2.4-dienals are detectable therein when they are exposed to thermal stressing episodes. Therefore, for this MOO product, the intensity of the total unsaturated aldehyde spectral region is largely but not exclusively ascribable to *trans*-2-alkenals. Moreover, because the oxygenated aldehydes 4,5-epoxy-, 4-hydroxy- and 4-hydroperoxy-*trans*-2-alkenals can also arise from the thermally induced transformation of *trans,trans*-alka-2,4-dienals [4], even higher levels of these secondary LOPs are generated in PUFA-rich rather than PUFA-deplete culinary oils when exposed to such high-temperatures.

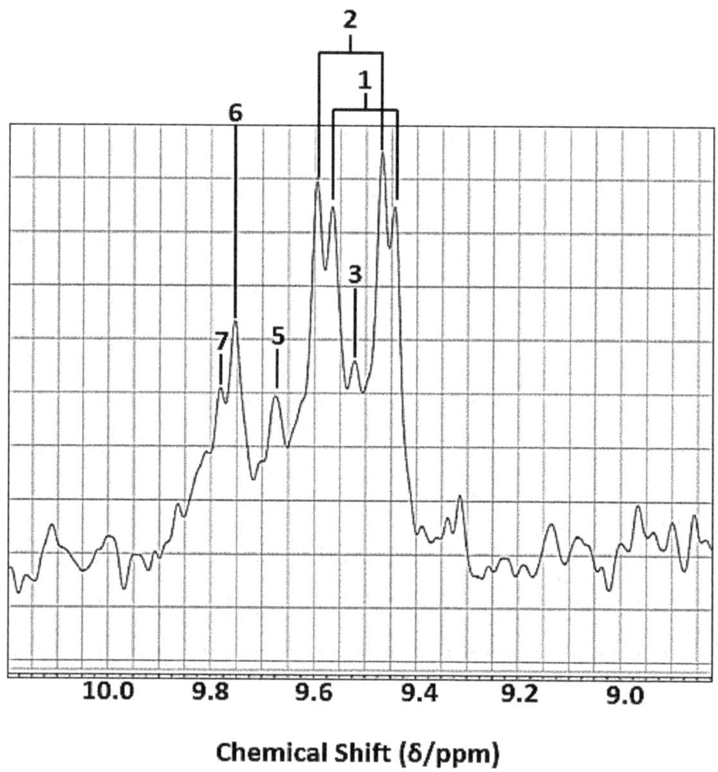

Figure 2. High-frequency (8.10–10.20 ppm) aldehydic proton region of the LF (60 MHz) ^{1}H NMR spectrum of SBO heated at 180 °C for a 90 min duration. Numbered label assignments correspond to those in Table 2. Aldehyde signals labelled 3 and 5 represent only one line from the doublet resonance usually observed for these α,β-unsaturated aldehydes at MF strength (400 MHz), as shown in Figure 3.

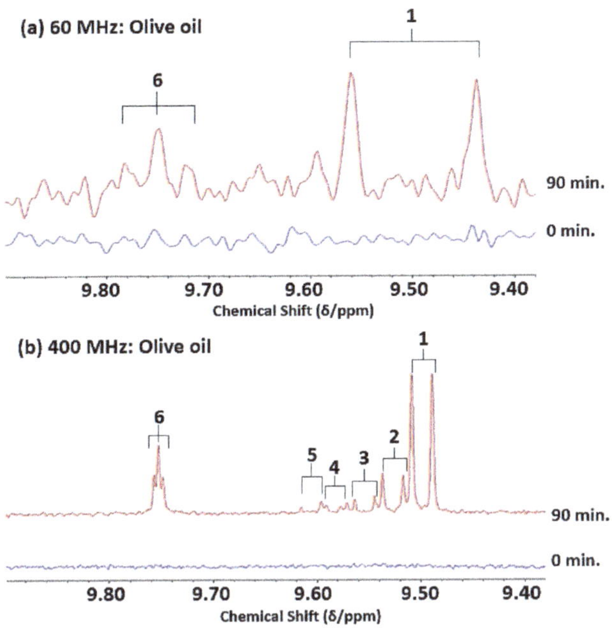

Figure 3. High frequency (8.10–10.20 ppm) aldehydic proton (-CHO) regions of the LF (60 MHz) and MF (400 MHz) ^{1}H NMR spectra of an olive oil (MOO) sample before and after heating for a period of 90 min at 180 °C. Numbered label assignments correspond to those available in Table 2.

Table 2. Assignments for the labelled aldehydic proton (-CHO) resonances present in the 60 and 400 MHz ^{1}H NMR profiles of culinary oils exposed to LSSFEs for periods of 60 or 90 min in Figures 2–4 and 8 (resonance 4 was undetectable at the 60 MHz operating frequency). These assignments were made by a consideration of those available in Refs. [2–4,12,13,15,25]. The spectral ranges and multiplicities of these signals are also provided in the second column. Abbreviations: *d*, doublet; *t*, triplet.

Resonance Assignment Code	δ/ppm (*Coupling Pattern*)	Assignment (-CHO Signal)
1	9.47–9.51 (*d*)	(*trans*)-2-Alkenals
2	9.51–9.55 (*d*)	(*trans,trans*)-Alka-2,4-dienals
3	9.54–9.58 (*d*)	4,5-Epoxy-(*trans*)-2-alkenals
4	9.57–9.61 (*d*)	Combined 4-Hydroxy-(*trans*)-/4-Hydroperoxy-(*trans*)-2-alkenals
5	9.62–9.65 (*d*)	(*cis,trans*)-Alka-2,4-dienals
6	9.73–9.76 (*t*)	*n*-Alkanals
7	9.78–9.82 (*t*)	Low-molecular-mass *n*-Alkanals

On consideration of the above aldehyde signal resolution problems encountered at 60 MHz, and for the clarity of comparative evaluations of 60 versus 400 MHz ^{1}H NMR quantitative determinations, we elected to group the resonances of aldehydic LOP species into only two major classification groups, specifically total α,β-unsaturated and saturated aldehydes, and electronically integrated their total grouped signal intensities at both 60 and 400 MHz operating frequencies. Spectral regions for these two aldehydic LOP classes were specified as the α,β-unsaturated aldehyde (δ = 9.40–9.70 ppm), and the saturated aldehyde integral domains (δ = 9.70–9.80 ppm). However, the *cis*-2-alkenal doublet resonance located at δ = 10.07 ppm in MF and above spectra [12] was excluded from the grouped α,β-unsaturated aldehyde resonances in order to avoid any complications arising from its lack of spectral response in the LF spectra acquired. Additionally, although only a minor issue, these ppm ranges were also used to further negate any issues of cross-species integration observed at MF (400 MHz), for example the partial superimposition of the 4-hydroxy- and 4-hydroperoxy-*trans*-2-alkenal signals observed at this higher operating frequency.

As noted in Section 2, total α,β-unsaturated and saturated aldehyde signals were electronically integrated and then normalised to the complete δ = 0.82–1.11 ppm region, which encompassed the bulk lipid chain terminal-CH_3 resonances of all acylglycerol FAs, including ω-3 acyl FAs. Therefore, each class of aldehyde integrals (total α,β-unsaturated and saturated) was computed to determine aldehyde concentration per mole of total FA (mmol aldehyde/mol FA).

3.3. *Analytical Calibration of Standard Authentic Aldehydes at 60 and 400 MHz Operating Frequencies*

Primarily, we explored the analytical calibration response of the ^{1}H NMR signals of one saturated and one α,β-unsaturated aldehyde calibration analytes; these aldehydes, *n*-hexanal and *trans*-2-octenal, represent major products arising from the peroxidation of PUFAs [12]. Both LF and MF NMR spectra acquired on series of *n*-alkanal and *trans*-2-octenal calibration standards in $CDCl_3$ solution containing unheated olive oil are shown in Figure 4.

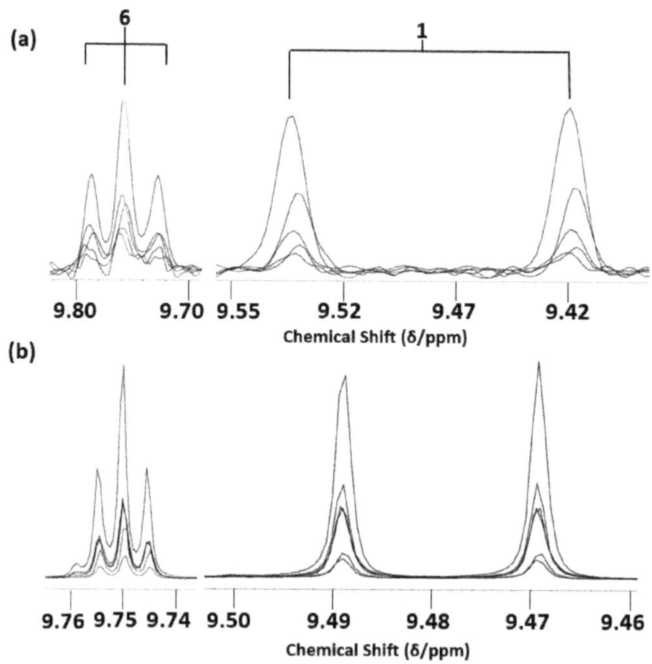

Figure 4. Partial aldehydic proton (-CHO) regions of the ^{1}H NMR spectra of n-hexanal (δ = 9.75 ppm, *t*) and *trans*-2-octenal (δ = 9.48 ppm, *d*) in a combined $CDCl_3$/unheated MOO medium at analyte solution concentrations of 5.40–54.98 and 2.42–40.40 mmol/L, respectively. Spectra acquired at (**a**) 60 and (**b**) 400 MHz operating frequencies are shown. The analyte media for these experiments were prepared according to one of the methods described in Section 2.3, and contained 0.20 mL of control (unheated) olive oil in a total volume of 0.76 mL. Numbered label assignments correspond to those available in Table 2.

Plots of the ratio of aldehydic proton (-C<u>H</u>O) resonance intensity to that of the terminal FA-C<u>H</u>$_3$ group of a fixed volume of an unheated (control) olive oil solution additive co-calibrant versus added *n*-hexanal calibrant concentration are shown in Figure 5a,b for spectra acquired at both 60 (blue datapoints) and 400 MHz (red datapoints) operating frequencies. These plots were clearly linear for this aldehyde, with Pearson correlation coefficients (r values) = 0.9757 ($p = 3.05 \times 10^{-7}$) and 0.9979 ($p = 5.78 \times 10^{-12}$) for the 60 and 400 MHz spectrometers, respectively, and 95% confidence intervals (CIs) for their regression coefficients (gradients) and ordinate (y-axis) intercepts were 4.88 to 6.87 $\times 10^{-4}$ and -2.65 to 1.88 $\times 10^{-3}$, respectively, for the 60 MHz, and 5.39 to 5.96 $\times 10^{-4}$ and -8.29 to 4.53 $\times 10^{-4}$, respectively, for the 400 MHz instruments (n = 10 datapoints). In view of the clear overlap of these CI values for both linear regression coefficients and y-intercept values, it may be confirmed that there were no major differences between *n*-hexanal concentrations determined by either the LF or MF NMR spectrometers employed for this analytical purpose. Moreover, the ordinate intercepts of both these calibration plots were not significantly different from zero. However, it should be noted that the 60 MHz facility's quantitative NMR (QNMR) linear calibrant response to this analyte was not as acceptable as that achieved at MF (400 MHz).

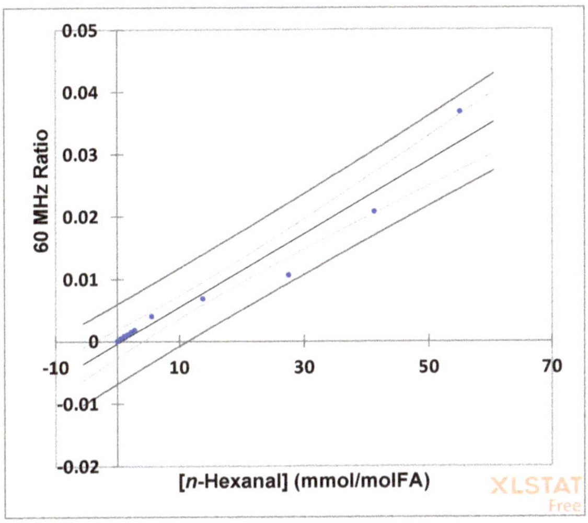

(a)

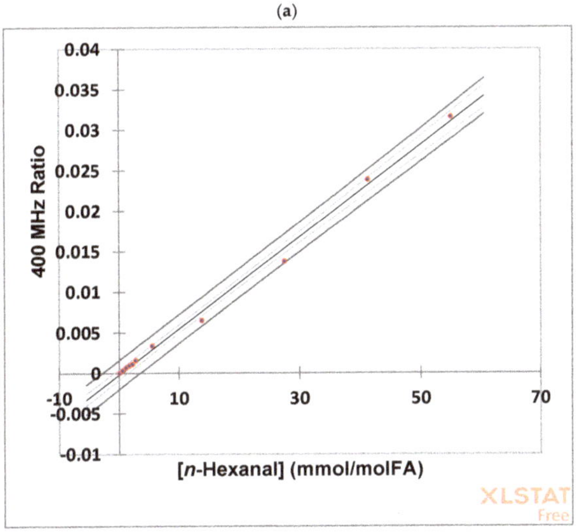

(b)

Figure 5. *Cont.*

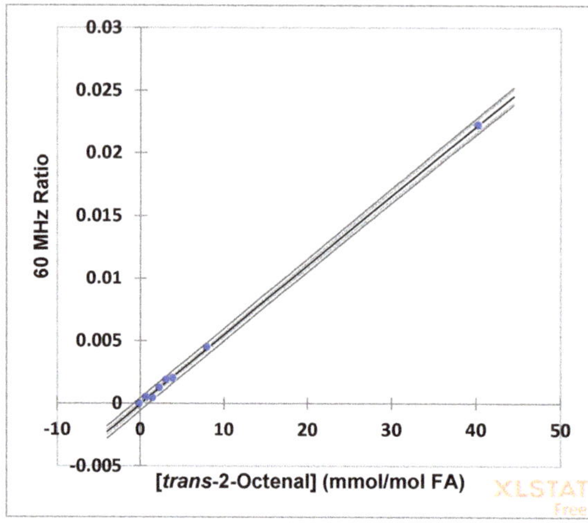

(c)

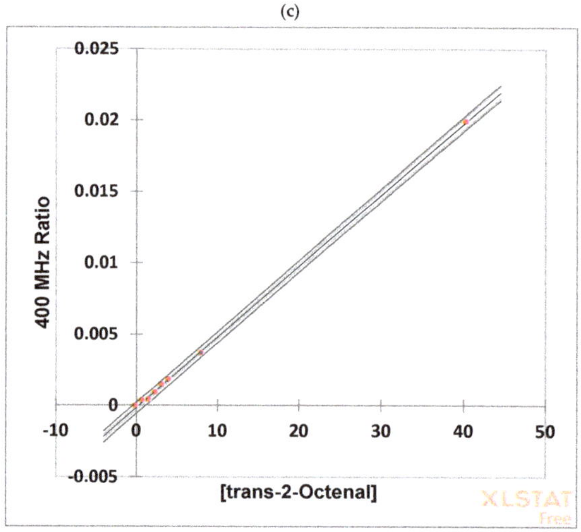

(d)

Figure 5. (**a**,**b**), Plots of the ratio of aldehydic proton (-CHO) intensity to that of the terminal FA-CH_3 group of a fixed volume of an unheated (control) olive oil solution co-calibrant additive versus *n*-hexanal calibrant concentration for spectra acquired at both 60 (blue datapoints) and 400 MHz (red datapoints) operating frequencies, respectively. (**c**,**d**), As (**a**,**b**), respectively, but for the *trans*-2-octenal calibration standard. Analytical calibrant solutions contained 0.20 mL of each aldehyde stock solution (0.00–54.98 mmol/L for *n*-hexanal, and 0.00–40.40 mmol/L for *trans*-2-octenal) in $CDCl_3$, 0.20 mL of the olive oil co-calibrant, a further 0.20 mL of the $CDCl_3$ solvent, and 0.10 and 0.06 mL aliquots of the 2,5-DTBHQ antioxidant (67.0 mmol/L) and the additional TCB ^{1}H NMR reference standard (19.8 mmol/L) solutions, also in $CDCl_3$. Abbreviations for the correlation plots shown in (**a**–**d**): - - - - -, grey, 95% confidence intervals (CIs) for means; --------, black, 95% CIs for observations.

Likewise, corresponding ratio versus calibrant concentration plots for *trans*-2-octenal yielded excellent linear relationships for both spectrometer operating frequencies (Figure 5c,d), with high r values of 0.9997 and 0.9998 (p = 5.34 and 2.21×10^{-11} for the 60 and 400 MHz instruments, respectively). The 95% CIs for their regression coefficients were 5.41 to 5.67×10^{-4} for the 60 MHz and 4.87 to 5.07×10^{-4} for the 400 MHz facilities, whereas for their ordinate intercepts, these CIs were −2.49 to 1.35 × 10^{-4} (60 MHz) and −3.28 to −0.31 × 10^{-4} (400 MHz) (n = 8 datapoints). Since the gradient parameter for the 60 MHz calibration plot was significantly greater than that observed at 400 MHz, it appears that the LF benchtop NMR instrument may overestimate this aldehydic LOP by a factor of ca. 11% when monitored as a single analyte standard in this manner, specifically calibration line gradients of 5.54 versus 4.97 × 10^{-4} for the LF and MF spectrometers, respectively. Although the 95% CIs for the 400 MHz calibration plot's ordinate-intercept did not quite cover the zero value, the upper limit for this was very close to this expected value. Reasons for this marginally negative ordinate-intercept observed are not simply explicable However, they may arise from analyte standard preparations which are of a concentration which is slightly lower than those expected, but this is very unlikely since the same solution calibration was conducted on the 60 MHz spectrometer, and the ordinate-intercept for that plot was not significantly different from its expected zero value (Figure 5c). Despite this observation, the effect on *trans*-2-octenal's linear calibration at 400 MHz can be considered to be largely negligible.

3.4. LLOD and LLOQ Values for Saturated and α,β-Unsaturated Aldehydes Monitored in Thermo-Oxidised Culinary Oils

Plots of STN ratio versus standard aldehyde calibrant concentration for *n*-hexanal and *trans*-2-octenal in either 'neat' $CDCl_3$ or olive oil/$CDCl_3$ analytical matrices are shown in Figure 6, and Table 3 lists simple linear regression parameters for these four plots. These results clearly show that there were no significant differences between the regression coefficients (gradients), nor ordinate- (y)-axis intercepts of these plots 'between-aldehyde nature' found, nor for samples which were analysed following the introduction of 0.20 mL of a culinary oil medium into the analytical matrix, as for all oil samples analysed. However, it should be noted that there were very marginally slimmer 95% CIs, together with slightly improved r values, obtained for the $CDCl_3$-only analyte medium, as might be expected. Estimated LLOD and LLOQ values for these two analytes were represented by STN values of 3.00 and 10.00, respectively, and these are shown in Table 4. These estimated LLOD and LLOQ values were 0.18–0.19 and 0.62–0.65 mmol/mol FA for both classes of aldehydes analysed in $CDCl_3$ media containing a 0.20 mL aliquot of olive oil.

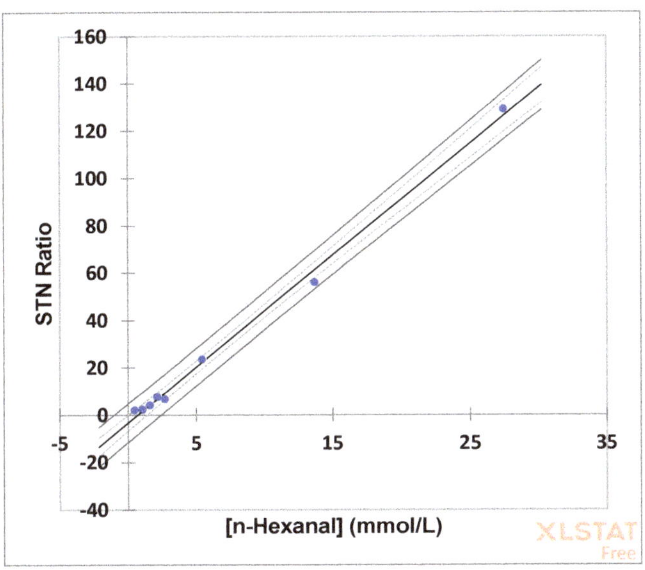

(a)

Figure 6. *Cont.*

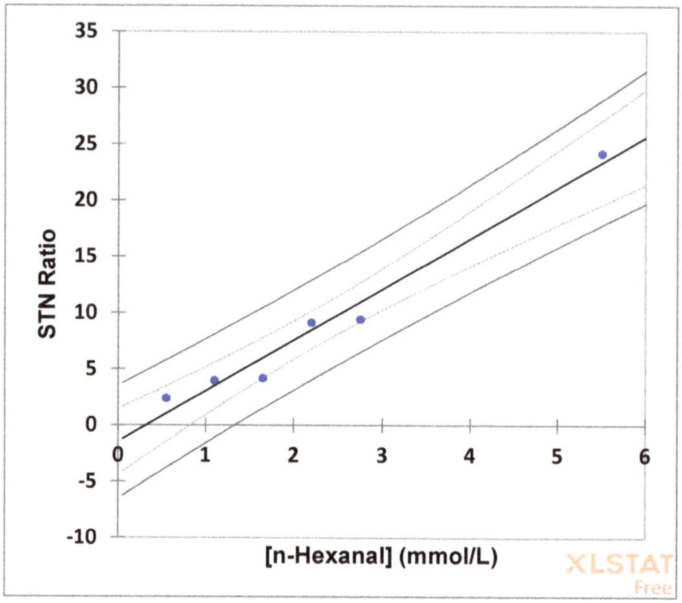

(b)

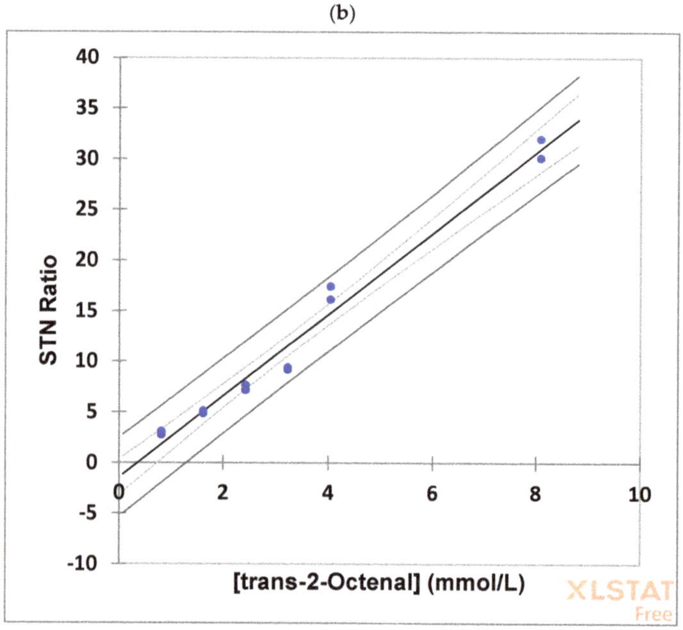

(c)

Figure 6. *Cont.*

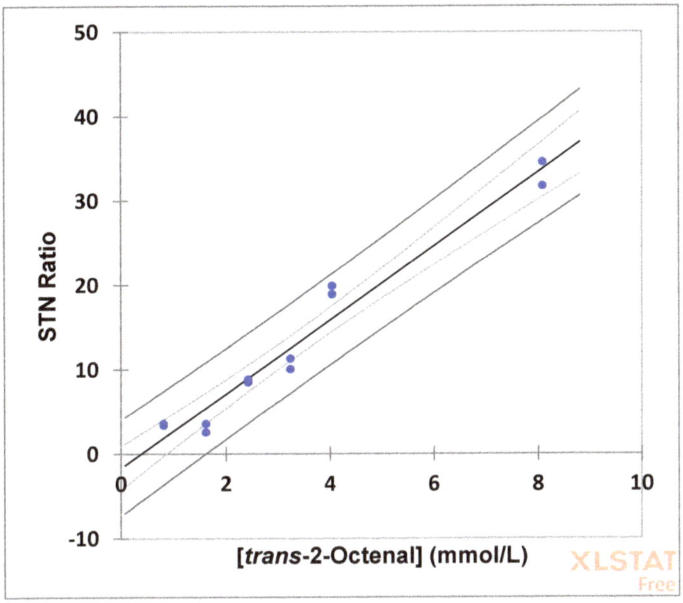

(d)

Figure 6. Plots of signal-to-noise (STN) ratio versus concentration of aldehyde calibrants. (**a**,**b**), *n*-Hexanal in 'neat' $CDCl_3$ and olive oil/$CDCl_3$ analytical solutions, respectively; (**c**,**d**), as (**a**,**b**), respectively, but for *trans*-2-octenal calibrant solutions. Abbreviations for the correlation plots shown in (**a**–**d**): as described in Figure 5.

Table 3. Linear regression parameters for plots of analytical STN ratio values versus added *n*-hexanal and *trans*-2-octenal calibrant concentrations in media including and excluding olive oil added to the analyte media at an operating frequency of 60 MHz. The 95% Confidence intervals (CIs) are provided for the estimated regression coefficients and ordinate intercepts of these plots. Abbreviation: r, Pearson correlation coefficient.

Aldehyde	Solution Medium	r	Regression Coefficient ± 95% CIs	Ordinate Axis Intercept ± 95% CIs
n-Hexanal	$CDCl_3$	0.9980	4.73 ± 0.30	−3.34 ± 3.35
n-Hexanal	Olive Oil/$CDCl_3$	0.9860	4.52 ± 1.04	−1.47 ± 2.92
trans-2-Octenal	$CDCl_3$	0.9886	4.01 ± 0.43	−1.41 ± 1.77
trans-2-Octenal	Olive Oil/$CDCl_3$	0.9801	4.38 ± 0.62	−1.67 ± 2.57

Table 4. Estimated LLOD and LLOQ values for n-hexanal and *trans*-2-octenal at an ^{1}H NMR operating frequency of 60 MHz. Units of mmol/L were employed for all analyte media, but mmol/mol FA units were also applicable to those prepared in a combined olive oil/$CDCl_3$ medium.

Aldehyde	Solution Medium	LLOD: 3(STN) (mmol/L)	LLOD: 3(STN) (mmol/mol FA)	LLOQ: 10(STN) (mmol/L)	LLOQ: 10(STN) (mmol/mol FA)
n-Hexanal	$CDCl_3$	0.63	n/a	2.10	n/a
n-Hexanal	Olive Oil/$CDCl_3$	0.66	0.19	2.21	0.65
trans-2-Octenal	$CDCl_3$	0.75	n/a	2.49	n/a
trans-2-Octenal	Olive Oil/$CDCl_3$	0.63	0.18	2.10	0.62

Notably, the -C<u>H</u>O function resonances of both of these aldehydic calibrant species did not suffer from the issue of signal collapse, and fully maintained their multiplicities at an operating frequency of 60 MHz.

3.5. Analytical Consistency of Aldehyde Calibrant Analyses at 60 and 400 MHz ^{1}H NMR Operating Frequencies

The analytical consistency of aldehyde analysis conducted at the two different ^{1}H NMR operating frequencies (60 versus 400 MHz) was also checked through an examination of Bland–Altman style dominance plots (BAPs) for both the *n*-hexanal and *trans*-2-octenal calibration standards (Figure 7). These BAP plots show the 400 MHz-determined concentrations subtracted from the corresponding 60 MHz ones (analytical deviations) versus the mean values of these matched sample analyses, and these provided valuable supporting information regarding the nature of operating frequency-dependent deviations between estimations of these two aldehydic analytes. For *n*-hexanal in (a), there were no apparent concentration-dependent trends for such deviations, although one datapoint laid outside the $mean_d \pm 1.96s_d$ limits, and hence overall, this analysis indicated an acceptable agreement between the two operating frequencies. For *trans*-2-octenal in plot (b), however, there appeared to be a positive deviational trend with increasing calibrant concentrations, although with only a single datapoint laying outside the $\pm 1.96s_d$ limits. This plot also confirmed a small but nevertheless significant analytical overestimation of this aldehyde at 60 MHz, but only at the higher concentrations investigated. In view of this observation, it is proposed that a benchtop facility operating at only 60 MHz should be employed for reliable aldehyde determinations only when the aldehyde-C$\underline{H}$O:FA terminal acylglycerol-C$\underline{H}_3$ signal intensity ratio values are ≤ 0.003 (which correspond to estimated aldehyde concentrations of $\leq$ca. 5.0 mmol/mol FA), so that the analytical deviations observed for measurements made at LF strength remain small.

Paired *t* tests performed on the datasets displayed in Figure 7 also verified that there was no significant 'between-operating frequencies' difference found for estimated concentrations of *n*-hexanal. Notwithstanding, it should be noted that *trans*-2-octenal concentrations estimated at 60 MHz were significantly greater than those determined at 400 MHz operating frequency ($p < 10^{-4}$), as shown in Figure 7d; however, as noted above, this was the case only for values with relatively high abscissa axis values, which represent the mean of the integration ratios determined at both operating frequencies. Nevertheless, for this unsaturated aldehyde, the BAP showed a significant increase in the 60–400 MHz difference value with these increasing abscissa axis values. Nevertheless, at low (60 MHz Ratio + 400 MHz)/2 ratios (i.e., <0.003), these difference values were actually lower than the mean difference value determined (Figure 7d), although all were higher than the zero-control optimum. Since this deviation from the optimal zero difference optimum is approximately 0.0005 when this ratio's values are <0.003, determinations of *trans*-2-octenal were considered acceptable within this abscissa axis limit.

3.6. Analytical Precision of Total Saturated and α,β-Unsaturated Aldehyde Classification Determinations in Control and Thermally Stressed Culinary Oils at 60 and 400 MHz Operating Frequencies

The analytical precision of the LF α,β-unsaturated aldehyde concentration data acquired was comparable to the results acquired at MF, with 'between-replicate' SD and 95% CI values for total saturated and α,β-unsaturated aldehydes being similar at these two operating frequencies. Indeed, 'between-replicate' coefficient of variation (CV) values for this class of aldehydes ranged from 0.4 to 13.4% (albeit with the exception of two values which were >20%) at LF, whereas those for the MF spectrometer varied from 1.2 to 12.1% (data not shown). However, for the LF data available for total saturated aldehydes, these values were less precise (CV 8.7 to 22.2%, with one value being >30%); those for the MF spectrometer employed, however, ranged from 3.0 to 12.2%, with the majority of these values being <7%.

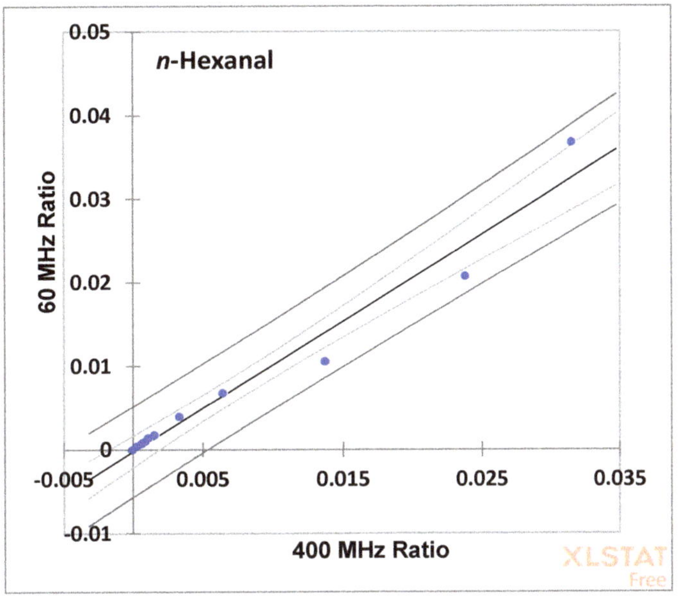

(a)

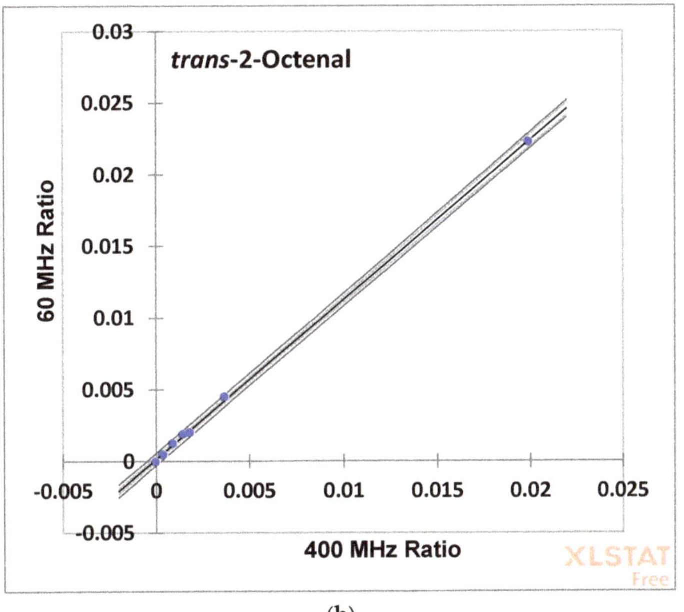

(b)

Figure 7. *Cont.*

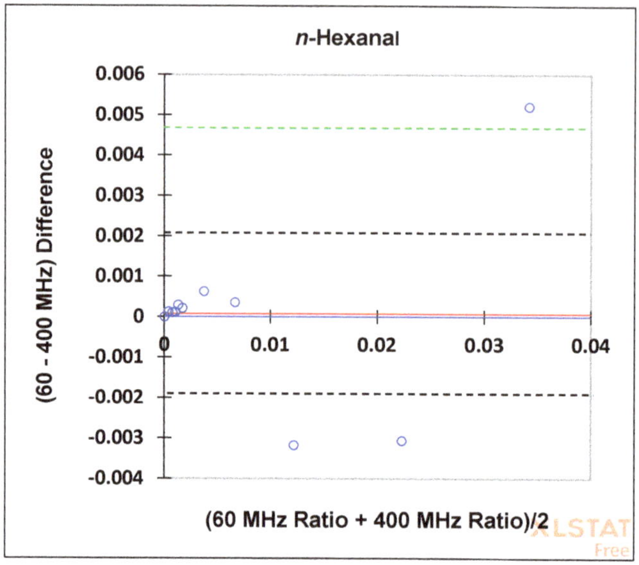

(c)

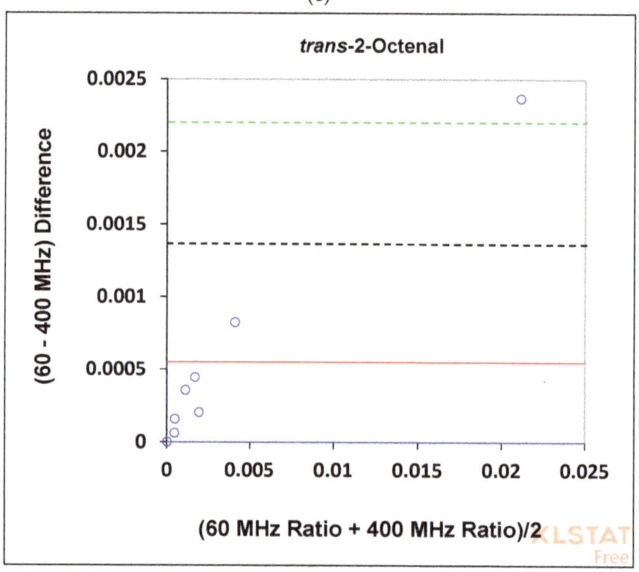

(d)

Figure 7. Plots of the concentrations of (**a**) *n*-hexanal (*t*, δ = 9.74 ppm) and (**b**) *trans*-2-octenal (*d*, δ = 9.48 ppm) determined at a 60 MHz operating frequency on a LF benchtop NMR spectrometer versus those obtained on a conventional MF 400 MHz NMR facility. Abbreviations for the correlation plots shown in (**a**,**b**): - - - - -, grey, 95% confidence intervals (CIs) for means; ——, black, 95% CIs for observations. (**c**,**d**), Corresponding Bland–Altman style dominance plots (BAPs) for comparisons of analytical results generated by the 60 and 400 MHz NMR facilities, respectively. For the BAP plots, the dark blue and red lines represent the null hypothesis zero difference and determined mean difference values, respectively. The dotted black lines represent 95% CIs for the mean difference (depicted as the red line), and the dotted green lines display the $\pm 1.96s_d$ limits for the latter; s_d represents the standard deviation of the difference values for each plot.

3.7. LF and MF ^{1}H NMR Analysis of Aldehydic LOPs in Culinary Oils Collected during Their Time-Dependent Exposure to Thermal-Stressing Episodes at 180 °C

For thermally stressed culinary oil samples collected at the 60 and 90 min LSSFE time-points, all the expected α,β-unsaturated aldehydes were readily detectable and quantifiable in thermally stressed culinary oils in spectra acquired at an operating frequency of 400 MHz (Figures 3 and 8), and with the exception of the partially superimposed 4-hydroxy- and 4-hydroperoxy-trans-2-alkenal signals, all their -C<u>H</u>O function ^{1}H NMR signals were sufficiently resolved. Importantly, at an operating frequency of only 60 MHz, the lowest determined concentration of the total α,β-unsaturated class of aldehyde at 60 MHz was found to be 0.95 mmol/mol FA (at a LSSFE time-point of 60 min), and therefore, with the exception of one missing replicate value, all 60 MHz determinations were included in the parametric ANOVA statistical analysis performed (Section 3.8).

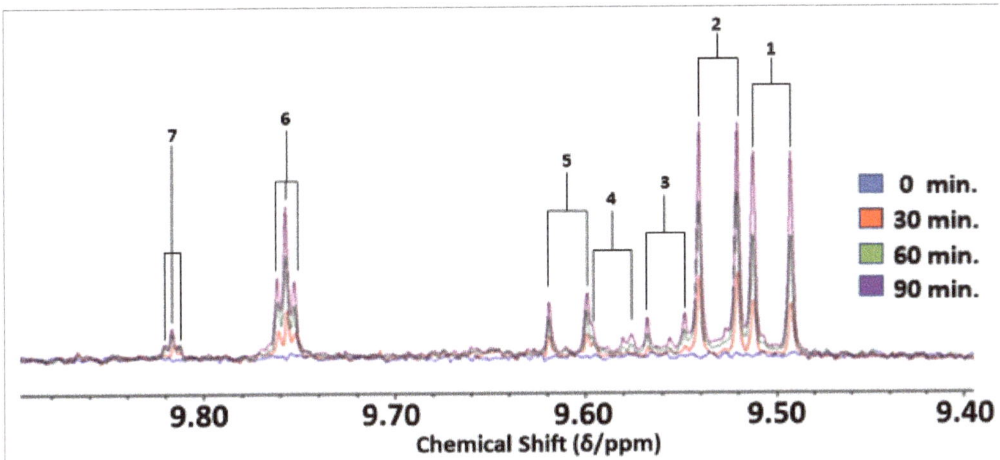

Figure 8. Aldehydic proton (-CHO) region (9.40–9.90 ppm) of the MF (400 MHz) ^{1}H NMR spectrum of a soybean oil (SBO) sample heated for periods of 0, 30, 60 and 90 min at 180 °C according to our LSSFEs. Numbered label assignments correspond to those available in Table 2. The triplet resonance located at δ = 9.817 ppm is this spectrum is attributable to low-molecular-mass *n*-alkanal species, which arise from the thermo-oxidative deterioration of ω-3 FAs, and higher levels of these are generally characteristic of oils containing significant contents of these highly unsaturated FAs (≥5% (*w/w*)) when exposed to high-temperature thermal stressing episodes [28]. Colour key code: blue, 0 min; red, 30 min; green, 90 min; violet, 90 min.

However, unfortunately this was not the case for the total saturated aldehyde variable. Indeed, for the MOO samples, all determinations made (with the exception of a small number at the 90 min time-point), were found to be below the specified LLOQ value (Table 4), and therefore these estimates were excluded from the statistical analysis performed. Likewise, two of the SBO (both at the 60 min time-point), and four of the RSO samples (three at the 60 min, and one at the 90 min time-points), were similarly excluded. However, three further values which were below the LLOQ threshold, but were ≥90% of this parameter, were permitted to remain in the saturated aldehyde analytical dataset for data analysis by ANOVA.

Therefore, although readily observable and detectable in the 60 MHz LF spectra acquired, the majority of *n*-alkanal concentrations determined in the MUFA-rich MOO product, especially those at the earlier LSSFE sampling time-point, remained below the LLOQ threshold. Indeed, in such cases the background noise and diminished resolution and sensitivity of the LF instrument hindered the direct observation of characteristic patterns of aldehydic LOP resonances.

As expected, an enhancement of total unsaturated aldehyde formation was also observed in the PUFA-rich oils analysed on increasing the thermal stressing duration from 0 to 60 and 60 to 90 min, although the ability to distinguish between each of the aldehydic species was compromised by the low resolution at an operating frequency of 60 MHz, as noted above for the overlapping *trans*-2-alkenal and *trans*-*trans*-2,4-alkadienal signals (Figure 2). Comparatively, when analysed at an operating frequency of 400 MHz, *trans,trans*- and *cis,trans*-alka-2,4-dienals, 4,5-epoxy-*trans*-2-alkenals, 4-hydroxy-/4-hydroperoxy-*trans*-2-alkenals, *cis*-2-alkenals and low-molecular-mass *n*-alkanals (i.e., propanal and *n*-butanal) were all detectable and quantifiable, in addition to *trans*-2-alkenals and *n*-alkanals (Figures 3 and 8). These secondary LOPs were also all observable at the 30 min heating time-point. Therefore, accurate determination of the contents of virtually all classes of both saturated and α,β-unsaturated aldehydes would require a spectrometer operating frequency of at least 400 MHz. Minor level low-molecular-mass saturated *n*-alkanal species were also observable and quantifiable at this operating frequency, However, this type of saturated aldehyde was also detectable, albeit not quantifiable, in ^{1}H NMR profiles obtained on the LF spectrometer (Figure 2).

For replicate MUFA-rich MOO samples exposed to 60 or 90 min thermal-stressing periods, estimates of the total α,β-unsaturated aldehyde concentrations made at an operating frequency of 400 MHz were 2.06 $\pm$ 0.13 and 2.82 $\pm$ 0.02 mmol/mol FA, respectively (mean $\pm$ SD values), whereas at 60 MHz, these values were only 1.12 $\pm$ 0.13 and 1.70 $\pm$ 0.01 mmol/mol FA, respectively. Therefore, for this oil, total α,β-unsaturated aldehyde levels determined at LF strength were only ca. 60% of those found at MF, and this is readily explicable by the inability of the 60 MHz instrument to detect and quantify lower levels of the more structurally complex minor aldehydic LOPs generated in cooking oils exposed to LSSFEs (specifically 4,5-epoxy- and 4-hydroxy-/4-hydroperoxy-*trans*-2-alkenals, and *cis*-2-alkenals), although the major *trans*-2-alkenal class is also derived from the thermally induced peroxidation of PUFAs [12].

Although largely unquantifiable in the MOO product at a LSSFE time-point of 60 min when analysed at an operating frequency of 60 MHz, at 400 MHz mean $\pm$ SD levels of total saturated aldehydes were found to be 0.64 $\pm$ 0.06 and 0.79 $\pm$ 0.05 mmol/mol FA at the 60 and 90 min sampling time-points, respectively.

Heated PUFA-rich soybean, rapeseed and chia oils were also included in this investigation to determine the ability of LF NMR analysis to provide some level of analytical distinction between unsaturated aldehyde species generated therefrom. In view of the inability of the 60 MHz instrument to fully resolve *trans*-2-alkenal and *trans*-*trans*-2,4-alkadienals resonances (Figure 2), the potential application of quantitative two-dimensional (2D) ^{1}H-^{1}H COSY or TOCSY determinations may be valuable. Indeed, such 2D spectra are readily acquirable on LF benchtop NMR facilities, may indeed facilitate the independent quantitative determination of these aldehydes in used or fried PUFA-rich edible oils, albeit only if present at concentrations sufficiently above the specified LLOQ value (Table 4).

3.8. ANOVA of Total Saturated and α,β-Unsaturated Aldehyde Concentration Datasets in Culinary Oils Exposed to LSSFEs for Periods of 60 and 90 min

An extensive ANOVA model was applied to further explore the level of analytical agreement between the LF and MF spectrometers employed for the determination of both total saturated and α,β-unsaturated aldehydic LOPs, in addition to evaluations of the statistical significance of concentration differences observed 'between oil products' and 'between LSSFE sampling time-points' (i.e., at the 60 versus 90 min sampling time-points).

This model demonstrated that for total saturated aldehyde levels, there were highly significant differences between all the main factors investigated, with $p = 0.036$ for oil type, 0.00030 for heating time (60 versus 90 min), and 0.005 for spectrometer operating frequencies employed for the analysis; the latter observation represents higher levels observed at 400 than at 60 MHz in view of its significantly higher STN and lower LLOQ values, most notably those for certain classes of saturated aldehydes such as low-molecular-mass n-alkanals (*t*, $\delta = 9.78$ ppm) which remain non-quantifiable at 60 MHz. Corresponding mean $\pm$ 95% confidence intervals (CIs) for this saturated aldehyde analysis protocol are shown in Figure 9. Figure 9a shows that the patterns of total saturated aldehydes were significantly lower for the 60 min heating time-point than that found at 90 min, and Figure 9c reveals that the levels of these aldehydes generated were in the product order RSO > CSO ≈ SBO > MOO.

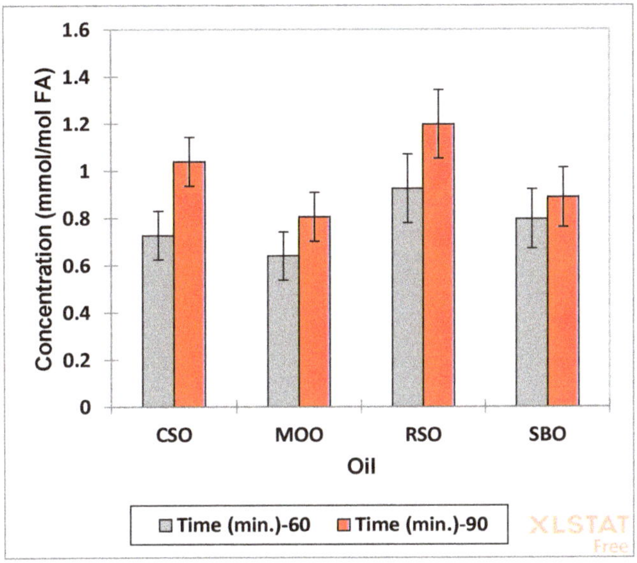

(a)

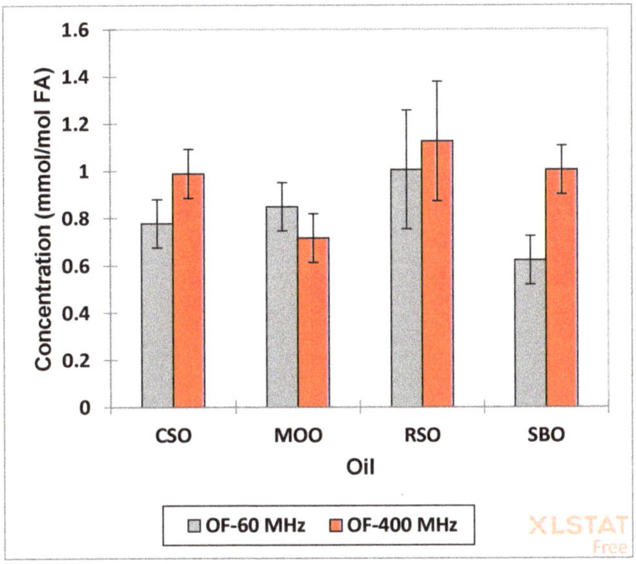

(b)

Figure 9. *Cont.*

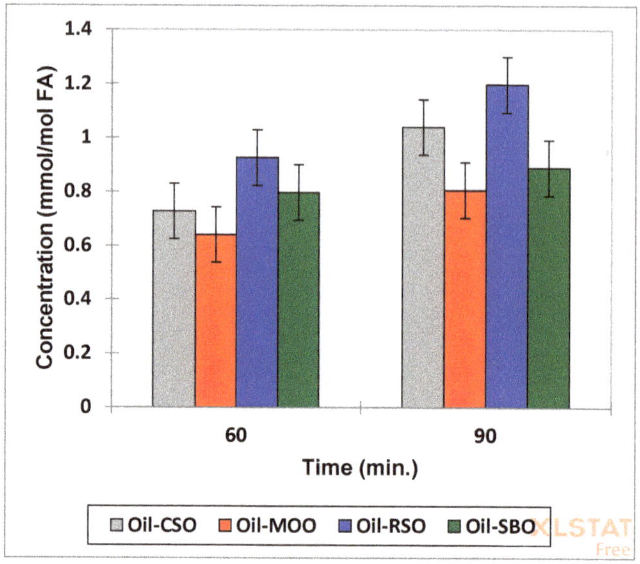

(c)

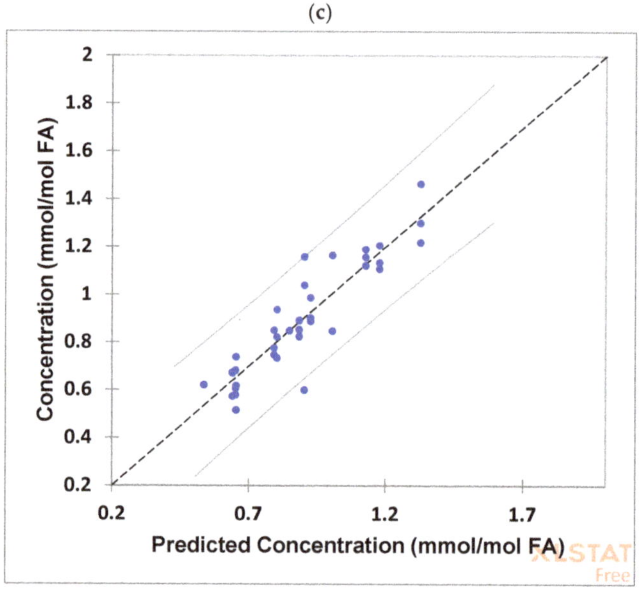

(d)

Figure 9. ANOVA of saturated aldehydic LOP concentration dataset from experiments involving the exposure of four different culinary oil products to LSSFEs. (**a**–**c**) Bar diagrams showing mean ± 95% CI values for saturated aldehydic LOP concentrations generated in culinary oils heated according to our LSSFEs at a temperature of 180 °C for periods of 60 and 90 min, and showing differences between heating time-points, NMR spectrometer field strength and oil products evaluated, respectively. (**d**) Plot of observed saturated aldehyde concentration versus those predicted from this ANOVA model (Equation (2)), which included a statistically significant first-order oil product x spectrometer operating frequency interaction effect (R^2 = 0.828). Abbreviations: OF, NMR spectrometer operating frequency; CSO, chia seed oil; MOO, refined olive oil; RSO, rapeseed oil; SBO, soybean oil.

Additionally, the oil type x spectrometer operating frequency first-order interaction effect was also statistically significant ($p = 0.027$), and this is best rationalised by again considering limits for the detection and quantification of different classes of saturated aldehyde LOPs. Indeed, Figure 9b shows that although there were higher mean values for the 400 MHz spectrometer saturated aldehyde concentrations found for CSO, RSO and SBO (those for CSO and SBO being significantly so), the extent of these differed in each case. Indeed, there was notably a quite high deviation of this parameter for the SBO product, which served as a major contribution to the statistical significance of this interaction effect. Nevertheless, this deviation is not simply explicable, although it should be noted that this oil contains a significant ω-3 FA (linolenoylglycerol) content (5% (*w/w*)), which can lead to the generation of low-molecular-mass alkanal species such as propanal and *n*-butanal following thermo-oxidation and fragmentation. It is therefore feasible that the levels of such secondary LOPs formed and remaining in the oil medium during LSSFEs are too low to be detectable and quantifiable in this oil, and hence are not included as a contribution towards its total saturated aldehyde concentration determined at both the 60 and 90 min heating time-points. This may also explain why the 'between-operating frequency' differences observed between the saturated aldehyde concentrations of the CSA oil product were lower than those observed for SBO, since the ω- FA content of the former was as high as 65% (*w/w*), and hence in principle it is possible that the total saturated aldehyde contents found for this oil included a significantly higher proportion of such low-molecular-mass species. However, if this was the case, a similar significant 60 MHz deviation might also be expected for the RSO product (ω-3 FA content 7% (*w/w*)), but this was not the case (Figure 9b). Moreover, an examination of both 60 and 400 MHz spectra acquired on the high ω-3 content CSA oil showed that the low-molecular-mass aldehyde resonance ($\delta = 9.80$ ppm, *t*) had an intensity that was always <10% of that of the more intense $\delta = 9.75$ ppm one.

A further possibility is that the LLOD and LLOQ values calculated for SBO may differ somewhat from those determined from an analyte solution containing a combined unheated olive oil/$CDCl_3$ medium according to our analytical protocol described in Sections 2.6 and 2.7; experiments to explore this are currently in progress.

The overall R^2 value for this model with this interaction effect included was 0.828, whereas that for the model excluding all first-order interactions was significantly lower (0.716), which demonstrates the importance of including it in the model. A plot of observed total saturated aldehyde level versus that predicted from this interaction-incorporated ANOVA model is shown in Figure 9d, and this confirms an acceptable concurrence between them.

Similarly, ANOVA results for the total unsaturated aldehyde levels are displayed in Figure 10. This analysis revealed that all the main effects were again very highly statistically significant ($p = 3.17 \times 10^{-14}$, 1.17×10^{-12} and 1.49×10^{-7} for the 'between-oil classes', 'between-heating time-points' and 'between-spectrometer operating frequencies' sources of variation, respectively), whereas the mean square estimates for all second-order interactions tested were not. Therefore, all these interaction effects were removed from the model, and mean squares for the main effects were recalculated following supplementation of that for the error term with these removed, albeit insignificant contributions. The R^2 value for the model which excluded all interaction effects was 0.876, but this value only increased to 0.907 following incorporation of these contributions. The lack of significance of these effects is clearly visible in the mean ±95% CI plots shown in Figure 10a–c, which showed similar patterns for each of the time-point, spectrometer operating frequency and oil type, respectively, with clearly higher total unsaturated aldehyde levels for the 90 min time-point, the 400 MHz operating frequency, and oil products in the order CSO > RSO ≈ SBO > MOO, as expected. The observation of higher levels of α,β-unsaturated aldehydes in the heated CSO product arose from its very high content of ω-3 FAs. Figure 10d shows a plot of observed total α,β-unsaturated aldehyde concentration versus that predicted from this interaction-free ANOVA model, and this shows a good agreement between these values.

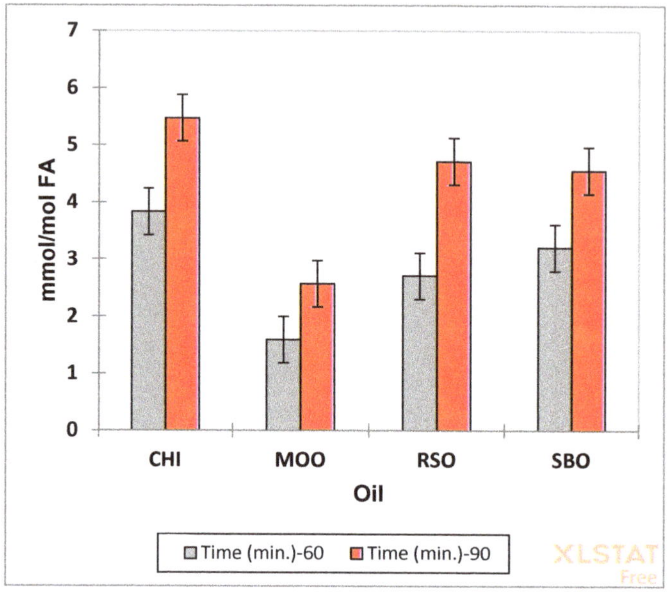

(a)

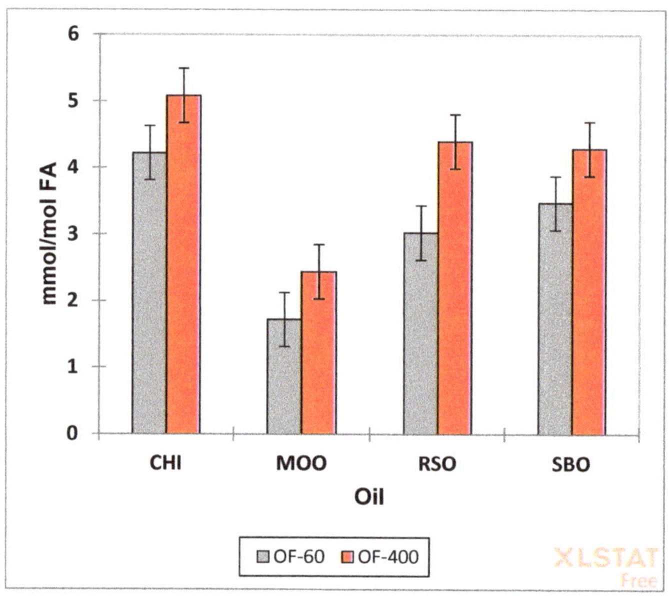

(b)

Figure 10. *Cont.*

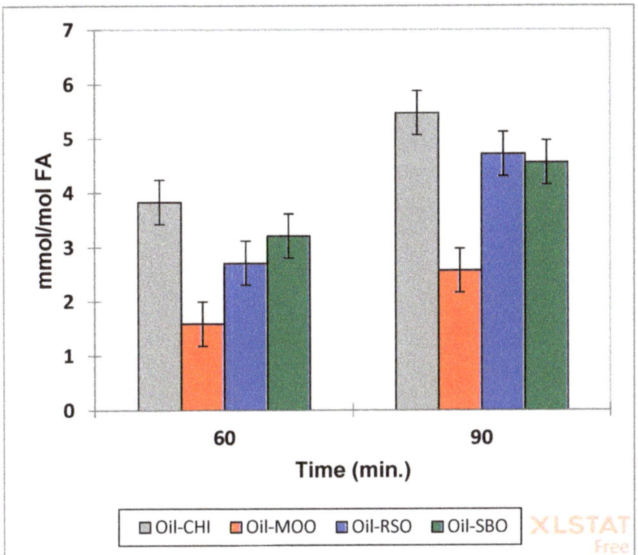

(c)

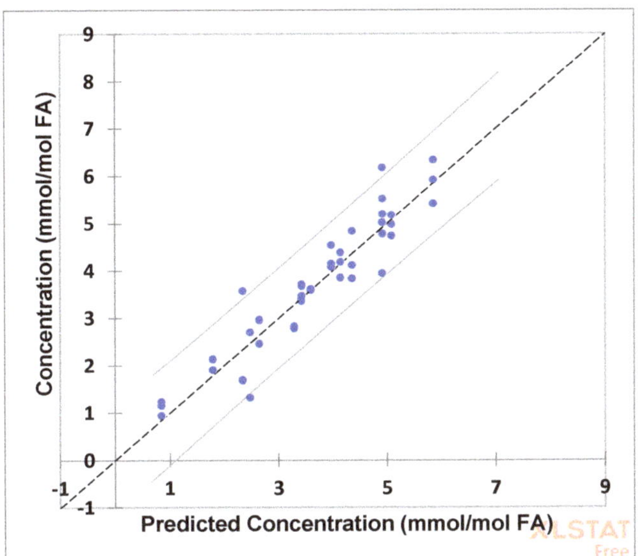

(d)

Figure 10. ANOVA of α,β-unsaturated aldehydic LOP concentration dataset from experiments involving the exposure of four different culinary oil products to LSSFEs. (**a**–**c**) Bar diagrams showing mean ±95% CI values for α,β-unsaturated aldehydic LOP concentrations generated in culinary oils heated according to our LSSFEs at a temperature of 180 °C for periods of 60 and 90 min, and showing differences between heating time-points, NMR spectrometer operating frequencies and oil products evaluated, respectively. (**d**) Plot of observed concentration versus those predicted from this ANOVA model (Equation (3)), which included no interaction effects (R^2 = 0.876). Abbreviations: OF, NMR spectrometer operating frequency; CSO, chia seed oil; MOO, refined olive oil; RSO, rapeseed oil; SBO, soybean oil.

The mean percentages of saturated and α,β-unsaturated aldehydes fromed at the 90 min LSSFE time-point for the CSO, MOO, RSO and SBO products evaluated were 16 and 84%, 23 and 77%, 20 and 80%, and 19 and 81%, respectively. Therefore, it appears that the MUFA-rich MOO product generated slightly higher proportions of saturated aldehydic LOPs over those of the other oil products when exposed to thermal stressing episodes. Moreover, the ω-3 FA-rich CSO oil appeared to produce marginally higher proportions of α,β-unsaturated aldehydes.

3.9. *LF ^{1}H NMR Detection of Aldehydic Lipid Hydroperoxide Precursors*

As previously observed in experiments conducted at both MF and HF operating frequencies, many of the LF ^{1}H NMR spectra acquired here on oils exposed to LSSFEs were also found to contain broad, relatively intense signals ascribable to exchangeable lipid hydroperoxide-OOH functions, i.e., the aldehyde precursors conjugated hydroperoxydienes (CHPDs) and/or unconjugated hydroperoxymonoenes (HPMs) derived from the peroxidation of PUFAs and MUFAs, respectively (Figure 11). Indeed, these signals are reported to have solution-state chemical shift values ranging from 8.0 to 8.9 ppm in $CDCl_3$ solution [2–4], although these values are highly dependent on analytical conditions such as analyte solution viscosity, and NMR acquisitional parameters such as NMR probe temperature for the sample, etc. Figure 11 demonstrates that 60 MHz ^{1}H NMR spectra acquired on thermo-oxidised SFO samples contain one major and one more minor -OOH function resonances centred at δ = 8.08 and 8.63 ppm, respectively. As previously observed [2,3], such resonances were found to completely disappear following the addition of a small μL aliquot to the sample after thorough mixing.

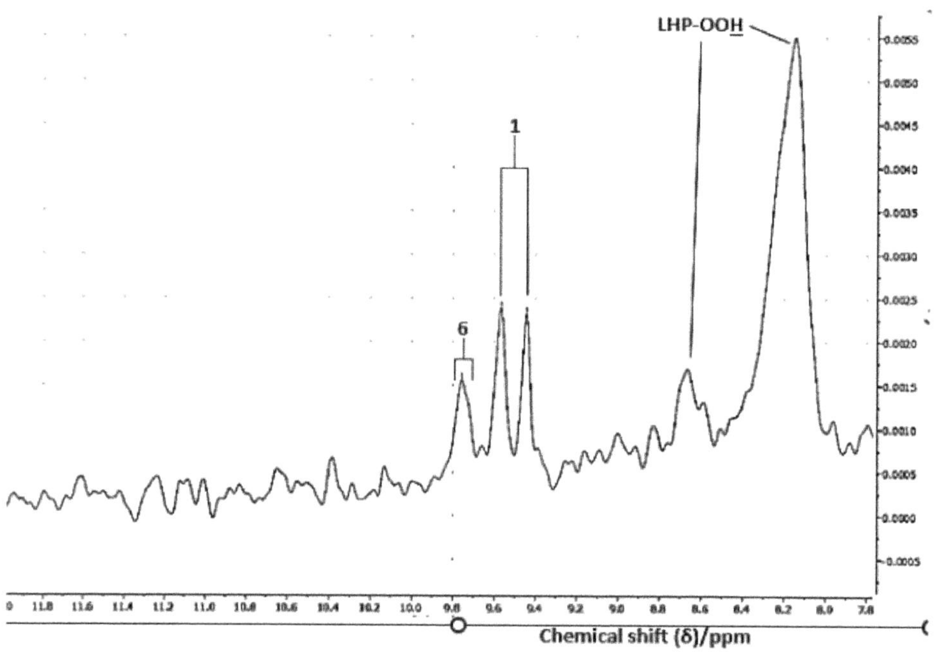

Figure 11. Expanded 7.80–12.00 region of the ^{1}H NMR spectrum of a sample of sunflower oil thermally stressed according to our LSSFEs (180 °C for a period of 90 min). Abbreviations: LHP-OOH, broad lipid hydroperoxide signals attributable to CHPDs and/or HPMs. Numbered label assignments correspond to those in Table 2.

Therefore, the LF ^{1}H NMR monitoring of concentrations of these hydroperoxide species offers a valuable addition to the information provided by this technique on the degradation of UFAs in cooking oils, and which is a major requirement for the quality monitoring of such products [29]. Data acquired regarding changes in the ^{1}H NMR-detectable levels of these primary LOPs with increasing LSSFE time-points (0–90 min) will be reported in detail in a follow-up paper.

4. Discussion

This study demonstrated that LF (60 MHz) ^{1}H NMR analysis could detect and quantify secondary total saturated and unsaturated aldehydic LOP concentrations in thermally stressed cooking oils as a function of high-temperature LSSFE exposure time, albeit most especially at thermal stressing time-points of 60 and 90 min However, in the case of MUFA-rich oils, such as the MOO product explored here, results obtained at LF were hampered by limitations of such analysis using this operating frequency, predominantly by the lowered sensitivity of this technique and the superimposition of resonances usually resolved at MF strength, along with increased levels of spectral noise. Indeed, for the purpose of making comparative analytical evaluations between results obtained on the 60 and 400 MHz spectrometers in this study, we had no option but to combine aldehyde contents within two classification groups: total saturated and total α,β-unsaturated types. This approach was adopted despite some partial resolution between *trans*-2-alkenal and *trans,trans*-alka-2,4-dienal resonances in all oils investigated (Figure 2). However, the absolute concentrations of these two different α,β-unsaturated aldehydes could, at least in principle, be quantified on LF spectrometers through the application of suitable spectral deconvolution strategies.

As previously reported, thermally stressed MUFA-rich oils produce greater relative amounts of *trans*-2-alkenals and only minor relative levels of all other α,β-unsaturated aldehyde species [1–4,12]. However, in MF spectra acquired, resonances assignable to virtually all these unsaturated aldehydes, including isomeric alka-2,4-dienals, are observed in spectra acquired on such heated oils, and may be individually included in the resonance integration values for quantification purposes, as shown in Figure 3. However, in LF spectra, the *trans,trans-* isomer signal (δ = 9.52 ppm, *d*) is not sufficiently resolved from that of *trans*-2-alkenals (δ = 9.48 ppm, *d*), whereas levels of the *cis,trans*-isomer (δ = 9.63 ppm, *d*) generally fall below the LLOQ threshold value specified, and this served as a limitation of the technique. Therefore, these more minor aldehydic products are not accurately represented by LF ^{1}H NMR data, and therefore a difference in total α,β-unsaturated aldehyde content between MF and LF analyses would be expected. This is indeed the case, with total α,β-unsaturated aldehyde concentrations in heated culinary oils ranging from 10 to 30% higher at the 400 MHz operating frequency (Figures 9b and 10b). Also expected was the evolution of a more diverse range of α,β-unsaturated aldehyde compound signals in the MF spectra acquired at the extended heating periods featured (>60 min). Notably, these resonances, e.g., those of oxygenated aldehydes such as 4,5-diepoxy- and 4-hydroxy/4-hydroperoxy-trans-2-alkenals, and that of malondialdehyde, may also arise from the thermally induced oxidation and/or degradation of pre-formed alka-2,4-dienal species [12].

The higher levels of isomeric alka-2,4-dienals observed in thermally stressed PUFA-rich oils arise from the thermo-oxidative fragmentation of higher concentrations of their specific CHPD precursors, and not of HPMs from the primary MUFA peroxidation stage (the latter yielding only *trans*-2-alkeanls and *n*-alkanals on degradation). Indeed, higher total concentrations of α,β-unsaturated aldehydic species derived from PUFA-rich oils are more readily observable at LF than only *trans*-2-alkenals arising from MUFAs (Figure 10b).

Interestingly, the analytical precision of the LF α,β-unsaturated aldehyde concentration data was found to be similar to those acquired at MF operating frequency. Indeed, 'between-replicate' CV values determined at 60 MHz predominantly ranged from 0.4 to 13.4%, and those at 400 MHz ranged from 1.2 to 12.1%. However, for saturated aldehyde species, estimated LF CV values were found to be not as precise, whereas those at MF were predominantly <7.0%. Additionally, the CSO product displayed a less than favourable level of reproducibility for both total saturated and α,β-unsaturated aldehydes determined at the 90 min heating time-point using the LF spectrometer.

However, CSO is certainly not recommended for cooking or frying purposes anyway in view of its very high ω-3 FA content. Indeed, this oil would be expected to lead to higher levels of aldehydic LOPs than those observed in oils with high linoleoylglycerol contents (e.g., natural sunflower and corn oils); this appears to be coupled with proportionate increases in their 'within- and between-assay' analytical variabilities. Furthermore, the potentially higher volatilities of aldehydic LOPs arising in linolenoylglycerol-rich oils may also contribute towards this higher variance (the inclusion of this unusual oil in this study was primarily to determine the capability of LF NMR analysis to identify and determine differential classes of aldehydic LOPs arising from the specific peroxidation of omega-3 rather than omega-6 PUFAs).

The STN ratios and LLOQ values of aldehydic-CHO function proton resonances determined here are, unfortunately, restricted by the fact that only a single proton gives rise to their signals observed at 60 MHz (δ = 9.4–9.9 ppm), which are split into doublets or triplets for α,β-unsaturated and saturated *n*-alkanals, respectively. Unfortunately, the use of alternative ^{1}H NMR resonances for these aldehydes is not possible, since that of the terminal-CH_3 group of all aldehydes is obscured by very intense, major bulk lipid ones, as are those of their chain-$(CH_2)_n$- groups; the STN values of the latter are also limited by their complex coupling patterns. Similarly, the use of olefinic proton resonances of α,β-unsaturated aldehydes for QNMR purposes at LF strength is also restricted by such complex coupling patterns.

As previously described, the potential toxicological effects of dietary LOPs are likely expended by their uptake by foods fried at high-temperatures in cooking oils, followed by the consumption of such fried foods, e.g., potato chips and fried chicken, etc., by humans [7,12]. Analytical estimates of the three major classes of aldehydic LOPs in potato chips were reported to be 121 ± 33, 157 ± 43 and 126 ± 25 μmol/kg for total n-alkanals, *trans*-2- alkenals and alka-(*trans,trans*)-2,4-dienals, respectively, using ^{1}H NMR analysis [12]. The mean *trans,trans*-alka-2,4-dienal level found was comparable to that determined for the single *trans,trans*-deca-2,4-dienal analyte alone by Boskou et al. [30] in French fries fried in a domestic deep-fryer at 170 °C (65 μmol/kg). It has been shown that fried food aldehyde levels are predominantly dependent on their uptake of LOP-containing frying oils during frying practices; this uptake may range from a few % to as high as 30% (*w/w*), however [7]. If we assume that this uptake level is 10% (*w/w*), a typical value for UK fast-food restaurants [13], then the above values would be ca. 10-fold higher in the oil itself, and would be equivalent to contents of 0.37, 0.49 and 0.39 mmol Aldehyde per mol Of oil FA for *n*-alkanals, *trans*-2-alkenals and *trans,trans*-alka-2,4-dienals, respectively, all of these values being lower than, but approaching, our LLOQ values for the analysis of cooking oil aldehydes by the LF benchtop NMR spectrometer employed here (Table 4). However, the total level of unsaturated aldehydes in these potato chip samples would be 0.88 mmol/mol of uptaken FA, a value which is indeed above our LLOQ threshold limit for determinations made at 60 MHz; although saturated aldehydes would not be validly quantifiable at this operating frequency, they would be detectable since the value of 0.37 mmol/mol FA remains above our LLOD limit, which is 0.19 mmol/mol FA for *n*-hexanal (Table 4). Our previously reported MF or HF ^{1}H NMR analysis of aldehydic LOPs in French fry samples relies on a $CDCl_3$ extraction process, and modifications to this protocol, such as the use of larger quantities of food samples for this purpose, a process giving rise to greater volumes of oils extracted therefrom, would, of course, yield lower LLOQ values for analysis at LF strength. However, it has been reported that the fried food aldehyde contents are markedly lower than those anticipated from the percentage oil uptake value alone [14], and this is likely to be attributable to their chemical consumption through Maillard or Michael addition reactions with food amino acids, peptides and proteins, and/or acetal and ketal formation on reaction with certain food carbohydrates. Indeed, it is well known that such aldehydes have a high level of reactivity with a range of biomolecules present in many biosystems. Fried food contents of aldehydic LOPs will, of course, also be strikingly dependent on a wide range of further factors, for example frying temperatures and fried food exposure times, the potential recycling of reused oils, frying type (i.e., shallow versus deep-frying episodes), frying oil MUFA, PUFA and ω-3 FA contents, and potentially also their lipid-soluble antioxidant contents, etc.

Therefore, in summary, LF NMR analysis has the capacity to monitor *n*-alkanals (saturated aldehydes) in thermally stressed culinary oils at detectable levels of 0.19 mmol/mol FA (equivalent to 0.66 mmol/L), and quantifiable at levels of 0.65 mmol/mol FA (equivalent to 2.21 mmol/L) when analysed in an olive oil co-calibrant placed in $CDCl_3$ solution. Moreover, it was determined that *trans*-2-alkenal species were detectable at levels of 0.18 mmol/mol FA (equivalent to 0.63 mmol/L), with quantifiable levels of 0.62 mmol/mol FA (equivalent to 2.10 mmol/L) in an olive oil/$CDCl_3$ medium. Similar LLOD and LLOQ estimates were found for both aldehyde calibrants when analysed in a 'neat' $CDCl_3$ solution medium alone (Table 4).

Of further interest, this technological LF NMR development was also found to be valuable for the detection and quantification of precursors of the aldehydic LOPs monitored, specifically CHPDs and HPMs, as demonstrated in Figure 11. Results arising from the LF NMR assessment of the generation of these precursors, and the dependence of their culinary oil concentrations on LSSFE heating time, will be made available in a follow-up paper shortly.

This LF NMR analysis option is also valuable for determining the FA and unsaturation status of culinary oils in general, as shown in Figure 1. One interesting feature of these major lipid component H NMR profiles was differences in the spectral appearance of selected triacylglycerol multiplet resonances, and this is ascribable to the previously reported 'roofing' effect which occurs when scalar *J*-coupled signals appear just a few Hz away from each other in spectra acquired, as indeed they are in LF (60 MHz) spectra.

In principle, near-portable, non-stationary LF NMR spectrometers could be employed for the purpose of directly monitoring toxic aldehydic LOPs, together with their lipid hydroperoxide precursors, in a range of culinary frying oils at commercial food production and manufacturing sites, or alternatively at large outlets of global fast-food restaurant chains. Indeed, analysis is rapid, and the skillset required for the operation and management of such devices is far simpler and appealing than that required for institutionally based MF or HF spectrometers; indeed, with the exception of a succinct level of training, no previous specialist operator knowledge is required.

5. Limitations of the Study

Although results obtained herein have demonstrated the ability of LF benchtop NMR spectrometers to detect several different classes of aldehydic LOP species in oil samples, unfortunately quantification of these species is limited to total saturated and total α,β-unsaturated aldehydes in view of an insufficient resolution of the ^1H NMR resonances of these LOPs, most notably that involving the latter class of these analytes. Veritably, LF NMR analysis does suffer from a number of limitations, including signal convolution which leads to issues of both identification and quantification for individual unsaturated aldehyde species. Furthermore, because of the reduced resolution and increased background noise of such instruments, the magnitude of and demands for satisfactory LLOD and LLOQ parameters were somewhat greater than those which are usually mandatory for QNMR approaches employed on MF and HF NMR facilities. Moreover, the analytical sensitivity threshold of the LF NMR technique for our analysis protocol (Table 4) also served as a further limitation; this involved the analysis of only a 0.20 mL aliquot of culinary oil in a total NMR tube volume of 0.760 mL. However, pilot experiments have revealed that we may increase this analytical volume of oil sample up to 0.40 mL, and therefore the LLOD thresholds determined here would be decreased 2-fold if that were the case. Unfortunately, larger increments in proportionate sample volume restrict signal resolution and hence the analysis of aldehydes and other LOPs in view of the resulting high viscosity of the analyte solution medium.

Data available in Figures 9b and 10b revealed that for both the 60 and 90 min heating time-points, mean total saturated aldehyde concentrations determined in PUFA-rich oils at an operating frequency of 60 MHz were 79 and 89% of that determined at 400 MHz for the CSO and RSO products, respectively, but only 62% for the SBO one (these 'between-oil' differences in analyte receptivity were also manifested by the observation of a statistically significant first-order oil × spectrometer operating frequency interaction effect for this aldehyde classification). For total α,β-unsaturated aldehyde levels, however, corresponding values were 83, 69 and 81% for the CSO, RSO and SBO products, respectively (MOO 71%). Therefore, it is important that this current restriction should be recognised by analysts and researchers, and perhaps the application of a suitable 'correction factor' should be considered to overcome this issue.

It should also be noted that saturated aldehydic LOPs (mainly two types of NMR-distinguishable *n*-alkanals) are generated at much lower concentrations than the α,β-unsaturated aldehyde classifications, and for this study, it was found that the former class were only ca. 20% of the total aldehyde levels determined by the LF spectrometer at the 90 min heating time-point. These diminished concentrations will also restrict their determination at an operating frequency of only 60 MHz.

Another consequence of the LF ^{1}H NMR monitoring of aldehydes in culinary oils, heated or otherwise, is that the total amounts of aldehydic LOPs liberated from the peroxidation of UFAs is not a direct reflection of those found in these oil products following exposure to thermal stressing periods. Indeed, a significant proportion of these toxins have boiling-points below standard frying temperature (now ranging from 160 to 180 °C), with some, such as acrolein (the lowest α,β-unsaturated aldehyde homologue), exceedingly so [14]. Hence, a quite substantial fraction of the total amount of aldehydic LOPs evolved from these frying oil 'reaction' media are in the volatile, i.e., gaseous form, and due caution should be exercised by humans to avoid the inhalation of these toxins during frying or cooking exercises, be they commercial or domestic. Therefore, oil aldehyde contents only serve as incomplete reflections of the total amounts generated.

Since our LF ^{1}H NMR-determined aldehyde concentrations were segmented into total saturated and α,β-unsaturated classifications, variance contributions towards these from structurally more complex LOPs would be expected to expand with increasing complexity level, most notably for PUFA-rich oils which have been exposed to prolonged heating durations. Ultimately, saturated aldehydes were only observed at quantifiable levels in the LF NMR spectra of heated oil samples from the 60 min, but not the 30 min, time-point, although not all oils thermally stressed in this manner for a 60 min period gave sample aldehyde levels higher than the LLOQ threshold level determined here. Despite this, LF-NMR analysis shows promise as an appropriate approach for detecting and quantifying total aldehydes in oils that have been heated for durations which are ≥60 min, and putatively less so for samples with higher oil contents present in analyte media solutions.

6. Conclusions

In this study, we have demonstrated the applications of LF NMR analysis for the identification and quantification of both saturated and unsaturated aldehydic LOP species in heated culinary oils. This investigation was also valuable for monitoring the abilities of LF and MF NMR spectrometers to distinguish between differential classes of these toxic products.

The investigations conducted demonstrated that this technology was valuable for determining the concentrations of LOP toxins in such culinary oil products exposed to LSSFEs, but only for those which were thermally stressed for minimal durations of 60 min or so. Although distinctive patterns of aldehydes were observed for different oil products investigated, unfortunately resonance superimposition and the limited sensitivity of the technique precluded the determination of individual molecular classes of aldehyde toxins in these products. However, LF NMR analysis remains a very promising technique, and in some thermally stressed PUFA-rich oils, a total of four or more unsaturated aldehydes, and two saturated aldehydes, along with their CHPD and/or HPM precursors, were detectable, although not all were directly quantifiable (Figures 2 and 11). However, the LF NMR determination of total saturated and α,β-unsaturated aldehyde oil contents, along with those of their hydroperoxide precursors, remains a relevant and acceptable avenue for aldehydic toxin analysis in commonly employed frying oils.

In principle, the non-stationary LF NMR technique could be employed for the 'on-site' monitoring of aldehydic toxins present in thermally stressed cooking oil samples at food manufacturing sites, or even at fast-food take-out restaurants.

Author Contributions: Conceptualization, M.G. (Martin Grootveld) and M.G. (Miles Gibson); validation, M.G. (Martin Grootveld), B.C.P., M.G. (Miles Gibson) and M.E.; formal analysis, M.G. (Martin Grootveld), B.C.P., M.G. (Miles Gibson) and ME; resources, M.G. (Martin Grootveld); data curation, M.G. (Martin Grootveld), M.G. (Miles Gibson), B.C.P. and M.E.; writing—original draft preparation, M.G. (Miles Gibson), B.C.P., M.G. (Martin Grootveld) and M.E.; writing—review and editing, M.G. (Martin Grootveld); visualization, M.G. (Martin Grootveld), B.C.P., M.G. (Miles Gibson) and M.E.; statistical analyses, M.G. (Martin Grootveld) and M.G. (Miles Gibson); supervision, M.G. (Martin Grootveld). All authors have read and agreed to the published version of the manuscript.

Funding: This research received no external funding.

Institutional Review Board Statement: Not applicable.

Informed Consent Statement: Not applicable.

Data Availability Statement: All study data will be provided to those who reasonably request this information from the correspondence author.

Acknowledgments: The authors thank Magritek GmbH for the loan of a Spinsolve 60 Ultra benchtop NMR spectrometer. All authors wish to thank Katy Woodason for useful discussions, and Benita Claire Percival would like to thank De Montfort University, Leicester, UK for the provision of a fees-waiver PhD scholarship to her.

Conflicts of Interest: None of the authors declare any conflicts of interest.

List of Abbreviations

Analysis-of-variance (ANOVA); Analysis-of-covariance (ANCOVA); conjugated hydroperoxydiene (CHPD); chia seed oil (CSO); fatty acid (FA); unconjugated hydroperoxymonoene (HPM); laboratory-simulated shallow frying episodes (LSSFEs); lower limit of detection (LLOD); lower limit of quantification (LLOQ); lipid oxidation products (LOPs); mixed refined olive oil (MOO); monounsaturated fatty acid (MUFA); nuclear magnetic resonance (NMR); NMR spectrometer operating frequency (OF); polyunsaturated fatty acid (PUFA); quantitative NMR (QNMR); rapeseed oil (RSO); saturated fatty acid (SAT); signal-to-noise (STN); soybean oil (SBO); standard deviation (SD); unsaturated fatty acid (UFA).

References

1. Percival, B.C.; Savel, E.; Ampem, G.; Gibson, M.; Edgar, M.; Jafari, F.; Frederick, K.; Wilson, P.; Grootveld, M. Molecular composition of and potential health benefits offered by natural East African virgin sunflower oil products: A 400 MHz ^{1}H NMR analysis study. *Int. J. Nutr.* **2019**, *3*, 22–43. [CrossRef]
2. Haywood, R.M.; Claxson, A.W.D.; Hawkes, G.E.; Richardson, D.P.; Naughton, D.P.; Coumbarides, G.; Hawkes, J.; Lynch, E.J.; Grootveld, M.C. Detection of aldehydes and their conjugated hydroperoxydiene precursors in thermally-stressed culinary oils and fats: Investigations using high resolution proton nmr spectroscopy. *Free Radic. Res.* **1995**, *22*, 441–482. [CrossRef]
3. Claxson, A.W.D.; Hawkes, G.E.; Richardson, D.P.; Naughton, D.P.; Haywood, R.M.; Chander, C.L.; Atherton, M.; Lynch, E.; Grootveld, M.C. Generation of lipid peroxidation products in culinary oils and fats during episodes of thermal stressing: A high field ^{1}H NMR study. *FEBS Lett.* **1994**, *355*, 81–90. [CrossRef]
4. Martínez-Yusta, A.; Goicoechea, E.; Guillén, M.D. A review of thermo-oxidative degradation of food lipids studied by ^{1}H NMR spectroscopy: Influence of degradative conditions and food lipid nature. *Compr. Rev. Food Sci. Food Saf.* **2014**, *13*, 838–859. [CrossRef]

5. Grootveld, M.; Atherton, M.D.; Sheerin, A.N.; Hawkes, J.; Blake, D.R.; Richens, T.E.; Silwood, C.J.L.; Lynch, E.; Claxson, A.W.D. In vivo absorption, metabolism, and urinary excretion of α,β- unsaturated aldehydes in experimental animals. Relevance to the development of cardiovascular diseases by the dietary ingestion of thermally stressed polyunsaturate-rich culinary oils. *J. Clin. Invest.* **1988**, *101*, 1210–1218. [CrossRef]
6. Ko, Y.C.; Cheng, L.S.; Lee, C.H.; Huang, J.J.; Huang, M.S.; Kao, E.L.; Wang, H.Z.; Lin, H.J. Chinese food cooking and lung cancer in women nonsmokers. *Am. J. Epidemiol.* **2000**, *151*, 140–147. [CrossRef] [PubMed]
7. Grootveld, M.; Percival, B.C.; Grootveld, K.L. Chronic non-communicable disease risks presented by lipid oxidation products in fried foods. *HepatoBiliary Surg. Nutr.* **2018**, *7*, 305–312. [CrossRef] [PubMed]
8. Weng, M.; Lee, H.W.; Park, S.H.; Hu, Y.; Wang, H.T.; Chen, L.C.; Rom, W.N.; Huang, W.C.; Lepor, H.; Wu, X.R.; et al. Aldehydes are the predominant forces inducing DNA damage and inhibiting DNA repair in tobacco smoke carcinogenesis. *Proc. Natl. Acad. Sci. USA* **2018**, *115*, E6152–E6161. [CrossRef] [PubMed]
9. Qin, P.; Zhang, M.; Han, M.; Liu, D.; Luo, X.; Xu, L.; Zeng, Y.; Che, Q.; Wang, T.; Chen, X.; et al. Fried-food consumption and risk of cardiovascular disease and all-cause mortality: A meta-analysis of observational studies. *Heart* **2021**, *107*, 1567–1575. [CrossRef] [PubMed]
10. Lippi, G.; Mattiuzzi, C. Fried food and prostate cancer risk: Systematic review and meta-analysis. *Int. J. Food Sci. Nutr.* **2015**, *66*, 587–589. [CrossRef]
11. Zhong, L.; Goldberg, M.S.; Parent, M.E.; Hanley, J.A. Risk of developing lung cancer in relation to exposure to fumes from Chinese-style cooking. *Scand. J. Work Environ. Health* **1995**, *25*, 309–316. [CrossRef]
12. Moumtaz, S.; Percival, B.; Parmar, D.; Grootveld, K.L.; Jansson, P.; Grootveld, M. Toxic aldehyde generation in and food uptake from culinary oils during frying practices: Peroxidative resistance of a monounsaturate-rich algae oil. *Sci. Rep.* **2019**, *9*, 4125. [CrossRef] [PubMed]
13. Le Gresley, A.; Ampem, G.; De Mars, S.; Grootveld, M.; Naughton, D.P. 'Real-world' evaluation of lipid oxidation products and legislated metals in French fries from two chain fast-food restaurants. *Front. Nutr.* **2021**, *9*, 620952. [CrossRef]
14. Grootveld, M. Evidence-based challenges to the continued recommendation and use of peroxidatively-susceptible polyunsaturated fatty acid-rich culinary oils for high-temperature frying practices: Experimental revelations focused on toxic aldehydic lipid oxidation products. *Front. Nutr.* **2021**, *8*, 711640. [CrossRef] [PubMed]
15. Silwood, C.J.L.; Grootveld, M. Application of high-resolution, two-dimensional ^{1}H and ^{13}C nuclear magnetic resonance techniques to the characterization of lipid oxidation products in autoxidized linoleoyl/linolenoylglycerols. *Lipids* **1999**, *34*, 741. [CrossRef]
16. Percival, B.C.; Grootveld, M.; Gibson, M.; Osman, Y.; Molinari, M.; Jafari, F.; Sahota, T.; Martin, M.; Casanova, F.; Mather, M.L.; et al. Low-field, benchtop NMR spectroscopy as a potential tool for point-of-care diagnostics of metabolic conditions: Validation, protocols and computational models. *High-Throughput* **2018**, *8*, 2. [CrossRef]
17. Grootveld, M.; Percival, B.C.; Gibson, M.; Osman, Y.; Edgar, M.; Molinari, M.; Mather, M.L.; Casanova, F.; Wilson, P.B. Progress in low-field benchtop NMR spectroscopy in chemical and biochemical analysis. *Anal. Chim. Acta* **2019**, *1067*, 11–30. [CrossRef] [PubMed]
18. Silva Elipe, M.V.; Milburn, R.R. Monitoring chemical reactions by low-field benchtop NMR at 45 MHz: Pros and cons. *Magn. Reson. Chem.* **2016**, *54*, 437–443. [CrossRef]
19. Edgar, M.; Percival, B.C.; Gibson, M.; Masania, J.; Beresford, K.; Wilson, P.B.; Grootveld, M. Benchtop NMR spectroscopy and spectral analysis of the *cis*- and *trans*-stilbene products of the Wittig reaction. *J. Chem. Educat.* **2019**, *96*, 1938–1947. [CrossRef]
20. Miloushev, V.Z.; Di Gialleonardo, V.; Salamanca-Cardona, L.; Correa, F.; Granlund, K.L.; Keshari, K.R. Hyperpolarized 13C pyruvate mouse brain metabolism with absorptive-mode EPSI at 1 T. *J. Magn. Reson.* **2017**, *275*, 120–126. [CrossRef]
21. Parker, T.; Limer, E.; Watson, A.D.; Defernez, M.; Williamson, D.; Kemsley, E.K. 60 MHz ^{1}H NMR spectroscopy for the analysis of edible oils. *TrAC—Trends Anal. Chem.* **2014**, *57*, 147–158. [CrossRef] [PubMed]
22. McDowell, D.; Defernez, M.; Kemsley, E.K.; Elliott, C.T.; Koidis, A. Low vs. high field 1h Nmr spectroscopy for the detection of adulteration of cold pressed rapeseed oil with refined oils. *LWT* **2019**, *111*, 490–499. [CrossRef]
23. Gouilleux, B.; Marchand, J.; Charrier, B.; Remaud, G.S.; Giraudeau, P. High-throughput authentication of edible oils with benchtop Ultrafast 2D NMR. *Food Chem.* **2018**, *244*, 153–158. [CrossRef] [PubMed]
24. Grootveld, M.; Ruiz-Rodado, V.; Silwood, C.J.L. Detection, monitoring, and deleterious health effects of lipid oxidation. *AOCS Inform* **2014**, *25*, 614–624.
25. Grootveld, M.; Percival, B.C.; Moumtaz, S.; Gibson, M.; Woodason, K.; Akhtar, A.; Wawire, M.; Edgar, M.; Grootveld, K.L. Iconoclastic reflections on the 'safety' of polyunsaturated fatty acid-rich culinary frying oils: Some cautions regarding the laboratory analysis and dietary ingestion of lipid oxidation product toxins. *Appl. Sci.* **2021**, *11*, 2351. [CrossRef]
26. Wishart, D.S.; Guo, A.; Oler, E.; Wang, F.; Anjum, A.; Peters, H.; Dizon, R.; Sayeeda, Z.; Tian, S.; Lee, B.L.; et al. HMDB 5.0: The Human Metabolome Database for 2022. *Nucleic Acids Res.* **2022**, *50*, D622–D631. [CrossRef]
27. Miller, J.N.; Miller, J.C. *Statistics and Chemometrics for Analytical Chemistry*; Pearson Higher Education: London, UK, 2018; ISBN -13: 9781292186719.
28. Wann, A.I.; Percival, B.C.; Woodason, K.; Gibson, M.; Vincent, S.; Grootveld, M. Comparative ^{1}H NMR-based chemometric evaluations of the time-dependent generation of aldehydic lipid oxidation products in culinary oils exposed to laboratory-simulated shallow frying episodes: Differential patterns observed for omega-3 fatty acid-containing soybean oils. *Foods* **2021**, *10*, 2481. [CrossRef] [PubMed]

29. Hwang, H.-S.; Winkler-Moser, J.K. Oxidative stability and shelf-life of frying oils and fried foods. In *Oxidative Stability and Shelf Life of Foods Containing Oils and Fats*; Hu, M., Jacobsen, C., Eds.; Academic Press: Cambridge, MA, USA; AOCS Press: Urbana, IL, USA, 2016; pp. 251–285.
30. Boskou, G.; Salta, F.N.; Chiou, A.; Troullidou, E.; Andrikopoulos, N.K. Content of *trans,trans*-2,4-decadienal in deep-fried and pan-fried potatoes. *Eur. J. Lipid Sci. Technol.* **2006**, *108*, 109–115. [CrossRef]

foods

MDPI

Article

Quantitative Analysis of Oat (*Avena sativa* L.) and Pea (*Pisum sativum* L.) Saponins in Plant-Based Food Products by Hydrophilic Interaction Liquid Chromatography Coupled with Mass Spectrometry

Anastassia Bljahhina [1,2,*], Dmitri Pismennõi [1,2], Tiina Kriščiunaite [1], Maria Kuhtinskaja [2] and Eeva-Gerda Kobrin [1]

[1] Center of Food and Fermentation Technologies (TFTAK), Mäealuse 2/4, 12618 Tallinn, Estonia
[2] Department of Chemistry and Biotechnology, Tallinn University of Technology, Akadeemia Tee 15, 12618 Tallinn, Estonia
* Correspondence: anastassia.bljahhina@tftak.eu

Abstract: This work presents the sample extraction methods for solid and liquid sample matrices for simultaneous quantification of oat (*Avena sativa* L.) and pea (*Pisum sativum* L.) saponins: avenacoside A, avenacoside B, 26-desglucoavenacoside A, and saponin B and 2,3-dihydro-2,5-dihydroxy-6-methyl-4H-pyran-4-one (DDMP) saponin, respectively. The targeted saponins were identified and quantified using a hydrophilic interaction liquid chromatography with mass spectrometric detection (HILIC-MS) method. The simple and high-throughput extraction procedure was developed for solid oat- and pea-based food samples. In addition, a very simple extraction procedure for liquid samples, without the need to use lyophilisation, was also implemented. Oat seed flour (U-^{13}C-labelled) and soyasaponin Ba were used as internal standards for avenacoside A and saponin B, respectively. Other saponins were relatively quantified based on avenacoside A and saponin B standard responses. The developed method was tested and successfully validated using oat and pea flours, protein concentrates and isolates, as well as their mixtures, and plant-based drinks. With this method, the saponins from oat- and pea-based products were separated and quantified simultaneously within 6 min. The use of respective internal standards derived from U-^{13}C-labelled oat and soyasaponin Ba ensured high accuracy and precision of the proposed method.

Keywords: oat; pea; avenacosides; saponin B; DDMP saponin; plant-based protein

Citation: Bljahhina, A.; Pismennõi, D.; Kriščiunaite, T.; Kuhtinskaja, M.; Kobrin, E.-G. Quantitative Analysis of Oat (*Avena sativa* L.) and Pea (*Pisum sativum* L.) Saponins in Plant-Based Food Products by Hydrophilic Interaction Liquid Chromatography Coupled with Mass Spectrometry. *Foods* **2023**, *12*, 991. https://doi.org/10.3390/foods12050991

Academic Editor: Gianfranco Picone

Received: 1 February 2023
Revised: 21 February 2023
Accepted: 24 February 2023
Published: 26 February 2023

1. Introduction

The demand for sustainable protein sources in food production is continuously growing [1,2]. Oat (*Avena sativa* L.) and pea (*Pisum sativum* L.) proteins in the form of concentrates or isolates can act as an alternative to animal proteins due to their potential ability to provide desirable technological properties in plant-based meat and dairy substitutes [3,4]. Pea protein is an insufficient source of methionine but, on the other hand, has a high content of the essential amino acid lysine and branched-chain amino acids-leucine, isoleucine, and valine [4]. In contrast to pea, oat contains enough methionine but a scarce amount of lysine [3]. Blending oat and pea proteins in products is one way to achieve a complete essential amino acid profile [5], and such products are already available on the market. However, one of the main obstacles in the application of plant-based proteins in food production is their bitter and astringent off-taste [6–8]. It has been suggested that saponins might be the main cause of this sensation [9–14] influencing consumer acceptance.

Saponins are a diverse group of secondary defence metabolites widely spread in plant species [15]. Saponins investigated in this study are amphiphilic molecules, with polar water-soluble sugar moieties attached to a nonpolar, water-insoluble steroid or triterpene core [16]. Oats, as the only cereals capable of accumulating saponins, contain bisdesmosidic steroidal saponins avenacoside A and B, and monodesmosidic 26-desglucoavenacoside A in their leaves and grains (Figure 1) [9,10,12,13,17]. Saponin B and 2,3-dihydro-2,5-dihydroxy-6-methyl-4H-pyran-4-one (DDMP) saponin are monodesmosidic triterpenoid saponins found in peas [18,19]. Besides being taste-active bitter compounds, saponins have also been reported as antinutrients. As such, they may affect nutrient absorption by inhibiting metabolic and digestive enzymes [20] and by binding to minerals such as zinc and iron [21]. High concentrations of saponins in the diet may lead to hypocholesterolemic effect [22], hypoglycemia [23], inefficient protein digestion, vitamin and mineral uptake in the gut, and the development of a leaky gut [24]. Despite the reported negative nutritional impact, some studies have also shown positive cholesterol-lowering [25] and anticancerogenic [26] effects of saponins.

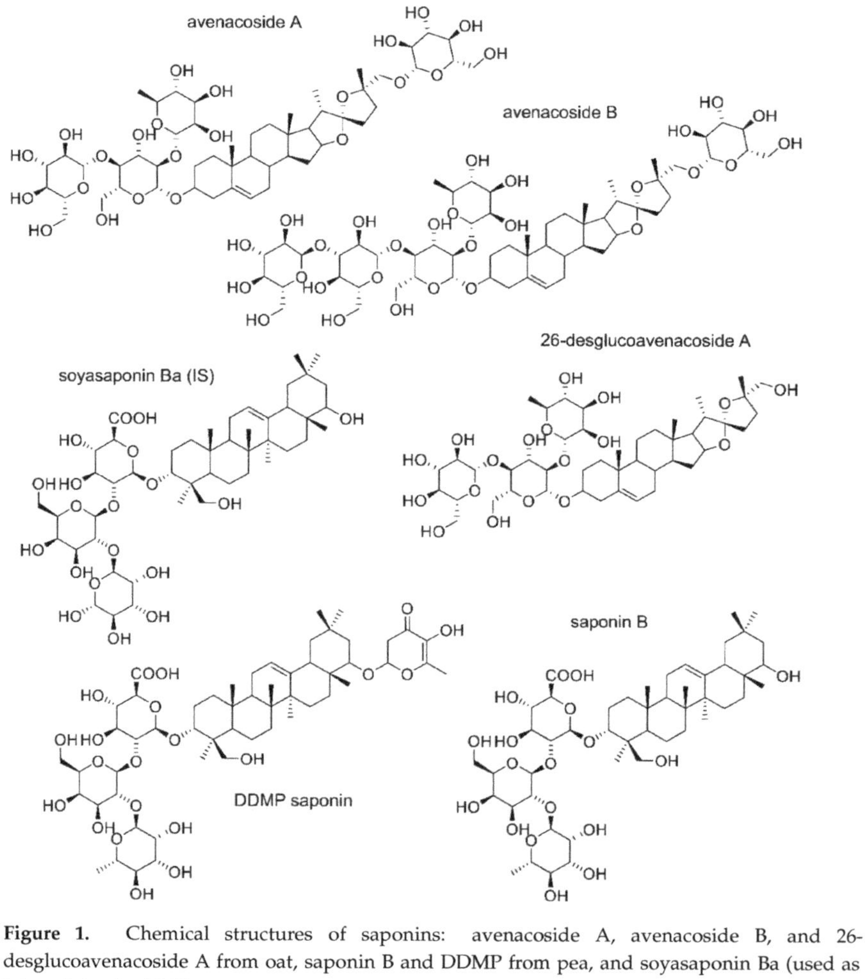

Figure 1. Chemical structures of saponins: avenacoside A, avenacoside B, and 26-desglucoavenacoside A from oat, saponin B and DDMP from pea, and soyasaponin Ba (used as internal standard [IS]).

The analysis of saponins could be performed using a wide range of classical methods such as gravimetry [15,27], hemolysis [28], bioassays [29], and spectrophotometry [30]. In addition, different saponins could be separated and analysed using chromatographic methods, e.g., thin-layer chromatography [31,32], gas chromatography [33], and high-performance liquid chromatography (HPLC) [19]. The detection of the saponin class compounds could be carried out using the simplest optical detection methods [34,35], but these methods usually lack the selectivity and sensitivity of more advanced analytical techniques such as mass spectrometry [9,17,36–38]. The saponin extraction and the pre- and post-extraction sample clean-up before the LC analysis [9,17,19,39] are required to obtain a clean extract which would minimise matrix effect in mass spectrometric measurements. However, these sample preparation procedures are time-consuming and unsuitable for routine analysis of large amounts of samples. This creates the need for an improved, efficient, sensitive, more selective, and reproducible extraction method of saponins prior to the analysis. The use of liquid chromatography coupled with mass spectrometry (LC-MS) allows more precise and selective determination of the contents of different types of saponins in various plant species: oat [9,13,17], pea [19,32], and soya [39,40]. Although, the amounts of saponins have been quantified mainly from the seeds or husks of numerous oat and pea varieties [9,19,32], there is a lack of data concerning the concentrations of saponins in processed food ingredients, and the half- and end-products produced therefrom. To the best of our knowledge, there is no versatile method for the determination of saponins derived from different plant species in various food matrices.

The objective of this study was to develop simple sample extraction methods for solid and liquid plant-based food sample matrices for the selective and quantitative determination of five oat and pea saponins: avenacoside A, avenacoside B, 26-desglucoavenacoside A, saponin B, and DDMP saponin, using hydrophilic interaction liquid chromatography with mass spectrometric detection (HILIC-MS). To our knowledge, there are no reports on simultaneous HILIC analysis of the above-mentioned saponins in solid and liquid samples containing concurrent oat and pea ingredients.

2. Materials and Methods

2.1. Chemicals and Materials

HPLC-grade acetonitrile (MeCN), methanol (MeOH), ethanol (EtOH), hexane, propan-2-ol (IPA), and formic acid (FA) (for MS, 98%) were purchased from Honeywell (Charlotte, NC, USA). The standard compounds avenacoside A, saponin B (soyasaponin I), and soyasaponin Ba phyproof® were purchased from Sigma-Aldrich (Darmstadt, Germany). Uniformly isotopically labelled oat seed flour (U-^{13}C oat seeds, *Avena sativa* 97 atom%) was obtained from IsoLife BV (Wageningen, The Netherlands). Ultrapure water (18.2 MΩ·cm) was prepared with MilliQ® HX 7040SD equipped with MilliQ LC-Pak (Merck KGaA, Darmstadt, Germany). Biotage Isolute® PLD+ and C18 columns (100 mg/1 mL) were purchased from Biotage Sweden AB (Uppsala, Sweden). Amicon Ultra-0.5 centrifugal filter units (3, 10, 30, 50 kDa) and Millex-LCR filters (pore size 0.2 µm, filter dimension 13 mm) were obtained from Merck KgaA (Darmstadt, Germany).

2.2. Food Samples

Yellow pea flour, whole-grain oat flour, and oat and pea drinks were purchased from a local supermarket. Pea protein isolate (Bang & Bonsomer Estonia OÜ, Tallinn, Estonia), pea protein concentrate (Aloja-Starkelsen Ltd., Limbažu novads, Latvia), and oat protein concentrate (Lantmännen, Stockholm, Sweden) were obtained from producers. The composition and nutritional information available on the product label of these products is available in Supplementary Table S1. Untreated and extruded blends of pea protein isolate, oat protein concentrate, and pea protein concentrate (52:28:20, *w*/*w*) were produced in-house by following a previously published protocol [41].

2.3. Extraction Method for Solid Samples and for Liquid Samples

Sample extraction methods 1A, 1B, 2A, 2B, and 2C, which were tested during solid sample extraction method development, are described in the supplementary information.

Solid sample extraction (method 2D) was performed according to Heng et al. [19] with some modifications. Powdered non-defatted solid sample (100 mg) was weighed into a 10 mL volumetric flask (n = 3), filled with aqueous EtOH (70%, *v*/*v*), mixed thoroughly, and ultrasonicated for 30 min (without additional heating). After ultrasonication, samples were centrifuged (14,000× *g* for 10 min at 10 °C) to remove insoluble matter. The supernatant (500 μL) was passed through PLD+ columns by applying positive pressure to remove proteins and phospholipids. The obtained filtrate was diluted to receive an aqueous MeCN (50%, *v*/*v*) solution. The diluted filtrate (100 μL) was transferred to the LC-MS vials, mixed with 50 μL soyasaponin Ba working solution and 50 μL U-^{13}C-oat extract working solution, and injected into the LC-MS.

A homogeneous liquid sample was weighed (0.25 g) into a 5 mL volumetric flask ($n = 3$), filled with ultrapure water, and mixed thoroughly. Diluted sample solutions were centrifuged (14,000× *g* for 15 min at 10 °C) to remove insoluble matter. Sample supernatant (200 μL) and 800 μL MeCN were transferred into the next tube, mixed thoroughly, and centrifuged (14,000× *g* for 15 min at 10 °C) to remove precipitated proteins. The supernatant (500 μL) was passed through PLD+ columns. The obtained filtrate (300 μL) was transferred into a clear tube and diluted with 180 μL ultrapure water to obtain an aqueous MeCN solution (50%, *v*/*v*). The diluted sample filtrate was combined with internal standard solutions as described for solid samples and injected into the LC-MS.

2.4. Preparation of Standard Solutions

The stock solution of avenacoside A (500 mg/L) was prepared in ultrapure water and the aliquots were stored at −80 °C. The stock solution of saponin B (500 mg/L) was prepared in aqueous EtOH (60%, *v*/*v*) and aliquots were stored at −80 °C. The internal standard stock solution of soyasaponin Ba (100 mg/L) was prepared in MeOH.

The stock solution of U-^{13}C-oat seed flour extract containing $^{13}C_{51}$-avenacoside A was prepared using the previously described solid sample extraction method 2D with some modifications. U-^{13}C-oat seed flour (150 mg) was weighed into a 50 mL volumetric flask, filled with EtOH (70%, *v*/*v*), and mixed thoroughly. The flask was ultrasonicated for 30 min (without additional heating) and the obtained solution was centrifuged (17,000× *g* for 10 min at 10 °C) to remove insoluble matter. The supernatant was passed through PLD+ columns using a vacuum manifold. The cleaned extract was aliquoted and stored at −80 °C.

The internal standard working solutions were prepared freshly before the analysis. The working solution of internal standard soyasaponin Ba was prepared by diluting stock solution in the aqueous MeCN (50%, *v*/*v*). The U-^{13}C-oat extract working solution was prepared by diluting the stock solution two-fold with neat MeCN.

2.5. Liquid Chromatography Mass Spectrometry

Samples were analysed using a Waters UPLC® system (Waters Corporation, Milford, MA, USA) coupled with a Waters Quattro Premier XE Mass Spectrometer equipped with ZSpray™ Source and controlled by Waters MassLynx™ 4.1 (V4.1 SCN805, Waters Corporation, Milford, MA, USA). Mobile phases were as follows: (A) 0.1% FA in ultrapure water, (B) 0.1% FA in MeCN. Weak needle wash was composed of MeCN in ultrapure water (90%, *v*/*v*), and strong needle wash consisted of IPA in MeCN (50%, *v*/*v*). The seal wash solution was aqueous MeCN (50%, *v*/*v*). Samples were stored in an autosampler which was set at 8 °C. The injection volume was 2 μL. Saponins were separated using BEH Amide column (1.0 × 50 mm, 1.7 μm) coupled with BEH Amide VanGuard Pre-column (2.1 × 5 mm) from Waters Corporation (Milford, MA, USA). The final gradient was as follows: 0–0.17 min at 10% A, 0.17–3.5 min linear gradient 10–70% A, 3.5–4.0 min at 70% A, 4.0–4.5 min linear gradient 70–10% A, 4.5–6.0 min at 10% A. The column temperature was held at 50 °C during all experiments. The flow rate was set at 200 μL/min.

The analytes were ionised under negative electrospray ionisation (ESI-) and optimised source conditions. The source temperature was set to 120 °C, and high-purity nitrogen was fed into the source at 25 L/h (cone) and 600 L/h (desolvation) and desolvation gas was heated to 350 °C. The capillary voltage was set to −1.5 kV, cone voltage to 80 V, and extractor voltage to 3 V. For measurement of analytes, a set of m/z values for single-ion-recording

(SIR) experiments was recorded simultaneously during one chromatographic run. For saponin quantification, deprotonated molecules [M-H]- were chosen based on a scan-type experiment. Mass-to-charge ratios ($m/z \pm 0.5$ Da) for SIR channels were set as follows: avenacoside A—m/z 1061.5; avenacoside B—m/z 1223; 26-desglucoavenacoside A—m/z 899.5; $^{13}C_{51}$-avenacoside A—m/z 1112.5 (internal standard); saponin B—m/z 941.5; DDMP saponin—m/z 1067; soyasaponin Ba—m/z 957.5 (internal standard). Data acquisition was performed in Waters MassLynx™ V4.1 (SCN805, Waters Corporation, Milford, MA, USA). Data analysis was performed in Waters QuanLynx™ V4.1 (SCN805, Waters Corporation, Milford, MA, USA) and Microsoft Excel® (Microsoft 365 Apps for enterprise).

2.6. Calibration and Quantification

The working solution was prepared by diluting standard stock solutions 100 times with MeCN:H_2O:EtOH solution (50:36:14, v/v). Internal standards, soyasaponin Ba and U-^{13}C-oat extract, were added before injection to the autosampler vial, and their concentration in the vial was set at 0.75 mg/L and 0.3 mg/L, respectively. Calibration curve standard solutions (100 μL) were mixed with internal standards working solutions (50 μL U-^{13}C-oat extract working solution and 50 μL soyasaponin Ba working solution). Calibration curves were built for avenacoside A (0.01–2.44 mg/L) and saponin B (0.01–2.48 mg/L) using eight-point measurements of serially diluted standards, which were run in triplicate. The regression was found by fitting points to the linear equation. The external standard calibration curves were built by correlating the concentrations of external standards to the response factors, which were calculated according to Equation (1).

$$\text{response factor (RF)} = (\text{area of analyte})/(\text{area of internal standard}) \quad (1)$$

As only the avenacoside A standard was commercially available, other analytes of interest (avenacoside B and 26-desglucoavenacoside A) were quantified relatively using the avenacoside A calibration curve. Avenacoside B and 26-desglucoavenacoside A results are presented in avenacoside A equivalents. Avenacosides were quantified using isotopically labelled ^{13}C-avenacoside A as an internal standard. As DDMP saponin could not be sourced commercially, its quantification was based on the saponin B standard curve, and the results are given in saponin B equivalents. Both were quantified using soyasaponin Ba as an internal standard.

2.7. Validation of the Method

The following parameters were assessed during method validation: linearity, limit of detection (LOD), limit of quantification (LOQ), precision, specificity, sample extraction recoveries, and matrix effect (ME). Developed extraction methods for solid and liquid samples were validated separately. Oat protein concentrate and pea protein isolate were used to validate the solid sample extraction method. Saponin determination in liquid samples was validated using oat and pea drinks.

The linear range and linearity were evaluated via repeated measurements of standard solutions of avenacoside A and saponin B consisting of 8 individual points obtained from serial dilution of stock solutions. For the calculation of LOD and LOQ values for avenacoside A and saponin B compounds, the standard deviation (SD), obtained by analysing the peak areas of the lowest standard concentration point, was multiplied by three or ten, respectively [42].

To determine the intra-day precision of the instrumental method, oat protein concentrate and pea protein isolate extracts containing all analytes and internal standards were injected six times, and for inter-day precision, sample extracts were studied across three independent days to confirm the stability of the retention times and peak areas. The precision of the extraction methods was determined by repeatability (intra-day) and intermediate precision (inter-day). Repeatability was carried out by performing six repeated analyses of the samples on the same day, while the intermediate precision of the method was assessed using samples that were analysed on three different days over two months under the same experimental conditions.

The total recoveries for avenacoside A and saponin B were evaluated by spiking the solid and liquid samples with a known amount of avenacoside A and saponin B at three different concentration levels. For estimation of solid sample extraction method recovery, oat protein concentrate and pea protein isolate (100 mg) were weighed into a 10 mL volumetric flask (n = 3). Aliquots of avenacoside A and saponin B standard solutions (10 mL) at three different concentrations were prepared in aqueous EtOH (70%, v/v) separately. These solutions were added to oat protein concentrate and pea protein isolate, mixed thoroughly and subjected to the solid sample extraction method as described above. The recoveries of avenacoside A and saponin B in oat and pea liquid samples were determined by cross-matrix spiking both sample matrices. For estimation of liquid sample extraction method recovery, separate standard stock solutions of avenacoside A and saponin B were prepared (200 mg/L). These solutions were added in different volumes to 0.25 g of liquid sample (oat and pea drink) (n = 3) weighed into a 5 mL volumetric flask, mixed thoroughly, and subjected to the liquid sample extraction method as described above. The total recovery was calculated using Equation (2) [43],

$$\text{total recovery (\%)} = (C_{spiked}/(C_{unspiked} + C_{spike})) \times 100\% \quad (2)$$

where C_{spiked} is the amount of saponin determined in the spiked sample, $C_{unspiked}$ is the amount of saponin in the unspiked sample, and C_{spike} is the amount of saponins at three different concentration levels.

ME as one of the most problematic issues in LC-MS was evaluated for all four sample matrices (oat protein concentrate and pea protein isolate and plant-based drinks) by post-extraction sample spiking with calibration curve standard solutions, then constructing a calibration curve based on response factors and spiked standard concentrations, and comparing the matrix-matched calibration curve slope with the calibration curve slope in solvent (Equation (3)) [42]

$$\text{ME (\%)} = slope_{matrix\text{-}matched}/slope_{solvent} \times 100\%. \quad (3)$$

Statistical analysis was carried out using Excel® (Microsoft® 365 for enterprise). The results are presented as mean ± SD or relative standard deviation (RSD).

3. Results and Discussion

3.1. Development of Liquid Chromatography Method

The HPLC method was developed and assessed by analysing external standards and compounds available in oat and pea sample matrices. During development of the liquid chromatography method, two types of stationary phase chemistry were tested (C18 and HILIC) as well as different column dimensions. The best separation performance in terms of time of analysis, selectivity, and efficiency was achieved by the BEH Amide column (1.0 × 50 mm, 1.7 µm). Based on the literature [9,19] and scan-type experiments of oat flour and pea flour sample extracts, m/z values for SIR channels were chosen for the detection and relative quantification of targeted compounds without existing standard compounds in these sample matrices. Avenacoside B and 26-desglucoavenacoside A were found to be present in the oat sample matrix in addition to avenacoside A. DDMP saponin also occurred in the pea sample matrix besides saponin B. MRM experiments were conducted during development of a methodology but we have found that the MRM approach did not bring any more selectivity but significantly reduced sensitivity by not producing consistent fragments. The example of a chromatogram obtained by injecting the oat and pea flour extracts is shown in Supplementary Figure S1.

3.2. Development of Sample Extraction Methods

Two previously published extraction methods (avenacosides in grain and husks of oats [9] and saponins in peas [19]) were the starting points for the development of a method for simultaneous saponin extraction from oat and pea matrices. As both extraction methods were time-consuming, a more efficient sample preparation was proposed for saponin quantification. All samples were analysed using LC-MS method described in the Materials and Methods section.

Table 1 shows the main steps of extraction methods and saponin extraction yields obtained by reference methods (1A and 2A) and modified methods (1B, 2B, 2C, and 2D). To demonstrate the efficiency of the optimized methods, oat protein concentrate and pea protein isolate were analysed in duplicate.

Since both reference methods [9,19] started by fat elimination, defatted oat protein concentrate (fat 18.9%) and pea protein isolate (fat 4.7%) were extracted using methods 1A, 1B, 2A, and 2B. The oat protein concentrate extracted using method 1B gave 37% higher avenacoside A concentration compared to method 1A, and method 2B resulted in 50% higher yield than method 2A. Overall, the highest avenacoside A content in oat protein concentrate was achieved using extraction method 2B. Using method 1B, the pea protein isolate gave two times higher saponin B yield than using extraction method 1A, and method 2B gave a 76% higher yield than method 2A. Thus, the highest saponin B amount from pea protein isolate was extracted using method 2B. Although both improved methods 1B and 2B gave similar saponin yields in analysed matrices, it was decided to proceed with more process-efficient method 2B, as method 1B utilizing two-step methanol reflux extraction is very time-consuming.

The necessity for fat removal before saponin extraction from the matrix was determined. For this, saponins from four samples (oat flour and protein concentrate and pea flour and protein isolate) were extracted using extraction methods 2B and 2C, and lastly, the extracts were filtered through different filtering devices (the molecular weight cut-off filters with different membrane pore sizes (3, 10, 30, and 50 kDa), 0.2 µm syringe filter, and ISOLUTE® PLD+ Protein and Phospholipid Removal columns) before the LC-MS analysis. The results of this experiment are shown in Supplementary Figure S2.

No significant differences in avenacoside A, avenacoside B, saponin B, and DDMP saponin content were determined in Soxhlet-defatted and non-defatted oat and pea matrices. On the other hand, different molecular cut-off sizes had a significant impact on the recovery of saponins. The 3 kDa and 10 kDa cut-off filters showed inferior performance irrespective of the sample matrix and saponin type determined. The maximum recovery of analytes in the samples was achieved using 50 kDa and in some cases 30 kDa cut-off devices. In all sample matrices except oat protein concentrate, the application of PLD+ columns and syringe filters gave even better results than 30 kDa or 50 kDa cut-off filters. Although the PLD+ and 0.2 µm syringe filters gave quite similar analyte recovery, the application of PLD+ columns resulted in clearer MS chromatograms with a minimum number of interfering peaks in the chromatogram baseline. Moreover, filtering through the PLD+ column enables an easy transition of the procedure to a high-throughput workflow in the case of using 96-well PLD+ plates. The ISOLUTE® PLD+ proprietary multifunctional sorbent phase is optimised to selectively retain proteins and phospholipids [44]. The results indicated that pre-extraction fat removal is not necessary before saponin extraction and could be omitted and the application of PLD+ columns is the best solution for post-extraction clean-up of sample extracts. This resulted in a modified method 2C (described in Table 1).

Table 1. The extraction steps of reference methods 1A and 2A [9,19], their modified versions (1B, 2B, 2C, and 2D), and saponin yields obtained using these methods [a].

	Sample Extraction 1A	Modified Sample Extraction 1B	Sample Extraction 2A	Modified Sample Extraction 2B	Modified Sample Extraction 2C	Modified Sample Extraction 2D
Defatted sample	yes	yes	yes	yes	no	no
Sample and solvent amount (10 g/L)	0.5 g, 25 mL × 2 MeOH	0.5 g, 25 mL × 2 MeOH	0.5 g, 50 mL EtOH (70%, *v*/*v*)	0.1 g, 10 mL EtOH (70%, *v*/*v*)	0.1 g, 10 mL EtOH (70%, *v*/*v*)	0.1 g, 10 mL EtOH (70%, *v*/*v*)
Extraction	2-step reflux at boiling point	2-step reflux at boiling point	1 h at 25 °C in a shaking incubator	1 h at 25 °C in a shaking incubator	1 h at 25 °C in a shaking incubator	30 min ultrasonic bath
Clean-up	decant	centrifuge (17,000× *g* × 10 min at 10 °C)	ashless filter paper	centrifuge (17,000× *g* × 10 min at 10 °C)	centrifuge (14,000× *g* × 10 min at 10 °C)	centrifuge (14,000× *g* × 10 min at 10 °C)
Solvent evaporation	vacuum rotary evaporator	-	vacuum rotary evaporator	-	-	-
Resuspended	an aqueous MeOH (5%, *v*/*v*)	-	-	-	-	-
Centrifuge	17,000× *g* × 10 min at 10 °C	-	17,000× *g* × 10 min at 10 °C	-	-	-
Sample clean-up and concentration	SPE C18	-	SPE C18	-	-	-
Solvent evaporation	N_2 flow	-	N_2 flow	-	-	-
Filtering	-	-	-	-	PLD+ column	PLD+ column
Reconstitution/dilution in an aqueous MeCN (50%, *v*/*v*)	yes	yes	yes	yes	yes	yes
Obtained results (mg/100 g ± SD)						
Avenacoside A [b]	19 ± 2 [c]	26 ± 1 [c]	18 ± 3 [c]	37 ± 2 [c]	36 ± 3	37 ± 2
Saponin B [d]	100 ± 3 [c]	214 ± 5 [c]	124 ± 10 [c]	219 ± 8 [c]	229 ± 22	240 ± 19

[a] Detailed description of extraction methods 1A, 1B, 2A, 2B, and 2C is available in the supplementary information. Each result represents mean ± SD (n = 2). [b] Measured in oat protein concentrate. [c] Result is presented on fat-containing sample. [d] Measured in pea protein isolate.

The influence of ultrasonic power on the saponin extraction yields was also investigated. Saponins from oat protein concentrate, pea protein isolate, and oat and pea flours were extracted using methods 2C and 2D (results are shown in Supplementary Table S2). The results showed that ultrasonication did not have a statistically significant effect on saponin yield but considering the extraction time the application of ultrasonication is preferable. It should be noted that heating taking place during sonication had no effect on the analytes. During this experiment, the ultrasonic bath heated itself from ambient temperature (23 °C) to 40 °C in 30 min. Previous research has shown that the exposure of DDMP saponin to a temperature higher than 40 °C has a profound effect on its degradation into saponin B [18]. However, in another study, it was reported that the pure DDMP saponin in methanolic solution started to decrease in concentration when heated at 65 °C [45].

Based on the obtained results and considering the extraction time and yield, method 2D was utilized for analysis and validation of all solid samples.

Liquid food samples were analysed without the need to use lyophilisation before the sample extraction. The sample preparation method was based only on the application of ISOLUTE® PLD+ cartridges for sample extract purification before LC-MS analysis, previously chosen as the most efficient for cleaning the extracts of the solid samples.

3.3. Validation of the Method

When the chromatographic methods and sample extraction methods were developed, validation was performed to evaluate the linear range, LODs and LOQs, precision, recoveries, and matrix effect of the proposed method. The linearity of response and other calibration parameters for avenacoside A and saponin B are presented in Table 2. Linearity for these two saponin standards was obtained in the concentration range of 0.01–2.5 mg/L. The LOQs were estimated from the lowest point of the calibration curve ranging from 0.015 mg/L for avenacoside A and 0.014 mg/L for saponin B. The obtained LOQ results were lower than or in accordance with previous research [9,13,17,39].

Table 2. The linear range, calibration curve, limits of detection (LODs), and limits of quantification (LOQs) of avenacoside A and saponin B.

Analyte	Linear Range (mg/L)	Calibration Curve [1]	R^2	LOD (mg/L)	LOQ (mg/L)
avenacoside A	0.01–2.44	$y = 0.2445x - 0.0197$	0.9998	0.004	0.015
saponin B	0.01–2.48	$y = 0.6350x - 0.0059$	0.9999	0.004	0.014

[1] Calibration curve: $y = ax + b$, where x is the response factor and y is the concentration in mg/L.

After linearity was found to be acceptable for avenacoside A and saponin B, the repeatability of the method was appraised. Repeatability of retention times and peak areas were studied first with six replicate injections of oat protein concentrate and pea protein isolate extract. Table 3 shows the repeatability of retention times, peak areas, and the precision of solid and liquid sample extraction methods. RSDs of peak areas for all saponins did not exceed 6%. Intra- and inter-day RSDs were at a similar level, indicating that the methods are reproducible to an acceptable extent for the routine analysis of oat and pea products. Intra-day and inter-day RSDs were determined by extracting oat protein concentrate, pea protein isolate, and plant-based drinks on different days. The RSD of the intra-day precision ranged from 6 to 13% and inter-day precision from 7 to 11% in powdered oat and pea samples. For oat and pea plant-based drinks, the intra-day precision ranged from 3 to 12% and inter-day precision from 7 to 16%. The precisions for the DDMP saponin pea drink were not evaluable despite multiple measurements (DDMP saponin content in this sample was <LOQ).

Table 3. Repeatability of retention times (RT) and peak areas of saponins, and precision of the whole method.

Analyte	m/z	RT, min	RT RSD (%)		Peak Area RSD (%)		Precision RSD (%)			
							Powdered Samples		Plant-Based Drinks	
			Intra-Day (n = 6)	Inter-Day (n = 18)	Intra-day (n = 6)	Inter-Day (n = 18)	Intra-Day (n = 6)	Inter-Day (n = 9)	Intra-Day (n = 6)	Inter-Day (n = 9)
Avenacoside A	1061.5	1.78	0.23 [a]	0.48 [a]	1.8 [a]	3.0 [a]	11	11	12	12
Avenacoside B	1223	1.93	0.20 [a]	0.27 [a]	4.1 [a]	4.5 [a]	13	9	3	8
26-desglucoavenacoside A	899.5	1.54	0.36 [a]	0.56 [a]	3.8 [a]	6.0 [a]	6	7	10	16
^{13}C-avenacoside A [b]	1112.5	1.78	0.28	0.54	2.9	4.2	n.i.	n.i.	n.i.	n.i.
Saponin B	941.5	1.42	0.29 [c]	0.98 [c]	2.6 [c]	3.1 [c]	6	7	6	7
DDMP saponin	1067	1.40	0.23 [c]	0.86 [c]	3.9 [c]	6.3 [c]	8	11	n.a.	n.a.
Soyasaponin Ba [d]	957.5	1.48	0.30	0.68	2.5	3.0	n.i.	n.i.	n.i.	n.i.

[a] Measured in oat protein concentrate. [b] Internal standard for avenacoside A, avenacoside B, and 26-desglucoavenacoside A. n.i. Not implemented, no RSD calculable. [c] Measured in pea protein isolate. n.a. Not available in this matrix. [d] Internal standard for saponin B and DDMP saponin.

The recoveries were determined in oat protein concentrate and pea protein isolate powder by spiking the oat matrix with avenacoside A and the pea matrix with saponin B. The recovery of analytes in the case of the liquid sample extraction method was investigated separately. Table 4 shows the recovery results of powdered and liquid samples. The recoveries of avenacoside A and saponin B ranged from 90 to 115% and from 82 to 100% in oat protein concentrate and pea protein isolate, respectively. In the oat drink, the recovery of avenacoside A ranged from 96 to 113% and saponin B from 98 to 113%. In the pea drink matrix, the recoveries of avenacoside A and saponin B were from 94 to 106% and from 89 to 98%, respectively. According to validation guidelines, the acceptable recovery range for this method should be in the range of 80 to 110% [46]. Thus, the mean values of obtained recoveries were acceptable for both matrices. The recovery results obtained with the current procedure were similar to ones reported for previously proposed methods [9,13].

Table 4. Recoveries of saponins in solid (oat protein concentrate and pea protein isolate) and liquid samples (oat drink and pea drink) [1].

	Spiking Level	Spike (mg/L)	Recovery (%)	Spike (mg/L)	Recovery (%)	Recovery (%)
			Oat Protein Concentrate [2]		**Oat Drink [3]**	**Pea Drink [4]**
Avenacoside A	L1	1.2	115 ± 7	0.24	104 ± 6	106 ± 1
	L2	0.9	90 ± 7	0.12	113 ± 7	99 ± 3
	L3	0.4	107 ± 12	0.06	96 ± 13	94 ± 1
			Pea protein isolate [5]		Oat drink [4]	Pea drink [6]
Saponin B	L1	1.1	82 ± 2	0.23	105 ± 2	98 ± 5
	L2	0.5	100 ± 1	0.11	113 ± 8	89 ± 2
	L3	0.3	96 ± 10	0.06	98 ± 15	93 ± 6

[1] Each result represents mean ± SD (n = 3). [2] Unspiked matrix initial analyte concentration 0.70 mg/L. [3] Unspiked matrix initial analyte concentration 0.45 mg/L. [4] Analyte-free sample matrix. [5] Unspiked matrix initial analyte concentration 0.55 mg/L. [6] Unspiked matrix initial analyte concentration 0.15 mg/L.

Oat protein concentrate ME on avenacoside A was 100%, and pea protein isolate ME on saponin B was 110%. Avenacoside A and saponin B ME were 107% and 105% in the oat drink and 105% and 102% in the pea drink, respectively. All measured ME were in the optimal range between 90 and 110% [47].

The stock solution of U-^{13}C-oat seed flour extract was analysed for purity. The unlabelled avenacosides were not detected; thus, isotopically labelled avenacoside A was regarded as fully labelled. The working solution of ^{13}C-oat flour was added into the LC-MS vial before the analysis to assess the quantity of analytes and take into account ME. Moreover, recovery experiments confirmed that the method could be used even with internal standards added post-extraction.

Overall, the method has demonstrated acceptable validation performance in terms of recovery, sensitivity, specificity, and precision, and could be characterised as robust and effective and could potentially be applied in a high-throughput environment. Thus, the developed sample extraction method and the LC-MS method are suitable tools for the analysis of oat and pea saponins in different matrices, e.g., flours, protein concentrates and isolates, mixed matrices, and liquid plant-based drinks.

3.4. Determined Concentrations of Saponins in Food Ingredients, Half- and End-Products

High sensitivity and reproducibility as well as very short analysis time make the developed method suitable for routine quality analysis of oat- and pea-based food ingredients and foods, as well as products containing oat and pea components. The results of saponin contents in various samples are shown in Table 5.

Table 5. Contents of saponins in different food samples [1].

Sample	mg/100 g				
	Avenacoside A	**Avenacoside B [2]**	**26-desglucoavenacoside A [2]**	**Saponin B**	**DDMP Saponin [3]**
Oat protein concentrate	42.3 ± 3.0	33.8 ± 0.7	5.1 ± 0.2	n.a.	n.a.
Whole-grain oat flour	23.4 ± 2.9	14.0 ± 1.5	<LOQ	n.a.	n.a.
Oat drink	4.6 ± 0.1	2.7 ± 0.2	<LOQ	n.a.	n.a.
Pea protein isolate	n.a.	n.a.	n.a.	243.8 ± 6.2	10.8 ± 0.7
Pea protein concentrate	n.a.	n.a.	n.a.	20.3 ± 1.6	107.6 ± 4.1
Pea flour	n.a.	n.a.	n.a.	6.2 ± 0.4	61.1 ± 2.0
Pea drink	n.a.	n.a.	n.a.	3.5 ± 0.2	<LOQ
Blend [4]	13.5 ± 1.0	10.9 ± 0.3	1.3 ± 0.3	123.9 ± 6.3	27.1 ± 3.5
Extruded blend [4]	10.6 ± 0.3	9.6 ± 0.9	1.1 ± 0.9	132.9 ± 12.4	11.4 ± 0.8

[1] Each result represents mean ± SD (n = 3). [2] Equivalent of avenacoside A. [3] Equivalent of saponin B. n.a. Not available in this sample matrix. [4] Blend: 52% pea protein isolate, 28% oat protein concentrate, and 20% pea protein concentrate.

In whole-grain oat flour, the contents of avenacoside A, avenacoside B, and 26-desglucoavenacoside A were 23.4 ± 2.9 mg/100 g, 14.0 ± 1.5 mg/100 g, and below LOQ, respectively. According to previous research, the concentrations of avenacosides and their ratios are different and depend largely on the variety of oats [9]. According to the latter study, the average avenacoside A content in oat grain in 16 analysed varieties was 36 ± 8 mg/100 g, avenacoside B content was in the range of 30 ± 4 mg/100 g, and 26-desglucoavenacoside A was 2.4 ± 0.8 mg/100 g [9]. Indeed, the contents of avenacoside A differed up to two-fold depending on the variety, and the ratios of avenacoside A to avenacoside B varied from 0.9 to 1.7 [9]. According to Günther-Jordanland et al. (2020), avenacoside A and avenacoside B content in oat flour has been reported to be 24.6 mg/100 g and 21.9 mg/100 g, respectively [13]. Thus, the concentration of avenacosides in the whole-grain oat flour determined in the present study is in a good correspondence with the results reported before [9,13]. In oat protein concentrate (53% protein; Table S1), avenacoside A content was 42.3 ± 3.0 mg/100 g, avenacoside B was 33.8 ± 0.7 mg/100 g, and 26-desglucoavenacoside A was 5.1 ± 0.2 mg/100 g. According to specification (Table S1), this product was manufactured from oat bran. Previous research has shown that the average content of avenacoside A and avenacoside B in three analysed oat bran products was 26 ± 7 mg/100 g and 8 ± 2 mg/100 g, respectively [17], which is similar to concentrations determined in the whole-grain flour in the current study. Thus, the increased content of avenacosides in oat protein concentrate should be ascribed to the partial concentration of the oat saponins together with the protein fraction during the production process of oat protein concentrate. In an oat drink, avenacoside A content was 4.6 ± 0.1 mg/100 g, avenacoside B was 2.7 ± 0.2 mg/100 g, and 26-desglucoavenacoside A was below LOQ. As it was a commercial liquid product with low dry matter content, it resulted in an apparently lower content of measured saponins. Nevertheless, according to specification (Table S1), the product contains only 1% of protein and the oat base is the only protein source in the oat drink. In this respect, considering the oat drink and, e.g., the whole-grain oat flour (12.5% of protein), the ratio of avenacosides to protein is much higher in the oat drink. One can suppose the considerable migration of saponins into the liquid phase when soaking the oats during the initial step of oat drink manufacture.

In pea flour (17.9% protein; Table S1), saponin B content was 6.2 ± 0.4 mg/100 g and relatively quantified DDMP saponin content was 61.1 ± 2.0 mg/100 g. In fact, our findings are inconsistent with the results of Reim and Rohn (2015), who analysed saponin B and DDMP saponin contents in hulls and peas in six different pea varieties using the HPTLC method [32]. They reported that saponin content in peeled peas was 10 to 40 mg/100 g of saponin B and 0 to 20 mg/100 g of DDMP saponin depending on pea variety [32]. Nonetheless, the present findings of high DDMP content in pea flour are comparable with the results of Heng et al. (2006): the DDMP saponin content varied from 70 to 150 mg/100 g DM, whereas saponin B varied from 0 to 40 mg/100 g DM [19]. Our results confirm that the DDMP saponin is the predominant naturally occurring saponin present in pea. The high level of DDMP saponin in pea flour was observed in the current study most likely because it has not been thermally treated and DDMP saponin has not been converted into saponin B. In pea protein concentrate (46.9% protein; Table S1), the saponin B content was 80.3 ± 1.6 mg/100 g and DDMP content was 107.6 ± 4.1 mg/100 g. Saponins are found in the cotyledons and are often associated with the protein bodies of legumes [4]. Therefore, saponin accumulation in pea concentrate produced by dry milling and air classification is evident [4], which is in accordance with at least twice higher levels of saponins in pea protein concentrate compared to pea flour determined in our study. In pea protein isolate (75% protein; Table S1), saponin B content was 243.8 ± 6.2 mg/100 g and DDMP content was 10.8 ± 0.7 mg/100 g. These results show that protein wet extraction and isoelectric precipitation, likely performed to achieve protein isolate, degrade unstable DDMP saponin naturally occurring in peas into saponin B. In the pea drink, saponin B content was 3.5 ± 0.2 mg/100 g and DDMP saponin was below LOQ. According to the product specification (Table S1), it contains 2% of protein, and the only protein source is pea. Although the exact production process of the pea drink is unknown, taking into account the content of saponin B per 1 g of pea drink protein (1.75 mg), the probable pea protein source should contain at least 175 mg of saponins (sum of saponin B and DDMP saponin, as DDMP saponin is converted into saponin B during drink pasteurization) per 100 g of pure pea protein.

To test the applicability of the developed method for simultaneous determination of oat and pea saponins from one matrix, the blend of pea isolate, oat protein concentrate, and pea protein concentrate was used. In addition, the part of the mixture was extruded according to the previously published article [41]. Results show that avenacoside A, avenacoside B, 26-desglucoavenacoside A, saponin B, and DDMP saponin content in the blend were 13.5 ± 1.0 mg/100 g, 10.9 ± 0.3 mg/100 g, 1.3 ± 0.3 mg/100 g, 123.9 ± 6.2 mg/100 g, and 27.1 ± 3.5 mg/100 g, respectively. Considering that this blend was composed of 52% pea protein isolate, 28% oat protein concentrate, and 20% pea protein concentrate, which were also analysed separately, the recoveries of avenacoside A, avenacoside B, 26-desglucoavenacoside A, saponin B, and DDMP saponin were 114%, 115%, 90%, 95%, and 100%, respectively. In the extruded blend, avenacoside B and 26-desglucoavenacoside A content did not change significantly, avenacoside A content decreased by 21%, and saponin B content increased from 123.9 to 132.9 mg/100 g, which could potentially happen due to DDMP saponin conversion into saponin B during extrusion cooking.

4. Conclusions

In conclusion, the HILIC-MS-based method for oat and pea matrices, with a relatively simple extraction procedure for solid and liquid samples, allowing the simultaneous quantification of avenacoside A and saponin B, and the relative quantification of avenacoside B, 26-desglucoavenacoside A, and DDMP saponin, was employed for analysis of saponins in various food ingredients and products. Oat protein concentrate, pea protein isolate, and oat- and pea-based drinks were chosen for development and validation of the sample extraction methods. The optimised HILIC-MS method was able to absolutely quantify avenacoside A and saponin B in the matrices; other compounds were quantified based on existing standard compounds. The validation of the improved methods for both sample types (solid and liquid) showed the acceptable linear range, LODs and LOQs, precisions, recoveries, and MEs. Generally, an inter-day precision was below 20%. The accuracy and the precision of quantification were achieved by using the labelled internal standard (^{13}C-avenacoside A) obtained from U-^{13}C-labelled oat flour and with soyasaponin Ba as internal standards. The content of saponins was measured in different plant-based oat and pea products (ingredients, half- and end-products). This method could be potentially extended for other plant-based sample matrices, and the absolute quantification of all analytes could be achieved if the missing saponin standards were to arrive on the market.

Supplementary Materials: The following supporting information can be downloaded at https://www.mdpi.com/article/10.3390/foods12050991/s1. Description of solid sample extraction methods (1A, 1B, 2A, 2B, and 2C) used during the method development; Table S1: Nutritional information of analysed products; Figure S1: LC-MS chromatograms of oat and pea flours (SIR and ESI-). In oat flour: avenacoside A, avenacoside B, 26-desglucoavenacoside A, and internal standard ^{13}C-avenacoside A. In pea flour: saponin B, DDMP saponin, and internal standard soyasaponin Ba; Figure S2: Saponin yield in (a) oat and (b) pea matrices. The effect of sample clean-up: the pre-extraction of fat and six post-extraction filtration possibilities. The results of avenacoside B are presented in equivalents of avenacoside A mg/g and DDMP saponin in equivalents of saponin B mg/g; Table S2: The effectiveness of ultrasonic bath extraction compared to reference extraction conditions using the tube rotator (extraction yield 100%).

Author Contributions: A.B.: conceptualization, methodology, design of the experiments, carried out the experiments, investigation, validation, data curation, visualisation, writing—original draft, review and editing. D.P.: design of the experiments, methodology, writing—review and editing. T.K.: supervision, investigation, formal analysis, funding acquisition, project administration, resources, writing—review and editing. M.K.: writing—review and editing, supervision. E.-G.K.: writing—review and editing. All authors have read and agreed to the published version of the manuscript.

Funding: The financial support for this research was provided by the European Union European Regional Development Fund (ERDF) and Estonian Research Council via projects RESTA16 and RESTA17.

Data Availability Statement: Data is contained within the article or supplementary material.

Acknowledgments: The authors thank Aleksei Kaleda for kindly providing samples, Georg Arju for assistance with chromatographic method development, and Kristiina Loit for helping with sample analysis (Center of Food and Fermentation Technologies).

Conflicts of Interest: The authors declare no conflict of interest. The financial supporters had no role in the design of the study; in the collection, analyses, or interpretation of data; in the writing of the manuscript; or in the decision to publish the results.

References

1. Alexander, P.; Brown, C.; Arneth, A.; Finnigan, J.; Moran, D.; Rounsevell, M.D.A. Losses, inefficiencies and waste in the global food system. *Agric. Syst.* **2017**, *153*, 190–200. [CrossRef] [PubMed]
2. FAO. *Dietary Protein Quality Evaluation in Human Nutrition: Report of an FAO Expert Consultation, 31 March–2 April 2011, Auckland, New Zealand*; FAO: Rome, Italy, 2013.
3. Mäkinen, O.E.; Sozer, N.; Ercili-Cura, D.; Poutanen, K. Chapter 6—Protein From Oat: Structure, Processes, Functionality, and Nutrition. In *Sustainable Protein Sources*; Nadathur, S.R., Wanasundara, J.P.D., Scanlin, L., Eds.; Academic Press: San Diego, CA, USA, 2017; pp. 105–119.
4. Tulbek, M.C.; Lam, R.S.H.; Wang, Y.C.; Asavajaru, P.; Lam, A. Chapter 9—Pea: A Sustainable Vegetable Protein Crop. In *Sustainable Protein Sources*; Nadathur, S.R., Wanasundara, J.P.D., Scanlin, L., Eds.; Academic Press: San Diego, CA, USA, 2017; pp. 145–164.
5. Kaleda, A.; Talvistu, K.; Vaikma, H.; Tammik, M.-L.; Rosenvald, S.; Vilu, R. Physicochemical, textural, and sensorial properties of fibrous meat analogs from oat-pea protein blends extruded at different moistures, temperatures, and screw speeds. *Future Foods* **2021**, *4*, 100092. [CrossRef]
6. Rackis, J.J.; Sessa, D.J.; Honig, D.H. Flavor problems of vegetable food proteins. *J. Am. Oil Chem. Soc.* **1979**, *56*, 262–271. [CrossRef]
7. Rochfort, S.; Panozzo, J. Phytochemicals for health, the role of pulses. *J. Agric. Food Chem.* **2007**, *55*, 7981–7994. [CrossRef] [PubMed]
8. Roland, W.S.U.; Pouvreau, L.; Curran, J.; van de Velde, F.; de Kok, P.M.T. Flavor Aspects of Pulse Ingredients. *Cereal Chem.* **2017**, *94*, 58–65. [CrossRef]
9. Pecio, Ł.; Wawrzyniak-Szołkowska, A.; Oleszek, W.; Stochmal, A. Rapid analysis of avenacosides in grain and husks of oats by UPLC–TQ–MS. *Food Chem.* **2013**, *141*, 2300–2304. [CrossRef] [PubMed]
10. Pecio, Ł.; Jędrejek, D.; Masullo, M.; Piacente, S.; Oleszek, W.; Stochmal, A. Revised structures of avenacosides A and B and a new sulfated saponin from *Avena sativa* L.: Steroidal saponins from *Avena sativa* L. *Magn. Reason. Chem.* **2012**, *50*, 755–758. [CrossRef]

11. Önning, G.; Asp, N.-G. Analysis of saponins in oat kernels. *Food Chem.* **1993**, *48*, 301–305. [CrossRef]
12. Günther-Jordanland, K.; Dawid, C.; Dietz, M.; Hofmann, T. Key Phytochemicals Contributing to the Bitter Off-Taste of Oat (*Avena sativa* L.). *J. Agric. Food Chem.* **2016**, *64*, 9639–9652. [CrossRef] [PubMed]
13. Günther-Jordanland, K.; Dawid, C.; Hofmann, T. Quantitation and Taste Contribution of Sensory Active Molecules in Oat (*Avena sativa* L.). *J. Agric. Food Chem.* **2020**, *68*, 10097–10108. [CrossRef] [PubMed]
14. Price, K.R.; Fenwick, G.R. Soyasaponin I, a compound possessing undesirable taste characteristics isolated from the dried pea (*Pisum sativum* L.). *J. Sci. Food Agric.* **1984**, *35*, 887–892. [CrossRef]
15. Price, K.R.; Johnson, I.T.; Fenwick, G.R.; Malinow, M.R. The chemistry and biological significance of saponins in foods and feedingstuffs. *Crit. Rev. Food Sci. Nutr.* **1987**, *26*, 27–135. [CrossRef] [PubMed]
16. Lásztity, R.; Hidvégi, M.; Bata, Á. Saponins in food. *Food Rev. Int.* **1998**, *14*, 371–390. [CrossRef]
17. Yang, J.; Wang, P.; Wu, W.; Zhao, Y.; Idehen, E.; Sang, S. Steroidal Saponins in Oat Bran. *J. Agric. Food Chem.* **2016**, *64*, 1549–1556. [CrossRef] [PubMed]
18. Heng, L.; Vincken, J.-P.; Hoppe, K.; van Koningsveld, G.A.; Decroos, K.; Gruppen, H.; van Boekel, M.A.J.S.; Voragen, A.G.J. Stability of pea DDMP saponin and the mechanism of its decomposition. *Food Chem.* **2006**, *99*, 326–334. [CrossRef]
19. Heng, L.; Vincken, J.-P.; van Koningsveld, G.; Legger, A.; Gruppen, H.; van Boekel, T.; Roozen, J.; Voragen, F. Bitterness of saponins and their content in dry peas. *J. Sci. Food Agric.* **2006**, *86*, 1225–1231. [CrossRef]
20. Cheeke, P.R. Nutritional and physiological implications of saponins: A review. *Can. J. Anim. Sci.* **1971**, *51*, 621–632. [CrossRef]
21. Southon, S.; Wright, A.J.; Price, K.R.; Fairweather-Tait, S.J.; Fenwick, G.R. The effect of three types of saponin on iron and zinc absorption from a single meal in the rat. *Br. J. Nutr.* **1988**, *59*, 389–396. [CrossRef]
22. Ikewuchi, C.C. Hypocholesterolemic effect of an aqueous extract of the leaves of *Sansevieria senegambica* Baker on plasma lipid profile and atherogenic indices of rats fed egg yolk supplemented diet. *EXCLI J.* **2012**, *11*, 346–356.
23. Barky, A.R.E.; Hussein, S.A.; Alm-Eldeen, A.-E.; Hafez, Y.A.; Mohamed, T.M. Saponins and their potential role in diabetes mellitus. *Diabetes Manag.* **2017**, *7*, 148.
24. Johnson, I.T.; Gee, J.M.; Price, K.; Curl, C.; Fenwick, G.R. Influence of Saponins on Gut Permeability and Active Nutrient Transport in Vitro. *J. Nutr.* **1986**, *116*, 2270–2277. [CrossRef] [PubMed]
25. Yamamoto, M.; Kumagai, A.; Yamamura, Y. Plasma lipid-lowering action of ginseng saponins and mechanism of the action. *Am. J. Chin. Med.* **1983**, *11*, 84–87. [CrossRef] [PubMed]
26. Dong, J.; Liang, W.; Wang, T.; Sui, J.; Wang, J.; Deng, Z.; Chen, D. Saponins regulate intestinal inflammation in colon cancer and IBD. *Pharmacol. Res.* **2019**, *144*, 66–72. [CrossRef] [PubMed]
27. Van Atta, G.R.; Guggolz, J.; Thompson, C.R. Plant Analysis, Determination of Saponins in Alfalfa. *J. Agric. Food Chem.* **1961**, *9*, 77–79. [CrossRef]
28. Von Buchi, J.; Dolder, R. Determining the hemolytic index of official medicinal drugs. *Pharm. Acta Helv.* **1950**, *25*, 179–188. [PubMed]
29. Morgan, M.R.A.; McNerney, R.; Matthew, J.A.; Coxon, D.T.; Chan, H.W.-S. An enzyme-linked immunosorbent assay for total glycoalkaloids in potato tubers. *J. Sci. Food Agric.* **1983**, *34*, 593–598. [CrossRef]
30. Thieme, H.; Hartmann, U. Comparison of various methods for the spectrophotometric determination of glycyrrhizic acid in *Radix liquiritiae* DAB 7-DDR. *Pharm* **1974**, *29*, 50–53.
31. Fenwick, D.E.; Oakenfull, D. Saponin content of soya beans and some commercial soya bean products. *J. Sci. Food Agric.* **1981**, *32*, 273–278. [CrossRef]
32. Reim, V.; Rohn, S. Characterization of saponins in peas (*Pisum sativum* L.) by HPTLC coupled to mass spectrometry and a hemolysis assay. *Food Res. Int.* **2015**, *76*, 3–10. [CrossRef]
33. Myoga, K.; Shibata, F. Ladino Clover Sapogenins and the Gas Chromatographic Assay Method. *Jpn. Soc. Grassl. Sci.* **1977**, *23*, 67–72.
34. Kesselmeier, J.; Strack, D. High Performance Liquid Chromatographic Analysis of Steroidal Saponins from *Avena sativa* L. *Z. Nat. C* **1981**, *36*, 1072–1074. [CrossRef]
35. Amarowicz, R.; Yoshiki, Y.; Pegg, R.B.; Okubo, K. Presence of two saponins in faba bean (*Vicia faba* L.) seeds. *Food Nahr.* **1997**, *41*, 352–354. [CrossRef]
36. Maillard, M.P.; Wolfender, J.-L.; Hostettmann, K. Use of liquid chromatography—Thermospray mass spectrometry in phytochemical analysis of crude plant extracts. *J. Chromatogr. A* **1993**, *647*, 147–154. [CrossRef]
37. Wolfender, J.L.; Hostettmann, K.; Abe, F.; Nagao, T.; Okabe, H.; Yamauchi, T. Liquid chromatography combined with thermospray and continuous-flow fast atom bombardment mass spectrometry of glycosides in crude plant extracts. *J. Chromatogr. A* **1995**, *712*, 155–168. [CrossRef] [PubMed]
38. Decroo, C.; Colson, E.; Lemaur, V.; Caulier, G.; De Winter, J.; Cabrera-Barjas, G.; Cornil, J.; Flammang, P.; Gerbaux, P. Ion mobility mass spectrometry of saponin ions. *Rapid Commun. Mass Spectrom.* **2019**, *33*, 22–33. [CrossRef]
39. Decroos, K.; Vincken, J.-P.; Heng, L.; Bakker, R.; Gruppen, H.; Verstraete, W. Simultaneous quantification of differently glycosylated, acetylated, and 2,3-dihydro-2,5-dihydroxy-6-methyl-4H-pyran-4-one-conjugated soyasaponins using reversed-phase high-performance liquid chromatography with evaporative light scattering detection. *J. Chromatogr. A* **2005**, *1072*, 185–193. [CrossRef]

40. Tian, J.; Fu, C.; Xu, L.; Lu, M. Extraction and identification of group B soybean saponins based on adsorption chromatography and HPLC-MS. In Proceedings of the 2010 3rd International Conference on Biomedical Engineering and Informatics, Yantai, China, 16–18 October 2010; pp. 2049–2051.
41. Kaleda, A.; Talvistu, K.; Tamm, M.; Viirma, M.; Rosend, J.; Tanilas, K.; Kriisa, M.; Part, N.; Tammik, M.-L. Impact of Fermentation and Phytase Treatment of Pea-Oat Protein Blend on Physicochemical, Sensory, and Nutritional Properties of Extruded Meat Analogs. *Foods* **2020**, *9*, 1059. [CrossRef]
42. Kruve, A.; Rebane, R.; Kipper, K.; Oldekop, M.-L.; Evard, H.; Herodes, K.; Ravio, P.; Leito, I. Tutorial review on validation of liquid chromatography–mass spectrometry methods: Part II. *Anal. Chim. Acta* **2015**, *870*, 8–28. [CrossRef]
43. Guidelines for Standard Method Performance Requirements AOAC. Appendix F. 2016, p. 18. Available online: https://www.aoac.org/wp-content/uploads/2019/08/app_f.pdf (accessed on 14 April 2022).
44. Biotage ISOLUTE®PLD+. Available online: https://www.biotage.com/phospholipid-removal (accessed on 15 January 2023).
45. Hu, J.; Lee, S.-O.; Hendrich, S.; Murphy, P.A. Quantification of the Group B Soyasaponins by High-Performance Liquid Chromatography. *J. Agric. Food Chem.* **2002**, *50*, 2587–2594. [CrossRef]
46. *Guidelines for the Validation of Chemical Methods for the FDA FVM Program*, 3rd ed.; U.S. Food and Drug Administration Foods Program: White Oak, MD, USA, 2019. Available online: https://www.fda.gov/media/81810/download (accessed on 31 March 2022).
47. Grant, R.P.; Rappold, B.A. Development and Validation of Small Molecule Analytes by Liquid Chromatography-Tandem Mass Spectrometry. In *Principles and Applications of Clinical Mass Spectrometry*, 1st ed.; Rifai Horvath, A.R., Wittwer, C.T., Eds.; Elsevier: London, UK, 2018; pp. 115–179. [CrossRef]

foods

MDPI

Article

Development and Method Validation of Butyric Acid and Milk Fat Analysis in Butter Blends and Blended Milk Products by GC-FID

Arunee Danudol [1] and Kunchit Judprasong [2,*]

1 Department of Medical Sciences, Bureau of Quality and Safety of Food, Nonthaburi 11000, Thailand
2 Institute of Nutrition, Mahidol University, Nakhon Pathom 73170, Thailand
* Correspondence: kunchit.jud@mahidol.ac.th; Tel.: +66-2800-2380

Abstract: Butyric acid is a short-chain saturated fatty acid with four carbon atoms in its molecule. It is unique to butter made from cow's milk and is an indicator to evaluate the quality of butter and milk products as stated in their ingredient labels. This study determined the milk fat content of butter blends and blended milk products by analyzing the content of butyric acid prepared as derivatives of methyl esters prior to injection into a gas chromatography flame ionization detector (GC–FID). Results revealed that this method had specificity, a linear relationship for concentration in the range of 0.04–1.60 mg/mL, a coefficient of determination (R^2) > 0.999, an instrumental limit of detection (LOD) and a limit of quantitative analysis (LOQ) at 0.01% and 0.10% of total fat, respectively, and an instrumental working range of 0.10–3.60% of total fat. The results of a precision study using relative standard deviation (RSD) was 1.3%, while an accuracy study using the spiking method showed % recovery in the range of 98.2–101.9%. The method linearity range for milk fat analysis had a good linear correlation in the range of 3–100% of total fat (R^2 > 0.999). Results for method LOD and LOQ were 1% and 3% of total fat, respectively. This method also had good precision (1.3% RSD) and accuracy (99.6–100.1% recovery), which indicates reliability in terms of precision and accuracy. This method, therefore, can be used to check claims about the quality of blended butter and blended milk products to ensure consumer confidence in product quality.

Keywords: butyric acid; butter fat; butter blends; blended milk products; GC-FID

Citation: Danudol, A.; Judprasong, K. Development and Method Validation of Butyric Acid and Milk Fat Analysis in Butter Blends and Blended Milk Products by GC-FID. *Foods* **2022**, *11*, 3606. https://doi.org/10.3390/foods11223606

Academic Editor: Gianfranco Picone

Received: 5 October 2022
Accepted: 9 November 2022
Published: 12 November 2022

1. Introduction

Butyric acid is a clear, colorless liquid with a pungent odor, a boiling point of 163.5 °C, and a chemical formula of $CH_3CH_2CH_2COOH$ or $C_4H_8O_2$. Butyric acid is a type of saturated fatty acid containing only four carbon molecules (short-chain fatty acid, C4:0). These molecules occur naturally in the form of triglyceride compounds in the milk fat of cows and other ruminants but not in animal adipose or vegetable fats [1], which makes butyric acid an indicator or marker for milk fat blends [2].

Butter contains butyric acid at about 4% of total fatty acids [3]. It also contains up to 400 other different fatty acids [3,4]. Butyric acid is used in food and pharmaceutical industries as additives as well as in health foods. Studies report that butyric acid can inhibit the growth of cancer cells and fight against the formation of atherosclerosis [5–8]. Consequently, butter has a higher nutritional value than other fats and is classified within a group of important food products due to its nutritious nature, especially for infants, children, working-age people, pregnant women, and the elderly [9–11].

Global demand for butter in dairy products has continued to increase [9,12] since it is a component of many everyday foods, such as butter itself, blended butter, blended milk products, cheese, chocolate, flavored milk, cream, dietary supplements, baked goods, snacks, and other foods. Butter is also an important food trade item and, due to competition, has been imitated by using vegetable oils and other fats (e.g., margarine) that are classified as non-milk fats or non-dairy fat products [13]. These products have lower nutritional value, but their appearance is similar to butter. Moreover, their use in replacing real butter, in part or in whole, may not be clearly stated in a food product's ingredient label, which can cause consumer confusion and misunderstanding. In addition, potential health risks arise because these substitutes are more difficult to digest in terms of their fat compared to butterfat products.

Producers and importers of blended butter and blended milk products [13] in countries such as the United States, New Zealand, Japan, the People's Republic of China, and the European Union (EU) have recognized the problem of unfair trade competition in the price and quality of blended butter and blended milk products [14]. The International Food Standards Program, Codex Stan 256-2007, requires disclosure of ingredients, including those of butterfat products, on food product ingredient labels to make it clearer for consumers and reduce potential economic disadvantages. Consequently, each country has developed analytical methods for quality inspection.

A variety of methods exist to determine butyric acid and fat content that differ in terms of chemical use, materials, procedures, and experimental conditions of the methods, and the tools used for analysis. Analyzing butyric acid and fat content using a gas chromatography flame ionization detector (GC-FID) is the preferred method for laboratory testing [1,3–5,8,9,15–18], with most laboratories using it in a general laboratory setting [18]. Use of advanced instrumentation for butyric acid analysis, such as LC-APCI-MS/MS [17], high-resolution gas–liquid chromatograph (HR-GLC) [16], and Carbon-13 nuclear magnetic resonance (^{13}C NMR spectroscopy), also have been reported [8]. Results using these instruments have relied on additional indicators other than butyric acid, such as triacylglycerols (TAGs, TGs). However, results have not been significantly different between GC-FID and advanced instrumentation.

As a producer and importer of blended butter and blended milk products, Thailand has taken steps to prevent the imitation of food products through the Thai Notification of the Ministry of Public Health (No. 348), B.E. 2555 Re: Margarine, blended butter, margarine products, and mixed milk products [19]. This Notification is based on the CODEX Alimentarius international food standards (CXS 256-2007 Standard for fat spreads and blended spreads [20]). It states that the quality of margarine and margarine products (fat spreads) must have milk fat not exceeding 3% of the total fat content and have a total fat content of 80–90% and 10–80% by weight, respectively. Blended butter and blended fat spreads must have a milk fat content of more than 3% of total fat and have a total fat content of 80–90% for blended butter and 10%, but less than 80% for blended fat spreads.

Milk fat content is an important indicator for butter blends and blended milk products with different fat contents depending on the type of product. Consequently, this present study centered on developing a method to determine fat content by analyzing butyric acid content from a butyric acid standard solution and a butter fat standard solution by adding valeric acid (internal standard) and preparing all solutions as a derivative of methyl ester. The reaction esterification allows easy vaporization of these fatty acids and is suitable for GC-FID analysis. The fat content is then analyzed by analogy with the content of butyric acid analyzed by standard curves. To our knowledge, the results of this new research concept have not been published previously. In addition, method validation of the new method was determined. This study's output is to support the country's laboratory mission to monitor the quality of fat in butter blends and blended milk products in order to facilitate consumer protection, prevent the adulteration of milk fat, and reduce potential damage to the country's population and economy.

2. Materials and Methods

2.1. Standards and Chemicals

Butyric acid ($C_4H_8O_2$, purity $\geq$ 99.6%) and valeric acid ($C_5H_{10}O_2$, purity $\geq$ 99.9%) were purchased from Sigma-Aldrich®, Switzerland. Anhydrous Milk Fat (AMF, purity $\geq$ 99.95%) was purchased from Fonterra®, New Zealand. Petroleum ether (40–60 °C), methanol, n-heptane, sodium hydroxide, anhydrous sodium sulfate, and sodium chloride were analytical grade from RCI Labscan®, Thailand. Boron trifluoride-methanol (20% solution in methanol) was obtained from Merck®, Germany.

2.2. Instruments and Allied Equipment

Gas chromatography (GC, Perkin Elmer Clarus 600, Flame Ionization Detector [FID]) was used in this study. The analytical capillary column was SP™ 2330 (30 m, 0.25 mm id, 0.2 μm film thickness). The GC condition for injection was: automatic liquid sampler injector, split mode 10:1, temperature set at 260 °C, and injection volume 1 μL. Oven temperature was programmed at initial 40 °C, held for 5 min, ramp rate 10 °C/min to 230 °C, and held for 3 min. Detector temperature was set at 260 °C with H_2 flow rate of 45 mL/min and air zero flow rate of 450 mL/min, which was used to ignite the vapor to ionization. Carrier gas (He) was set at velocity of 20 cm/sec. Analysis time required for all separation stages was about 27 min.

Before analysis, a system suitability test was performed by injection, at 5 repetitions, using a standard solution (butyric acid, C4) and internal standard (valeric acid, C5) by preparing derivatives of methyl ester at a concentration of 0.40 mg/mL. Percentage relative standard deviation (%RSD) of retention time (RT) and peak area were calculated with acceptance criteria at RSD $\leq$ 1% and $\leq$ 2.5%, respectively.

2.3. Preparation of Standard Solutions

An internal standard solution (0.4 mg/mL) was prepared by weighing 100 mg of valeric acid, dissolving, and then adjusting by volume with methanol to 250 mL. A stock standard solution (2 mg/mL) was prepared by weighing 100 mg of butyric acid, dissolving, and then adjusting by volume with methanol to 50 mL. Working standard solutions at different concentrations (0.04, 0.08, 0.40, 0.80, 1.20, and 1.60 mg/mL) were prepared to create a matrix calibration curve by pipetting stock standard solution (2 mg/mL) at 0.1, 0.2, 1, 2, 3, and 4 mL, respectively, into flasks containing 0.2 g of palm oil (matrix blank). A standard solution of valeric acid (0.4 mg/mL) was pipetted (5 mL) into each flask to obtain a final concentration of 0.4 mg/mL.

Milk fat standard solutions were prepared at 8 concentrations (3, 5, 10, 20, 40, 60, 80, and 100% of total fat) to analyze butyric acid content and create a standard curve. They were prepared by weighing a standard substance (butter fat or anhydrous milk fat, AMF) mixed with palm oil to concentrations of butter fat at 3, 5, 10, 20, 40, 60, 80, and 100% of total fat. The prepared solution (0.2 g) and standard valeric acid solution (0.4 mg/mL, 5 mL) were added into each flask to a final concentration of 0.4 mg/mL.

2.4. Sample Preparation

Butter blends and blended milk products were randomly purchased from department stores (2–3 containers) to represent a total sample of approximately 200–300 g. They were mixed thoroughly, taking care in terms of liquid separation. The reserved samples were stored in a refrigerated compartment at 2–8 °C. To prepare each sample, fat was extracted from the sample by weighing 2 g of the prepared sample in a centrifuge tube (50 mL polypropylene) according to the standard method (ISO 17189/IDF 194, 2003: Butter, edible oil emulsions and spreadable fats—Determination of fat content [Reference method] [21]). Petroleum ether (20 mL) was added, mixed well with a vortex mixer, and shaken by hand for 1–2 min. Each sample was centrifuged at 3000 rpm for 5 min, after which the clear solution was poured into a beaker and evaporated in a water bath. The extraction processes were performed 2–3 times. The combined extracted fat was evaporated by blowing with N_2 before drying in a hot air oven at 102 $\pm$ 2 °C for 2 h (to evaporate the residual moisture) and weighing to a constant weight (dried another 1 h in a hot air oven, weight difference less than 0.002 g). The lower weight value was used to calculate the amount of extracted fat.

Method blank was analyzed according to the above method without samples added. Matrix blank (palm oil) was also analyzed using all chemicals and processes. Due to the major constituents of both palm oil and butter being triglycerides of oil or fat, the analysis of butyric acid in the sample was analyzed only for oil/fat extraction. The prepared solutions were injected into the GC-FID machine to check for interferences. If the chromatogram had no interference peaks (only the peaks of butyric acid and valeric acid), the palm oil samples were used as matrix blanks for further testing of the method's accuracy.

2.5. Analytical Method and Method Validation

To analyze butyric acid content, the extracted fat (0.2000–0.2005 g) was weighed into a 50 mL flat bottom flask. The standard solution of valeric acid (0.4 mg/mL) was added at 5 mL and then 4 mL of 0.5 N NaOH in methanol. The solution was connected to a condenser and refluxed for 8 min. Boron trifluoride-methanol (5 mL) was added via the condenser and boiled for 2 min. Organic solvent (n-heptane, 5 mL) was added via the condenser and boiled for 1 min. The flask was removed from the condenser, and 15 mL of saturated sodium chloride was added. The stopper was closed and shaken vigorously for approximately 15 sec. The saturated sodium chloride was continuously added until the n-heptane layer formed at the neck of the flask. The n-heptane layer, at approximately 2–3 mL, was transferred via a dropper into a test tube containing approximately 2 g of anhydrous sodium sulfate. This solution was mixed with a vortex mixer and left to stand for 5–10 min. The clear solution was then transferred into a vial and injected into the GC-FID for butyric acid determination.

Butyric acid analysis was performed in standard solution, butyric acid (working standard solution 6 concentrations), milk fat standard solution (8 concentrations), and sample solution according to Equation (1) below. Milk fat content was determined by creating a standard graph relationship between butyric acid content (percent of total fat, y-axis) and fat content (percent of total fat, x-axis) and then calculating the fat content from the linear equation according to Equation (2) below.

$$\text{Butyric acid content (\% of total fat)} = \frac{C_S \times V_S}{W \times 10} \quad (1)$$

where C_S is the amount of butyric acid in the solution obtained from the standard curve (mg/mL); W is the weight of the extracted fat (g); V_S is the volume of the solution in the n-heptane layer (5 mL)

$$\text{Milk fat content (\% of total fat)} = \frac{y - b}{m} \quad (2)$$

From a standard calibration graph, the relationship between butyric acid content (% of total fat) and fat content (% of total fat) is shown in linear equation as y = mx + b, where y is the amount of butyric acid (% of total fat), b is intercept, m is slope.

The linearity and working range of the method were assessed using standard butyric acid (working standard solution) at 6 concentration levels: 0.04, 0.08, 0.40, 0.80, 1.20, and 1.60 mg/mL. Each concentration was analyzed 3 times. Standard calibration curve was created between the peak area ratio of butyric acid and valeric acid and the butyric acid concentration. The determination of coefficient (R^2) was calculated using the concentration at 8 levels (3, 5, 10, 20, 40, 60, 80, and 100% of total fat). Each concentration was analyzed 3 times. The determination of coefficient (R^2) calculated the relationship between the amount of butyric acid (% of total fat) and milk fat content (% of total fat).

Limit of detection (LOD) and limit of quantitation (LOQ) of butyric acid content analysis were evaluated by adding the lowest concentration of standard butyric acid into a matrix blank and analyzed for 10 repetitions. LOD and LOQ were calculated from 3 times and 10 times of the standard deviation (SD), respectively. The accuracy and reliability of the LOQ value of both butyric acid and milk fat were verified by analyzing 10 times of LOQ concentration. The percentage of recovery (demonstrate accuracy) and the relative standard deviation (demonstrate precision) were assessed.

Accuracy and precision tests were performed with standard butyric acid at concentrations of 0.1, 0.8, 1.5, 2.2, 2.9, and 3.6% of total fat. Each concentration was analyzed for 10 repetitions and calculated in terms of the % recovery and the relative standard deviation of each concentration. The acceptance criteria for accuracy (% recovery) was 95–105%, and the acceptance criteria for precision using Horwitz's equation. For milk fat at the concentrations of 3, 5, 10, 20, 40, 60, 80, and 100% of total fat, the acceptance criteria for accuracy (% recovery) was 98–102%, and the acceptance criteria for precision using Horwitz's equation. For intermediate precision (analysis between days) study, blended milk products were analyzed twice a day for 10 days, and results were calculated using Analysis of variance (ANOVA) [22] as an output shown in Table 1. The intermediate precision (S_I) is calculated from Equation (3).

$$S_I = \sqrt{S_r^2 + S_{between}^2} \quad (3)$$

where $S_r = \sqrt{MS_w}$, $S_{between} = \sqrt{\frac{MS_b - MS_w}{n}}$, MS_b = mean square of between-group standard deviation, MS_w = mean square of within-group standard deviation.

Table 1. One-way ANOVA output table, as example.

Source of Variations	Sum of Squares (SS)	ν	Mean Square (MS)	p	F_{crit}
Between groups	SS_b	p−1	$MS_b = SS_b/(p-1)$	MS_b/MS_w	
Within groups (residuals)	SS_w	N−p	$MS_w = SS_w/(N_p)$		
Total	$SS_{tot} = SS_b + SS_w$	N−1			

Measurement uncertainty of the analytical method was estimated according to the Eurachem/CITAC guideline [23] by taking into account all uncertainty sources. The expanded uncertainty was reported at the 95% confidence level (coverage factor, k = 2).

3. Results

For the specificity of the method, the peak methyl ester of butyric acid (C4) and valeric acid (C5) in blended milk products were well-separated and showed symmetrical peaks, as in Figure 1. The retention times (Rt) of these two compounds were 5.07 and 7.26 min, respectively. No other fatty acid peaks were found to interfere with the interested peaks of samples.

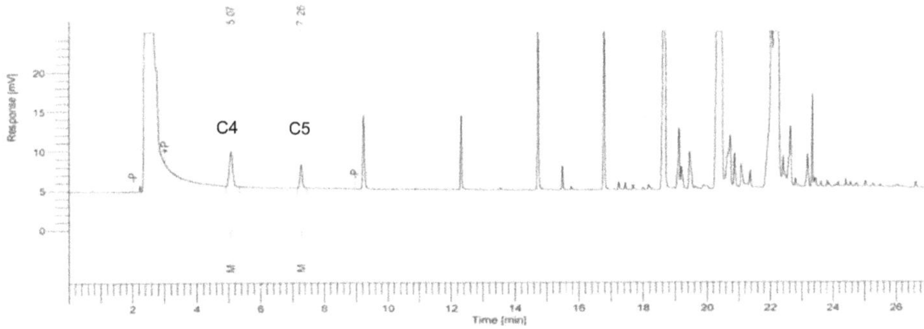

Figure 1. Chromatogram of butyric acid (C4) and valeric acid (C5) methyl esters in blended fat spreads analyzed by GC-FID.

For linearity and analytical range of the standard solution, all concentrations of butyric acid (0.04, 0.08, 0.40, 0.80, 1.20, and 1.6 0 mg/mL) provided a good relationship between the peak area ratio of butyric acid and valeric acid (C4/C5 ratio) and the concentration (Figure 2). A coefficient of determination (R^2) was 0.9995, which was confirmed by the residual plot. The linearity for milk fat standard solutions at 3, 5, 10, 20, 40, 60, 80, and 100% of total fat was obtained (Figure 3). A standard calibration graph showed a good relationship between the concentration of butyric acid and the concentration of milk fat, with an R^2 of 0.9993.

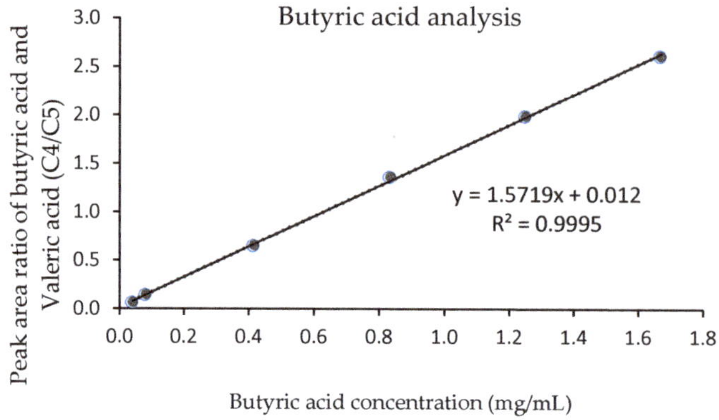

Figure 2. The standard calibration graph shows the relationship between the peak area ratio of butyric acid and valeric acid and the butyric acid concentration.

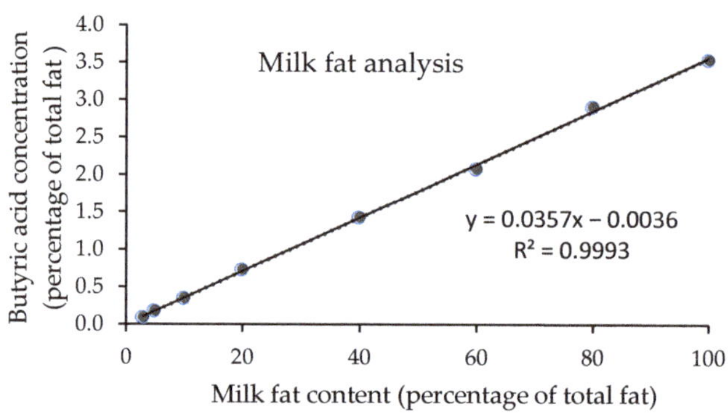

Figure 3. The standard calibration graph shows the relationship between butyric acid concentration and milk fat content.

For the detection limit with a peak signal higher than that of signal to noise at three times, the limit of detection (LOD) for butyric acid was 0.01% of total fat and for milk fat was 1% of total fat. The limit of quantitative (LOQ) of butyric acid was 0.1% of total fat, and milk fat was 3% of total fat. The % recovery of butyric acid at concentrations of 0.1, 0.8, 1.5, 2.2, 2.9, and 3.6% of total fat was in the range of 98.2–101.9%, which met the AOAC standard guideline (97–103%) [24]. For the precision study assessing % RSDr, the range was 0.70–1.33%, as shown in Table 2. For milk fat fraction at concentrations 3, 5, 10, 20, 40, 60, 80, and 100% of total fat, the % recovery of all eight concentrations was in the range of 99.6–100.1%, which met the AOAC standard guideline (98–102%) [24] as shown in Table 2. The precision was in the range of 0.71–1.31% RSDr.

Table 2. Recovery study of butyric acid and milk fat content (n = 10).

Standard	Concentration (% of Total Fat)	% Recovery (Min–Max)	Mean ± SD
Butyric acid	0.1	98.22–101.89	99.91 ± 1.24
	0.8	98.49–101.51	99.81 ± 0.98
	1.5	98.25–101.40	99.80 ± 0.95
	2.2	98.56–101.68	100.13 ± 1.02
	2.9	98.97–101.03	100.12 ± 0.70
	3.6	98.89–101.72	100.02 ± 0.94
Milk fat	3	98.13–101.90	99.77 ± 1.31
	5	98.00–102.08	99.55 ± 1.23
	10	98.11–101.43	99.81 ± 1.14
	20	98.60–101.51	99.82 ± 0.97
	40	98.32–101.40	99.78 ± 1.02
	60	98.56–101.68	100.13 ± 1.02
	80	98.97–101.03	100.09 ± 0.71
	100	98.89–101.72	100.02 ± 0.94

For intermediate precision (SI) from control sample analysis, two examples of mixed milk products with a concentration of butyric acid at 0.17 and 1.46% of total fat analyzed twice a day on 10 different days were analyzed by using one-way ANOVA statistic [22]. The results of the SI value presented as %RSD were 4.38 and 2.03%, respectively, which passed the acceptance criteria of the AOAC standard guideline [24] for reliability (Horwitz's ratio; HORRAT < 2 and p-value > 0.05). This result indicated no significant difference between the two samples, as shown in Table 3a–c.

Table 3. (a). Results of butyric acid in milk products, analysis on different days, twice a day for 10 days. (b). Results of intermediate precision (SI) of butyric acid in sample 1 assessed by one-way ANOVA. (c). Results of intermediate precision (SI) of butyric acid in sample 2 assessed by one-way ANOVA.

a

Date	Example 1 (g/100 g Fat)		Example 2 (g/100 g Fat)	
	Replicate 1	Replicate 2	Replicate 1	Replicate 2
1	0.165	0.162	1.494	1.455
2	0.184	0.173	1.435	1.438
3	0.171	0.165	1.469	1.542
4	0.169	0.185	1.406	1.399
5	0.172	0.168	1.432	1.427
6	0.185	0.178	1.420	1.455
7	0.153	0.167	1.514	1.470
8	0.175	0.182	1.458	1.503
9	0.163	0.157	1.434	1.478
10	0.186	0.170	1.511	1.465

b

Source of Variation	SS	df	MS	F	p-value	F Critical	S_I
Between Groups	0.0012	9	0.00013	2.663	0.0715	3.020	0.0074
Within Groups	0.0005	10	0.00005				
Total	0.0017	19					

Table 3. *Cont.*

c

Source of Variation	SS	df	MS	F	*p*-value	F Critical	S_I
Between Groups	0.0196	9	0.00218	2.699	0.0689	3.020	0.0296
Within Groups	0.0081	10	0.00081				
Total	0.0277	19					

All sources of measurement uncertainty were included for estimating measurement uncertainty; for instance, repeatability of measurement, uncertainties due to calibration curve, analytical balances, volumetric flask, pipettes, and purity of standards. They were calculated into combined uncertainty (u_c) and expanded uncertainty (U) at a 95% confidence level (k = 2). For example, the measurement uncertainty of butyric acid at 0.69% of total fat was 0.06% of total fat, and for milk fat, at 20% it was 2% of total fat (Table 4).

Table 4. Estimation of measurement uncertainty of butyric acid and milk fat in butter blends.

Components	Value, x_i	Unit	$u_{(xi)}$	$u_{(xi)}/x_i$	Contribution (%)
Standard:					
- P_{std} (purity) (butyric acid, C4)	99.6	%	0.11547	0.00116	1.61
- W_{std} (weight) (butyric acid	0.100	g	0.00021	0.00212	4.17
- W_{std} (weight) (valeric acid, C5)	0.100	g	0.00021	0.00212	
- W_{std} (weight) (butter, MF)	25 × 7	g	0.00056	0.000003	
Volume:					
- V_{C4} (final volume)	50	mL	0.11815	0.00236	12.95
- V_{C4} (pipette, std. curve)	0.1	mL	0.00025	0.00254	
- V_{C4} (pipette, std. curve)	0.2	mL	0.00047	0.00237	
- V_{C4} (pipette, std. curve)	1	mL	0.00592	0.00592	
- V_{C4} (pipette, std. curve)	2	mL	0.00714	0.00357	
- V_{C4} (pipette, std. curve)	3	mL	0.00881	0.00294	
- V_{C4} (pipette, std. curve)	4	mL	0.01072	0.00268	
- V_{C5} (final volume)	250	mL	0.58222	0.00233	
- V_{C5} (pipette, 5 mL)	5 × 18	mL	0.05416	0.00060	
Sample:					
- W_s (weight) (Blends)	2.0082	g	0.000212	0.000106	6.39
- W_{fat} (weight) (Fat)	0.4016	g	0.000300	0.000747	
- W_{fat} (weight) (Fat)	0.2000	g	0.000900	0.004500	
- V_s (pipette, 5 mL)	5 × 18	mL	0.054160	0.000602	
C_0 from calibration curve	0.2762	mg/mL	0.01233	0.044635	62.01
Precision (% RSD) *	100	%	0.92645	0.009264	12.87
Combined standard uncertainty (u_c/c) = 0.04569 (c = 0.68926 g/100 g)					
Standard uncertainty (u_c) = 0.03 g/100 g					
Expanded uncertainty (U), k = 2 = 0.06 g/100 g					

* % RSD was obtained from method validation (maximum RSD, 1.31%, standard uncertainty divided by 1.414).

4. Discussion

The analytical method based on ISO 17678/IDF 202: 2010 (milk and milk products–determination of milk fat purity by gas chromatographic analysis of triglycerides, reference method) [25] analyzes milk fat purity through the analysis of triglycerides. However, milk fat contains up to 400 different types of fatty acids and has more than 50 atoms of carbon, which leads to a complicated method for determining milk fat. Analysts must be highly skilled in analyzing data and interpreting results correctly, making this method difficult in practice and analytically time-consuming. The International Food Standard, Codex Stan 256-2007 [20] (standard for fat spreads and blended spreads) recommends the AOAC official method (990.27: butyric acid in fats containing milk fat) [26] using gas chromatography flame ionization detector (GC-FID) for measurement. However, using a glass column for the analysis of butyric acid limits this method due to separation difficulties and a long conditioning time. To improve the separation efficiency of butyric acid and solvent peaks, adding % H_3PO_4 acid must be applied, which adversely affects peak symmetry and leads to instability. This method also has limitations in terms of milk fat analysis. Consequently, this present study aimed to address these issues through the analysis of both butyric acid and milk fat in butter blends and blended milk products. This new method could make laboratory analysis easier, faster, more efficient and accurate, and achieve more reliable results.

Extraction of butyric acid from total fats by method ISO 17189/IDF 194:2003 (Butter, edible oil emulsions, and spreadable fats—determination of fat content, Reference method) [21] provided only fats without nutrients and other substances. The fat extraction of this method was not fully validated. To improve accuracy and precision, this present study's method used valeric acid as an internal standard for the analysis of butyric acid by GC-FID. Valeric acid had similar properties to the analyzed substance, provided a symmetry peak, clearly separated, and was stable, which can reduce variation and errors in the results. Some previous studies also chose valeric acid as the internal standard. Methyl esterification of butyric and valeric acids to methyl butyrate and methyl valerate provided efficiency of detection, no interference, and were well-separated with suitable retention times (5 and 7 min, respectively). In using palm oil as a matrix blank to prepare the working standard solution and milk fat standard solution, neither butyric acid nor valeric acid was found. All parameters of method validation of butyric acid content were performed according to the Eurachem Guideline [22] and Generation Accreditation Guidance [27]. Results showed that the developed method had a specificity and linearity test range from 0.04 to 1.60 mg/mL, which was sufficient for the analysis of samples. There was linearity in the standard calibration graph with the value of R^2 0.9995, a good limit of detection, and detection of quantitation. The results of this method also provided good precision and accuracy. Internal quality control using a control sample for analysis of each set of analyses provided good results. This method also participated in an external quality control (proficiency testing) program in Thailand, which achieved satisfactory results ($|z| \leq 2$).

The analysis of milk fat content in blended milk fat using the results of butyric acid at each milk fat concentration led to a straight line in the calibration curve (calculate the value of the variable X or the fat content from the linear equation $y = mx + b$). This technique reduced the discrepancy of the result by more than one point on the average or representative value of the butyric acid content [14]. Results of the accuracy test for milk fat content revealed linearity (R^2 0.9993) with concentration ranging from 3% to 100% of total fat. The LOD and LOQ of milk fat were 1% and 3% of total fat, respectively, which is comparable to the CODEX STAN 256-2007 standard and conforms to the Thai Notification of the Ministry of Public Health (No. 348, B.E. 2555) [19] (Re: Margarine, blended butter, margarine products, and blended milk products) which states that milk fat must contain more than 3% of the total fat. For accuracy and precision tests, the developed method analyzed milk fat between 3 to 100% of total fat; acceptable recovery was achieved (99.6–100.1%) as well as acceptable precision (relative standard deviation of repeatability, RSD_r = 0.71–1.31%).

An assessment of the measurement uncertainty (MU) of the developed method was calculated from all sources of uncertainty, which can fulfill the requirements of international standard of testing laboratories ISO/IEC 17025: 2017 (Topic 5.4.6.2, the laboratory shall at least attempt to identify all the components of uncertainty and make a reasonable estimation). This method also provided measurement traceability to the SI unit and unbroken chain. Based on the butyric acid content at a concentration of 0.69% of total fat, the MU was reported at 0.06% of total fat (at a 95% confidence level) with 8.7% relative uncertainty, which met the criteria for consideration [28] which is less than two times of the predicted Horwitz's relative standard deviation of reproducibility (RSD_R). For the milk fat in blended milk at a concentration of 20.0% of total fat, the MU was 2% of total fat with 8.7% relative uncertainty. Considering all sources of measurement uncertainty, the source of uncertainty due to the calibration curve (C_0) was a major part of the MU (up to 62%). The sources of uncertainty from repeatability (precision) and the volume were approximately the same percentage (about 13%). For the reduction of relative uncertainty of measurements, all three sources of uncertainty could be considered and reduced, especially uncertainty due to the calibration curve.

GC-FID is one of the instruments used in most analytical laboratories. The cost of a machine, running cost, and maintenance are approximately 8–10 times cheaper than high-technology instruments such as high-resolution gas–liquid chromatography (HR-GC) or gas chromatography–mass spectrometry. In this study, the analytical results of butyric acid showed no significant difference between GC-FID and HR-GC. A study by Joachim and Dietz [15] analyzed milk fat using the mean value of C4 content as a substitute for the milk fat calculation in the mixed fat sample that showed a relatively high error (± 10%). The result from actual C4 content showed a decrease in error (± 4%) at milk fat contents of 60% and 25%, and also did not study at a low amount of milk fat. In this study, the developed method determined the milk fat content from the actual milk fat contents as a calibration curve to control the tolerance to cover the milk fat content ranging from 3% to 100% of total fat. This method presented an error of less than 2% and can also analyze milk fat content as low as 3% that is consistent with or compatible with the CXS 256-2007 [20] benchmark with a milk fat content LOQ of 3%. Therefore, this study demonstrated the capability of GC-FID to analyze butyric acid and milk fat in butter blends and blended milk products with reliable results.

5. Conclusions

Development and method validation of butyric acid and milk fat analysis in butter blends and blended fat spreads by GC-FID is essential for laboratories that must conduct analyses for food production, quality control during production, and inspection tasks for the import and export of these food products. Due to incidents of food fraud and adulteration in both the quality of products and the prices of butter blends and blended milk products, the reliable measurement of butyric acid and milk fat is essential for all related stakeholders. This study demonstrated that the developed method provided reliable results according to international guidelines in terms of good specificity, linearity, LOD, LOQ, precision, accuracy, and measurement uncertainty. This method can also be used to analyze samples of other dairy products, such as butter, cheese, cream, and other fat products like margarine and margarine products. The outcome of this study could directly affect a country's economy and mediate harmful effects on consumer health.

Author Contributions: Conceptualization, A.D. and K.J.; methodology, A.D. and K.J.; laboratory analysis, A.D.; statistical analysis, A.D., and K.J.; writing original draft preparation, K.J. and A.D.; writing review and editing, K.J. and A.D.; project administration, A.D. and K.J. All authors have read and agreed to the published version of the manuscript.

Funding: A Thailand Science Study and Innovation grant (funding number DIG6280001) supported this research.

Institutional Review Board Statement: This study did not require ethical approval.

Informed Consent Statement: Not applicable.

Data Availability Statement: Data are contained within the article.

Conflicts of Interest: The authors declare that they have no conflict of interest.

References

1. Molkentin, J.; Precht, D. Precision of milk fat quantitation in mixed fats by analysis of butyric acid. *Chromatographia* **1998**, *48*, 758–762. [CrossRef]
2. Raffaele, S.; Antonello, P.; Nicola, C.; Gianluca, P.; Raffaele, R.; Francesco, A. Assessment of milk fat content in fat blends by ^{13}C NMR spectroscopy analysis of butyrate. *Food Contr.* **2018**, *91*, 231–236.
3. Helena, L.M. Fatty acids in bovine milk fat. *Food Nutr. Res.* **2008**, *52*, 1821.
4. Gustavo, A.; Malio, V. Total and free fatty acids analysis in milk and dairy fat. *Separations* **2019**, *6*, 14. [CrossRef]
5. Vaseji, N.; Mojgani, N.; Amirinia, C.; Iranmanesh, M. Comparison of butyric acid concentrations in ordinary and probiotic yogurt samples in Iran. *Iran J. Microbiol.* **2012**, *4*, 87–93. [PubMed]
6. Fei, T.; Peng, W.; Lin, Y.; Ying, M.; Li, D. Quantification of fatty acids in human, cow, buffalo, goat, yak, and camel milk using an improved one-step GC-FID method. *Food Anal. Methods* **2017**, *10*, 2881–2891.
7. Peter, W.P. Cows' milk fat components as potential anticarcinogenic agents. *J. Nutr.* **1997**, *127*, 1055–1060.
8. Parodi, P.W. Conjugated linoleic acid and other anticarcinogenic agents of bovine milk fat. *J. Dairy Sci.* **1999**, *82*, 1339–1349. [CrossRef]
9. Ha-Jung, K.; Jung-Min, P.; Jung-Hoon, L.; Jin-Man, K. Detection for non-milk fat in dairy product by gas chromatography. *Korean J. Food Sci. Anim. Resour.* **2016**, *36*, 206–214.
10. Cossignani, L.; Pollini, L.; Blasi, F. Authentication of milk by direct and indirect analysis of triacylglycerol molecular species. *J. Dairy Sci.* **2019**, *102*, 5871–5882. [CrossRef] [PubMed]
11. Robert, K.; Eva, S.; Lenka, P.; Oto, H.; Kestutis, S.; Dalia, R. An overview of determination of milk fat: Development, quality control measures, and application. *Acta Univ. Agric. Silvic. Mendel. Brun.* **2018**, *66*, 1055–1064.
12. Jung-Min, P.; Na-Kyeong, K.; Cheul-Young, Y.; Kyong-Whan, M.; Jin-Man, K. Determination of the authenticity of dairy products on the basis of fatty acids and triacylglycerols content using GC analysis. *Korean J. Food Sci. Anim. Resour.* **2014**, *34*, 316–324.
13. Lee, C.L.; Liao, H.L.; Lee, W.C.; Hsu, C.K.; Hsueh, F.C.; Pan, J.Q.; Chu, C.H.; Wei, C.T.; Chen, M.J. Standards and labeling of milk fat and spread products in different countries. *J. Food Drug Anal.* **2018**, *26*, 469–480. [CrossRef] [PubMed]
14. Glaeser, H. Determination of the milk fat content of fat mixtures. *Grasas Y Aceites.* **2002**, *53*, 357–358. [CrossRef]
15. Joachim, M.; Dietz, P. Validation of a gas-chromatographic method for the determination of milk fat contents in mixed fats by butyric acid analysis. *Eur. J. Lipid Sci. Technol.* **2000**, *102*, 194–201.
16. Manuela, B.; Simona, A.; Elke, A. Quantification of milk fat in chocolate fats by triacylglycerol analysis using gas-liquid chromatography. *J. Agric. Food Chem.* **2007**, *55*, 3275–3283.
17. Kazuaki, Y.; Toshiharu, N.; Hoyo, M.; Koichi, K.; Naohiro, G. Simple method for the quantification of milk fat content in foods by LC-APCI-MS/MS using 1,2-dipalmitoyl-3-butyroyl-glycerol as an indicator. *J. Oleo Sci.* **2013**, *62*, 115–121.
18. Hayoung, K.; Joseph, K.; So Young, C.; Yun Gyong, A. Method development for the quantitative determination of short chain fatty acids in microbial samples by solid phase extraction and gas chromatography with flame ionization detection. *Anal. Sci. Technol.* **2019**, *10*, 1–6.
19. Food Act, B.E. 2522. Notification of the Ministry of Public Health No. 348 (2012) Re: Margarine, mixed butter, margarine products and mixed milk products. In *Government Gazette*; Vol. 130, Special Part 11 Ngor, dated 25th January 2013; Food and Drug Administration: Nonthaburi, Thailand. Available online: https://food.fda.moph.go.th/law/data/announ_moph/V.English/No.%20348%20Margarine,%20Blends,%20Fat%20spread%20sand%20Blended%20fat%20spreads-fomat.pdf (accessed on 11 November 2022).
20. *CXS 256-2007*; Codex Alimentarius. International Food Standards. Standard for Fat Spreads and Blended Spreads. Available online: https://www.fao.org/fao-who-codexalimentarius/sh-proxy/zh/?lnk=1&url=https%253A%252F%252Fworkspace.fao.org%252Fsites%252Fcodex%252FStandards%252FCXS%2B256-1999%252FCXS_256e.pdf (accessed on 5 September 2022).
21. *ISO 17189/IDF 194*; International Organization for Standardization. Butter, Edible Oil Emulsion and Spreadable Fats—Determination of Fat Content (Reference Method). International Organization for Standardization (ISO): Geneva, Switzerland, 2003.
22. Eurachem Guide. *The Fitness for Purpose of Analytical Methods: A Laboratory Guide to Method Validation and Related Topics*, 2nd ed.; Eurachem: Torino, Italy, 2014.
23. Ellison, S.L.R.; Williams, A. *Eurachem/CITAC Guide: Quantifying Uncertainty in Analytical Measurement*, 3rd ed.; Eurachem/CITAC: Teddington, UK, 2012.
24. AOAC International. Guidelines for standard method performance requirements. In *Official Methods of Analysis of AOAC International*, 21st ed.; Latimer, G.W., Ed.; AOAC International: Rockville, MD, USA, 2019; Appendix F; p. 9.
25. International Organization for Standardization. *ISO 17678/IDF 202*; Milk and Milk Products-Determination of Milk Fat Purity by Gas Chromatographic Analysis of Triglycerides (Reference Method). International Organization for Standardization (ISO): Geneva, Switzerland, 2010.
26. AOAC International. AOAC International. AOAC official method 990.27 Butyric acid in fats containing butterfat. In *Official Methods of Analysis of AOAC International*, 21st ed.; Latimer, G.W., Ed.; AOAC International: Rockville, MD, USA, 2019; pp. 44–45, Chapter 41.

27. National Association of Testing Authorities. *Generation Accreditation Guidance-Validation and Verification of Quantitative and Qualitative Test Methods*; National Association of Testing Authorities: Rhodes, Australia, 2018.
28. Tippawan Ningnoi. *Practice Guidelines for Testing the Accuracy of Chemical Analysis Methods by a Single Laboratory*; Department of Medical Sciences Ministry of Public Health: Nonthaburi, Thailand, 2006.

foods

MDPI

Article

Cheese Whey Milk Adulteration Determination Using Casein Glycomacropeptide as an Indicator by HPLC

Ricardo Vera-Bravo [1,*], Angela V. Hernández [2], Steven Peña [2], Carolina Alarcón [3], Alix E. Loaiza [1] and Crispín A. Celis [1]

1 Chemistry Department, Pontificia Universidad Javeriana, Bogotá 110231, Colombia
2 Engineering Department, Fundación Universidad de América, Bogotá 110311, Colombia
3 Productos Naturales de la Sabana S.A.S. Bic, Cajicá 250247, Colombia
* Correspondence: verar@javeriana.edu.co

Abstract: Raw milk adulteration with cheese whey is a major problem that affects the dairy industry. The objective of this work was to evaluate the adulteration of raw milk with the cheese whey obtained from the coagulation process, with chymosin enzyme using casein glycomacropeptide (cGMP) as an HPLC marker. Milk proteins were precipitated with 24% TCA; with the supernatant obtained, a calibration curve was established by mixing raw milk and whey in different percentages, which were passed through a KW-802.5 Shodex molecular exclusion column. A reference signal, with a retention time of 10.8 min, was obtained for each of the different percentages of cheese whey; the higher the concentration, the higher the peak. Data analysis was adjusted to a linear regression model, with an R^2 of 0.9984 and equation to predict dependent variable (cheese whey percentage in milk) values. The chromatography sample was collected and analyzed by three tests: a cGMP standard HPLC analysis, MALDI-TOF spectrometry, and immunochromatography assay. The results of these three tests confirmed the presence of the cGMP monomer in adulterated samples with whey, which was obtained from chymosin enzymatic coagulation. As a contribution to food safety, the molecular exclusion chromatography technique presented is reliable, easy to implement in a laboratory, and inexpensive, compared with other methodologies, such as electrophoresis, immunochromatography, and HPLC-MS, thus allowing for the routine quality control of milk, an important product in human nutrition.

Keywords: casein glycomacropeptide; adulteration; raw milk; whey; HPLC

Citation: Vera-Bravo, R.; Hernández, A.V.; Peña, S.; Alarcón, C.; Loaiza, A.E.; Celis, C.A. Cheese Whey Milk Adulteration Determination Using Casein Glycomacropeptide as an Indicator by HPLC. *Foods* **2022**, *11*, 3201. https://doi.org/10.3390/foods11203201

Academic Editor: Gianfranco Picone

Received: 1 September 2022
Accepted: 8 October 2022
Published: 14 October 2022

1. Introduction

Milk of animal origin is a highly nutritional food, with 3.5% protein, 3% to 4% fat, and 5% lactose. It is an important product of the human diet, due to its essential nutrients, such as calcium, magnesium, selenium, riboflavin, vitamin B12, and pantothenic acid [1]. Statistics from the Food and Agricultural Organization of the United Nations (FAO) reveal Colombia is among the foremost milk producers in Latin America, with an annual volume of 6900 L. According to the Colombian Ministry of Agriculture, milk production represents 12% of agricultural gross domestic product (GDP) and generates 20% of farming jobs, where the cost of milk production per liter is around 0.25 US dollars. At the national level, the dairy sector averages contribute 1.5% of the total GDP, where 1.1% corresponds to milk production, with the remaining 0.4% corresponding to dairy products [1,2].

There are different types of food fraud, such as perception, adulteration, artificial enhancement, counterfeiting misuse of undeclared-unapproved or prohibited biocides, misrepresentation of nutritional content, fraudulent labeling, or removal of authentic constituents, etc.). This type of food fraud seeks financial gain for food manufacturers, retailers, or importers, which is of concern in the production of food and beverages, including milk [3]. In developing countries, milk is usually adulterated with formaldehyde, rice flour, glucose, water, turmeric, whey, cane sugar, neutralizers (caustic soda, caustic potash, sodium carbonate, lime water, etc.), and sodium and potassium nitrates [3]. In Colombia, influenced by the high demand for milk, the dairy industry faces the adulteration of milk with whey, altering its physicochemical properties and food quality. Reports of the indiscriminate use of whey protein indicate that, in high doses, or taken for prolonged periods, they can have detrimental effects on the body (stomach pain, cramps, reduced appetite, nausea, sore throat, headache, fatigue, acne, kidney and liver damage, and altered microbiota), that are aggravated by sedentary habits. Furthermore, from a nutritional point of view, it is strange to consume whey protein, and there is no natural equivalent [4,5]. These alterations in raw milk cannot be detected by the routine analyzes used in the dairy industry (pH, cryoscopy, total protein, total solids, specific gravity, etc.) [6], which particularly affects suppliers, distributors, and final consumers.

There are two methods to coagulate milk: one by lactic or acid coagulation and the other by enzymatic coagulation. In the first method, the caseins coagulate, due to a change in the milk pH (isoelectric point), depending on the amount of acid produced by lactic bacteria or directly added. The curd is partially demineralized, porous, disintegrable, and not very contractile. The second method uses enzymatic coagulation, where enzymatic proteolysis is carried out by chymosin or rennet. The curd obtained is highly mineralized, compact, flexible, contractile, elastic, and waterproof. Other available coagulants from animal, plant, or microbial sources are used less frequently, due to changes in the manufacturing method, costs, and finished product [7]. In Colombia, the dairy industry mainly uses enzymatic coagulation (chymosin). However, in the coastal region of the country, acid coagulation is very common [8].

During cheese manufacturing, the κ-casein protein present in milk is hydrolyzed by the rennet enzyme acting on the phenylalanine 105-Methionine 106 peptide bond, resulting in two fractions: a solid para-κ-casein fraction (cheese curd) and liquid casein glycomacropeptide (cGMP) fraction. In other words, the milk separates into the liquid whey from a solid curd. Furthermore, GMP in the liquid fraction shows a particular chemical structure with 64 amino acids, in which some threonine residues are attached to short carbohydrate chains, known as O-glycosidic bonds. Moreover, it is hydrophilic and remains in suspension in the whey fraction, while the remaining section (para-k-casein) precipitates to form cheese. In general, in bovine milk, cGMP should be present in very low concentrations, which allows it to be used as a marker of milk adulteration, since it is only found in whey in high concentrations. In cheese manufacturing, during the hydrolysis process, the cGMP released by casein is almost ten times higher than the cGMP present in cow's milk [4].

In Colombia, the milk industry is regulated by Decree 616 of 2006, stating the conditions that milk must comply with for human consumption, processing, transporting, bottling, commercialization, exporting, or importing to the country. In this decree, milk and dairy products have been considered priority foods for public health, so they must meet several requirements for consumption. Additionally, article 14 lists all the prohibitions, including adding whey to milk at any stage of the production chain. Although it is illegal to add cheese whey to milk, Colombia does not currently have a method to detect its addition, according to the standard [9].

Several methods are currently available for isolating and quantifying cGMP from cheese whey, such as protein precipitation with trichloroacetic acid (TCA) or ethyl alcohol. In addition, chromatographic techniques, such as molecular exclusion chromatography, affinity chromatography (AC), hydrophobic interaction chromatography (HIC), and ion exchange chromatography (IEC), can be used to separate and quantify cGMP. Other methods used for identification and quantification are colorimetric and fluorometric analysis, immunological methods (Elisa), western blot, immunochromatographic assay, polyacrylamide gel electrophoresis (SDS-PAGE), and biosensors. However, chromatographic techniques are preferred because of their accuracy, replicability, and repeatability [4,10–15].

The objective of this research is to present the molecular exclusion chromatography technique as a tool to detect milk adulteration with cheese whey, using whey from enzymatic coagulation of milk with chymosin as a natural standard and cGMP as a marker. The good results obtained will make it possible to generate milk quality control and regulate these control mechanisms at the governmental level. This fast, simple, accurate, and reliable technique allows us to determine the quality of milk and its derivatives, so it can be implemented by governmental and non-governmental entities (dairy industry) and contribute to improving the quality of a basic product in human nutrition.

2. Materials and Methods

2.1. Materials and Chemicals

Trichloroacetic acid, monopotassium phosphate, and dipotassium phosphate were purchased from Loba Chemie Pvt (Mumbai, India). Sodium sulfate was obtained from Schaurlau (Barcelona, Spain). Deionized water was purified using a Milli-Q system (Millipore; Bedford, MA, USA). Rennet (chymosin) enzyme was supplied by the CAL Group (Bogota, Colombia). A cGMP pattern was acquired from Sigma-Aldrich (St. Louis, MO, USA. An immunochromatographic test for casein glycomacropeptide detection in milk (Stick cGMP) was acquired from Operon (Zaragoza, Spain).

2.2. Sweet Cow Whey Production as a by Prooffrom Enzymatic Coagulation with Chymosin

Raw cow's milk samples were obtained from a dairy farm in Cundinamarca, Colombia. Samples were collected in clean containers, maintaining the cold chain during transport and storage until use. Sweet whey was prepared by adding 0.12 mg of commercial chymosin/liter of raw milk (following the manufacturer's specifications). Samples were heated to 36 °C and incubated for 45 min until curdled. The whey (liquid cheese residue or supernatant) was then centrifuged at 4000 rpm for 10 min at 4 °C and filtered through Whatman No. 1 paper. The whey was bottled and stored at 4 °C ± 2 °C until use.

2.3. Adulterated Milk Sample Preparation with Sweet Whey

Adulterated milk samples were prepared by mixing raw milk with whey obtained by enzymatic coagulation in the following percentages (% m/m (50 g of base as maximum established weight): 0% (fresh unadulterated milk), 2.5% (48.75 mL of adulterated milk with 1.25 mL of whey), 5%, 7.5%, 10%, and 12.5%. Milk and whey were mixed gently for 1 min and stored at 4 °C for 3 h. Subsequently, the majority of proteins were precipitated with 24% (*w/v*, aqueous) trichloroacetic acid (TCA) under constant agitation. After one hour, the mixture was centrifuged at 3500 rpm for 15 min at 4 °C. The supernatant was completely removed from the curd (precipitate) and filtered through a 0.22 µm membrane. The filtered samples were stored at 4 °C, until HPLC analysis was performed, following the methodology reported by the European Official Standard [10]. The analysis of the samples was performed in quadruplicate for each adulteration point (or mixture of raw milk and whey), which is expressed as an average of the chromatography area for each adulterated sample.

2.4. HPLC Analysis

Chromatographic analysis was performed on a Shimadzu Prominence liquid chromatograph (HPLC), equipped with a diode array detector at 205 nm and Shodex KW-802.5 molecular exclusion chromatographic column. A total of 20 µL of each sample (mixture of adulterated milk and sweet whey) was injected with a flow rate of 0.9 mL/min of the mobile phase (0.1 M phosphate buffer and 0.15 M sodium sulfate) at 40 °C. Chromatographic analyzes were performed for each adulterated sample in quadruplicates.

The presence of a distinct and increased signal in the analyzed chromatographic profile from whey adulterated samples, as well as its absence in unadulterated milk sample profile, indicates the chromatography molecular exclusion technique can be effectively implemented for the analysis of adulterated milk with cheese whey.

To confirm the presence of cGMP in the signal of interest, an additional assay was performed under the same HPLC device chromatographic conditions. A Sigma-Aldrich bovine cGMP standard (Catalog number, C7278) was used, and the retention times were compared with the target signal obtained at 10.8 min in the adulterated sample. The cGMP standard was dissolved in 0.1 M phosphate buffered saline at a concentration of 2 ppm. From this stock solution, dilutions were prepared (0.25, 0.5, 1.0, and 2 ppm) and analyzed by HPLC.

2.5. MALDI TOF Analysis and Immunochromatography

The fraction corresponding to the target signal at 10.8 min, observed by HPLC, was collected employing a fraction's collector and analyzed by MALDI-TOF-MS (matrix assisted laser desorption ionization-time of flight mass spectrometry) from Bruker Microflex LT Biotyper (Bruker, Bremen, Germany). For the analysis, 1 µL of the fraction collected at 10.8 min retention time was spotted onto a polished steel target plate, air-dried at room temperature, and overlaid with 1 µL of matrix solution (alpha-cyano-4-hydroxycinnamic acid, diluted in 50% acetonitrile and 2.5% trifluoroacetic acid, followed by air-drying). The mass spectrum for cGMP was analyzed using Flex Control software to verify its presence, based on its molecular weight.

Additionally, qualitative identification of cGMP was performed with Immunostick cGMP visual assay to selected samples of interest. To identify the presence of cGMP in each sample's fraction, an immunochromatographic strip (OPERON S.A.), containing monoclonal antibodies specific for cGMP, was introduced into the collected fraction, following the manufacturer's specifications.

3. Results and Discussion

The molecular exclusion chromatography technique, based on the separation of particles by weight and molecular size, was utilized to detect casein glycomacropeptide, as an indicator of the fraudulent addition of sweet whey obtained by coagulation with the enzyme chymosin in milk samples. Whey affects milk's physicochemical properties, thus decreasing its quality. This practice also generates problems throughout the supply chain because whey contains chymosin (a milk clotting enzyme), which, in its active state, can change the properties of milk, thus altering the final product and possibly deteriorating its microbiological properties. Furthermore, it can affect its protein content, thus diminishing its stability.

For many years, for certain populations, cheese whey was fed to pigs or even considered a waste product. Therefore, cheese whey wastewater was discarded without appropriate treatment, possibly affecting the environment. On the other hand, cheese whey is an important reservoir of food protein, which can be consumed by athletes, infants, and patients. Currently, a variety of products and ingredients are being developed and tested, such as fortified beverages and foods, which are becoming increasingly popular among young bodybuilders. In addition, it has been reported that whey proteins help in the prevention or treatment of obesity, type 2 diabetes mellitus, hypertension, oxidative stress, cancer, high blood pressure, hepatitis B, osteoporosis, and metabolic syndrome associated with metabolic complications [16,17].

Under Colombian law, and in many other countries, adding whey to milk is forbidden because it is a fraud involving milk adulteration. It involves every actor in the production chain: from the producer to the consumer, via the distributors, with nutritional and legal consequences. Therefore, it is necessary to implement, both for governmental control entities and the dairy industry, a precise, accurate, fast, reliable, accessible, and quantitative technique, such as the molecular exclusion chromatography technique, using casein glycomacropeptide as a marker of adulteration, to ensure milk's quality for the consumers.

3.1. HPLC Analysis

The basic physical–chemical parameters that many dairy plants receiving raw milk evaluate are the amount of water, fat, protein, pH, titratable acidity, presence of chlorides and neutralizers, and determination of total solids and non-fat solids [18]. However, these analyzes do not determine the presence of cheese whey in milk at the time it arrives at the plant. Hence, it is necessary to implement a reliable, quantifiable, fast, and easily accessible technique for quality control in dairy plants.

In this study, the isolation and separation of casein glycomacropeptide from cheese whey were achieved using of 24% TCA for protein precipitation, as well as a high efficiency liquid chromatography technique using a KW-802.5 Shodex size exclusion column for its separation. Adulterated sample (2.5%, 5%, 7.5%, 10%, and 12.5% (m/m)) HPLC analyses identified a signal in the chromatographic profile, with a retention time of 10.8 min (x-axis). This signal increased its intensity (area under the peak) as the percentage of adulteration with cheese whey increased (Figure 1). On the contrary, for the milk sample not adulterated with cheese whey (0% fresh milk, without adulteration), this signal was very low or not present in the chromatogram profile.

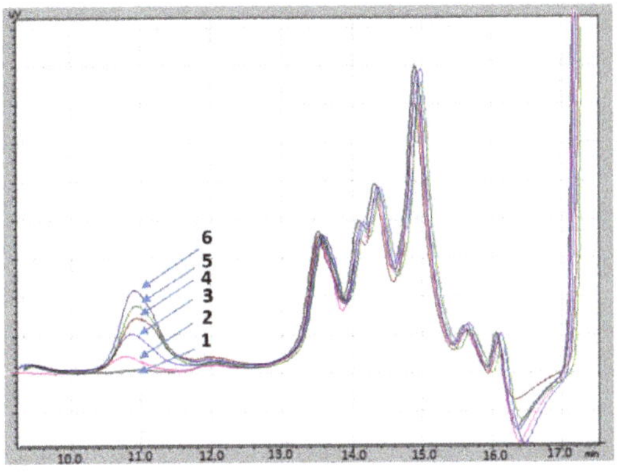

Figure 1. Chromatograms obtained from adulterated milk samples with cheese whey: (1) 0% unadulterated milk sample, (2) 2.5% adulterated sample, (3) 5% adulterated sample, (4) 7.5% adulterated sample, (5) 10% adulterated sample, and (6) 12.5% adulterated sample. A distinct single signal appeared at 10.8 min retention time. An increase in the chromatographic area was noticeable as the percentage of cheese whey increased in the adulterated milk.

Based on these results, we suggest that this obtained signal at 10.8 min is presumably due to the presence of cGMP. Table 1 illustrates the chromatographic area integration's average value for the quadruplicate analyses of every adulteration percentage. This may well be because, as the adulteration percentage increased, so did the presence of cGMP. Thus, the area of the chromatographic signal of interest increased in the profile.

Table 1. Data were obtained for the signal of interest chromatographic area, according to cheese whey adulteration percentage. High accuracy was achieved by this method, according to the results' standard deviations and coefficient of variance.

Cheese Whey Adulteration Percentage	Chromatographic Area Average	Standard Deviation	Coefficient of Variance
0	153,603	6690	0.044
2.5	756,269	26,076	0.034
5	1,381,484	20,389	0.015
7.5	1,845,755	11,828	0.006
10	2,401,283	45,091	0.019
12.5	2,942,586	46,265	0.016

Further more, unadulterated samples revealed a very small area value, corresponding to the small amounts of free cGMP naturally present in raw milk. cGMP can be produced by the degradation induced by bacteria, depending on the cow's stage of lactation and sanitary condition of the udder. However, its concentration will not achieve the high value observed in milk adulterated with cheese whey. According to Furlanetti et al. [19], the average content of free cGMP in mature milk is almost ten times less than the cGMP present in cheese whey or adulterated milk. This is because cGMP constitutes between 20–25% of total proteins in cheese whey [15].

Data from the chromatographic area and adulteration percentage, presented in Table 1, were plotted in a linear model, applying the minimum squares method. In the calibration curve generated with these values, we obtained an adjusted equation for the straight line, where Y = 221,077X + 198,433, Y represents the chromatographic area, and X represents the percentage of adulteration of milk with whey. The slope value was 221,077, and 198,433 was the intercept value. The correlation coefficient (R) calculation was 0.9992, which specifies a strong positive linear relationship between the two variables, as well as a determination coefficient of (R^2) 0.9984, thus indicating that 99.84 of the chromatographic area changes were due to changes in adulteration percentages (Figure 2). The R^2 value provides an estimate of the association strength between the proposed linear model and response variable, which corresponds to the chromatographic area [12]. The results obtained for the association between the chromatography area and cheese whey percentages demonstrate the data fit a linear regression model.

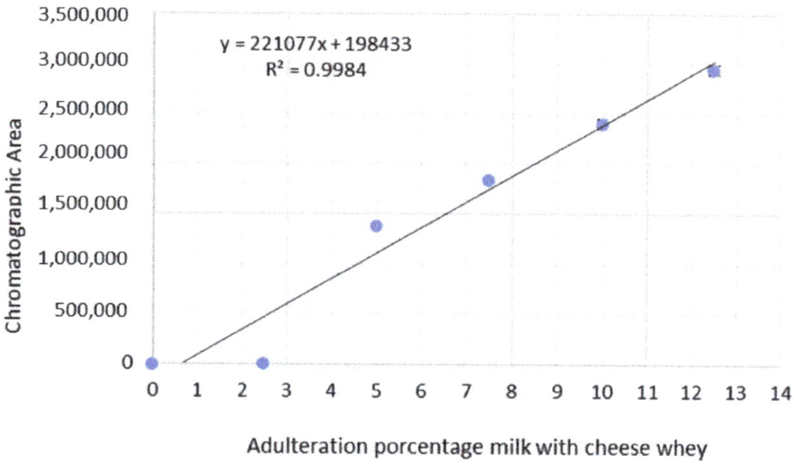

Figure 2. Calibration curve and linear regression model equation obtained for the chromatographic area of each adulterated milk with cheese whey percentage. The samples analyzed were prepared in quadruplicate for each of adulterate sample point (0%, 2.5%, 5%, 7.5%, 10%, 12.5%, 15%, and 20% (m/m)); the average and precision of data is presented for each adulteration point.

3.2. Assays to Confirm cGMP Present in the Chromatographic Profile

3.2.1. HPLC Analysis of a cGMP Standard

The samples corresponding to Sigma-Aldrich's lyophilized cGMP peptide powder pattern (Catalog number, C7278) were dissolved in PBS at 2 ppm; different concentrations were analyzed by HPLC under the same conditions used to separate and analyze adulterated cheese whey samples in order to compare the retention times between the standard and the fraction collected in 10.8 min. The retention times obtained in the chromatographic profile for the cGMP standard showed a single signal at 10.8 min. for each of the dilutions prepared, with a chromatographic area that increased as the concentration of the standard did (Figure 3). These results suggest that the signal of interest obtained in the chromatographic profile for adulterated milk samples with cheese whey may correspond to cGMP.

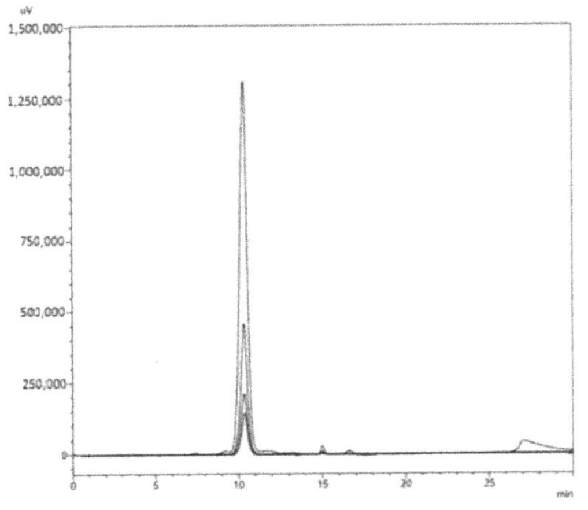

Figure 3. A standard of cGMP (Sigma-Aldrich) was dissolved in 0.1 M PBS at different concentrations and then analyzed by HPLC under the same conditions used to separate and analyze the samples adulterated milk with cheese whey. The dilutions of the standard presented a fine single signal at the same retention time as the adulterated milk samples.

3.2.2. MALDI TOF Analysis

To identify the presence of the cGMP marker in the fraction corresponding to the signal of interest at 10.8 min from adulterated samples, the fraction was collected in the chromatographic assays and then analyzed by MALDI-TOF MS. Figure 4 shows the results obtained by MALDI-TOF mass spectrometry for an adulterated sample, where a signal was observed for a compound with a molecular weight of 6780 Da.

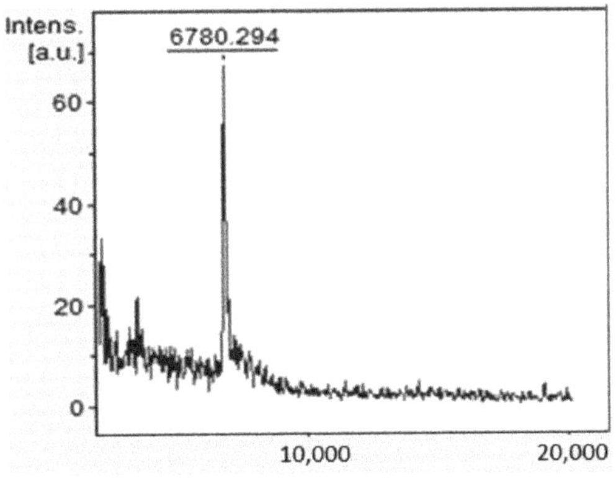

Figure 4. MALDI-TOF mass spectrum from recollected fraction corresponding to the signal of interest at 10.8 min by HPLC of an adulterated sample. The molecular weight result (Da) identified for a peptide in this fraction is similar to the molecular weight corresponding to cGMP monomer reported by other authors, between 6755 and 6787 Da [15,20].

The casein glycomacropeptide is a peptide that comprises 64 amino acids, corresponding to the C-terminal end of the hydrophilic region of k-casein protein. Some of its threonine residues may or may not be glycosylated (sialic acid-galactose-N-acetyl); additionally, this peptide is also phosphorylated. It is the result of the hydrolysis of the k-casein protein with the chymosin (rennet) enzyme, making the molecular weight of cGMP dependent on the κ-casein variant from which it derived, as well as the degree of glycosylation and phosphorylation of the molecule. Vreeman et al. reported that approximately 40% of cGMP is not glycosylated [20,21]. Using MS liquid chromatography, it was found that the non-glycosylated molecule had a molecular mass between 6755 and 6787 Da, while, for the glycosylated form, the molecular weight was around 9631 Da. Findings from other studies suggest that, under certain pH conditions, this monomer can aggregate or dissociate. The reported molecular weight varies, depending on the technique used to isolate it; for example, by using SDS–PAGE, the peptide is in a polymeric form, with a molecular weight ranging between 14 to 30 kDa, two or three times higher than the theoretical weight, due to the association of monomers [15,22,23]. According to our MALDI-TOF assay results for the interest signal fraction, this corresponds to the cGMP's reported molecular mass.

3.2.3. Immunochromatography

The immunochromatographic strip assay involves a recognition of antigen–antibody, specific on a strip that detects the presence or absence of a target analyte (cGMP) in the sample (milk) and enables the rapid detection of the analyte with a high sensitivity. For appropriate cGMP qualitative identification in the signal fraction of interest, with a 10.8 min retention time, an Immunostick cGMP visual assay was carried out, employing an immunochromatographic strip (OPERON S.A.) [24]. This strip contains monoclonal antibodies specific for cGMP that guarantee high specificity for cGMP detection, in addition to an automatic qualitative recognition. The results revealed the presence of GMP by the appearance of a red band on the strip for the collected and analyzed chromatographic fraction (signal at 10.8 min), as shown in Figure 5b, thus confirming cGMP presence in adulterated milk with cheese whey sample obtained from enzymatic coagulation. This red band was not present in unadulterated milk samples (0% fresh milk without adulteration), as shown in Figure 5a, because the milk was not adulterated. The results achieved in this assay are similar to those of previous assessments, obtained by Oancea, for the qualitative identification of cGMP as an adulteration marker using this type of immunochromatographic strip in adulterated milk with cheese whey [14]. Additionally, Martín-Hernández found that the results of immunostick strips correlate and coincide with the results obtained by HPLC, using a size exclusion column, and were more reliable than other methodologies [25].

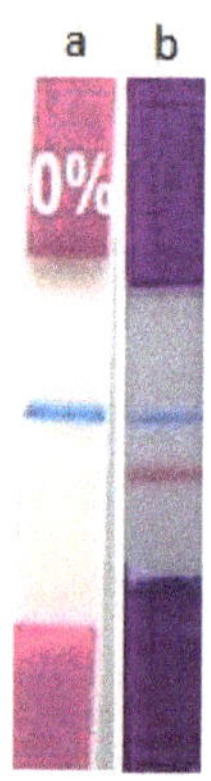

Figure 5. Immunochromatograms from the collected fraction corresponding to the signal of interest at 10.8 min by HPLC of an adulterated sample, (**a**) Absence of recognition of cGMP in the sample negative control, only the blue control line was observed; fresh milk without adulteration (0%) and (**b**) Presence of cGMP in the 5% cGMP sample; recognition of the monoclonal antibodies located on the strip (Immunostick) of cGMP present in the collected fraction for the signal of interest. The blue control line and the positive test red line were observed.

Two factors govern the quality of milk: The first one is the milk's physicochemical composition, where different parameters are evaluated, such as acidity, protein, fat, lactose, minerals, vitamins, non-fat solids, and total solids, i.e., parameters that determine its nutritional value. The second factor is the hygienic quality of the milk, related to the microbial content of raw milk, which directly affects the shelf life of the finished product. These factors are controlled through the application of good manufacturing practices by companies that produce dairy products. Raymond et al. found a correlation between the total bacteria count and cGMP content in milk. In these tests, a concentration of 33 mg/L of cGMP was obtained for a total bacterial count of 5.6×10^6 CFU/mL [26].

Many authors have concluded that there is a close relationship between the quality of milk and content of psychrotrophic bacteria (cold-tolerant bacteria). In raw milk, the predominant genera include *Pseudomonas, Achromobacter, Aeromonas, Serratia, Alcaligenes, Chromobacterium*, and *Flavobacterium* spp. And, to a lesser extent, *Bacillus, Clostridium, Corynebacterium, Streptococcus, Lactobacillus*, and *Microbacterium* spp. After arrival at the dairy plant and storage for 48 h at 6 °C, bacterial counts have shown *Pseudomonas* to be the predominant genus, to which milk and milk-product organoleptic deterioration have been largely attributed. It has been estimated that populations of psychrotrophs, in the order of 5×10^6 to 20×10^6 CFU/mL, are capable of generating detectable organoleptic changes in milk, such as gelling, unpleasant odors, and flavors [27,28]. According to the type of psychrotroph, Punch et al. [29] found unpleasant flavors in milk when the count was in the order of 5.2×10^6 to 200×10^6 CFU/mL for *Pseudomonas*, between 2.5×10^6 and 14×10^6 CFU/mL for *Alkaligens*, 8.3×10^6 to 120×10^6 CFU/mL for *Flavobacteria*, and 2.7×10^6 to 150×10^6 for *Coliforms* [27,30]. Tolle et al. determined that populations of 5×10^6 CFU/mL of *Pseudomona Fragi* and *Pseudomona fluorescens* cause organoleptic changes in milk at 6 °C [30]. However, it all depends on the amount and type of psychrotrophic bacteria present, storage temperature and time, and good practices for collecting, preserving, transporting, and processing milk in the collection center.

According to Raymond et al., high bacterial counts can generate a visible deterioration in raw milk ($>5 \times 10^6$ CFU/mL), thus allowing for its detection in the dairy plant, through routine microbiological tests in the production process [20–22,26]. Consequently, such a high value, in the order of 33 mg/L of cGMP in milk or a commercial dairy product, cannot be solely due to psychrotrophic bacteria. Therefore, it is important to establish a reliable methodology to detect food fraud, and chromatographic methods have been presented as a good alternative to detect milk adulteration with cheese whey, using cGMP as a marker, thus allowing for the precise control of milk quality.

The results of this methodological proposal show that the addition of 24% TCA for protein precipitation, followed by high-performance liquid chromatography using a molecular exclusion column (Shodex KW-802.5) of the supernatant obtained, allows for finding a signal in the chromatogram (10.8 min), whose intensity increases (area under the peak) with the concomitant increase in the percentage of adulterated milk with cheese whey. In addition, these data are fitted to a linear regression model, with an R^2 value (determination coefficient) of 0.9985, which suggests a strong correlation between the two variables (chromatogram area versus percentage of adulteration). This is a versatile model with a short analysis time and little interference in the determination process. In other countries, measures have already been taken to prevent food fraud, due to milk adulteration with cheese whey. In the European Union, the official method to determine adulteration with cheese whey is based on HPLC by molecular exclusion chromatography. Countries such as Spain and Brazil have also implemented this same methodology within their regulations [10,31–33].

To confirm the presence of cGMP in adulterated samples, the signal fraction of interest was collected and tested in different assays. The results indicate that this fraction contains the monomer of glycomacropeptide casein, which was recognized against cGMP-specific monoclonal antibodies, presented a molar mass of 6760 Da in agreement with that reported for cGMP, and presented a chromatographic run, similar to that obtained for the commercial standard of cGMP. The method presented in this research presents short analysis times and is quantitative, reliable, reproducible, precise, and exact, in comparison with other proposed methodologies, such as electrophoresis, immunochromatography, fluorometry, and even HPLC-Ms. This method is easy to develop and implement in the laboratory. Collectively, we propose that this assay can become a test for the detection of cheese whey adulteration of milk using cGMP as a marker. The method executed in this research requires a short analysis time and is quantitative, reliable, reproducible, precise, and exact, in comparison with other proposed methodologies, such as electrophoresis (long analysis times), immune-chromatography (it is a quantitative method), the fluorometry method (quantitates sialic acid to estimate cGMP content), and even HPLC-MS (a costly method to implement) [4,15]. The method proposed by HPLC and exclusion by size is easy to develop and implement in the laboratory. Collectively, we propose that this assay can become a test for the detection of milk adulteration with cheese whey obtained from enzymatic coagulation (chymosin), using cGMP as an adulterated marker to ensure milk's quality for consumers.

4. Conclusions

Casein glycomacropeptide (cGMP) is a peptide resulting from cleavage at residue 105-106 of k-Casein protein by hydrolysis of the enzyme rennet or chymosin during cheese production. It is present in whey with chemical and nutritional properties and high functional benefits for health. Some dairy product companies are already commercializing it to take advantage of its potential as a food supplement. cGMP is the marker for the adulteration of milk with cheese whey that has been used in the different analysis methods. This report presents the development of a method for determining the adulteration of milk with cheese whey using HPLC, in which 24% TCA was added to milk for protein precipitation; then, it was centrifuged and filtered prior to chromatographic analysis, employing a column (Shodex KW-802.5). The chromatograms obtained showed a signal of interest, whose chromatographic area increased concomitant with the percentage of adulteration with cheese whey. These results were adjusted to a linear regression model. This procedure is robust, shows very good precision, and is reproducible, which allows its use in qualitative and quantitative tests in milk. This is a versatile method with short analysis time and reduced interference in the determination. The presence of cGMP in the signal of interest collected in the HPLC analysis of adulterated samples was confirmed by three analyses: mass spectrometry, comparison of the chromatographic run with a commercial cGMP standard (retention time), and by the recognition of cGMP-specific monoclonal antibodies on an immunostick strip. The presented method is easy to implement and develop in the laboratory; it can be applied to routine tests of milk arriving at the dairy plant. It can also be used in finished products by small or large milk processing industries, distributors, and even government regulatory entities to promote the quality and protection of the authenticity of milk, as a product for daily consumption, which is considered a basic in the family diet.

Author Contributions: In this article the contributions of the authors were as follows: Conceptualization, A.V.H., S.P., C.A. and R.V.-B.; methodology, A.V.H., C.A.C. and R.V.-B.; formal analysis, A.V.H., C.A. and A.E.L.; writing—preparation of original draft, A.V.H., C.A., A.E.L., C.A.C. and R.V.-B.; project management and fundraising, R.V.-B. All authors have read and agreed to the published version of the manuscript.

Funding: This research was financed under the University-Company modality, between the Pontificia Universidad Javeriana-Bogotá and Productos Naturales de la Sabana S.A.S. Bic, ID project 007130, and The APC was founded by Pontificia Univeridad Javerian-Bogotá ID 10168.

Data Availability Statement: Data is contained within the article.

Acknowledgments: Authors would like to express their gratitude to numerous colleagues, i.e., Claudia Parra, Alejandro Reyes, Mario Rodríguez, Julian Contreras, María Lucía Gutiérrez, and Elizabeth Torres, for their help with the development of the research project, as well as to the anonymous reviewers for their insightful comments.

Conflicts of Interest: The authors declare no conflict of interest. Additionally, the funders had no role in the design of the study; in the collection, analyses, or interpretation of data; in the writing of the manuscript, or in the decision to publish the results.

References

1. Programa de transformación Productiva. *Propuesta de valor de la cadena Láctea colombiana*; Consejo Nacional Lácteo: Bogota, Colombia, 2011.
2. Unidad de Planificación Rural Agropecuaria-UPRA. *Cadena Láctea Colombiana, Analisis situacional cadena láctea*; Ministerio de agricultura Bogota: Bogotá, Colombia, 2020.
3. Tola, A. Global Food Fraud Trends and Their Mitigation Strategies: The Case of Some Dairy Products: A Review. *Food Sci. Qual. Manag.* **2018**, *77*, 30–42.
4. Poonia, A.; Jha, A.; Sharma, R.; Singh, H.B.; Rai, A.K.; Sharma, N. Detection of adulteration in milk: A review. *Int. J. Dairy Technol.* **2017**, *70*, 23–42. [CrossRef]
5. Damaris Jones Severino Vasconcelos, Q.; Paschoalette Rodrigues Bachur, T.; Frota Aragão, G. Whey protein supplementation and its potentially adverse effects on health: A systematic review. *Appl. Physiol. Nutr. Metab.* **2021**, *46*, 27–33. [CrossRef]
6. Agudelo Gómez, D.A.; Bedoya Mejía, O. Composición nutricional de la leche de ganado vacuno. *Rev. Lasallista Investig.* **2008**, *2*, 38–42.
7. Troch, T.; Lefébure, É.; Baeten, V.; Colinet, F.; Gengler, N.; Sindic, M. Cow milk coagulation: Process description, variation factors and evaluation methodologies. A review. *Base* **2017**, *21*, 276–287. [CrossRef]
8. Granados, R.A.C. *Producción y Comercialización de Quesos Frescos*; Universidad de los Andes: Bogotá, Colombia, 2011.
9. Ministerio de la Protección Social. *Decreto Numero 616 de 2006*; Ministerio de Proteccion Social: Bogotá, Colombia, 2006. Available online: https://www.ica.gov.co/getattachment/15425e0f-81fb-4111-b215-63e61e9e9130/2006D616.aspx (accessed on 26 July 2022).
10. European Union. *7.2.2001 ES Diario Oficial de las Comunidades Europeas I (Actos cuya publicación es una condición para su aplicabilidad) L 37/1 REGLAMENTO (CE) No 213/2001 DE LA COMISIÓN de 9 de enero de 2001*; European Union: Brussels, Begium, 2001; pp. 63–68.
11. Neelima; Rao, P.S.; Sharma, R.; Rajput, Y.S. Direct estimation of sialic acid in milk and milk products by fluorimetry and its application in detection of sweet whey adulteration in milk. *J. Dairy Res.* **2012**, *79*, 495–501. [CrossRef]
12. Thomä, C.; Krause, I.; Kulozik, U. Precipitation behaviour of caseinomacropeptides and their simultaneous determination with whey proteins by RP-HPLC. *Int. Dairy J.* **2006**, *16*, 285–293. [CrossRef]
13. Nakano, T.; Betti, M. Isolation of κ-casein glycomacropeptide from bovine whey fraction using food grade anion exchange resin and chitin as an adsorbent. *J. Dairy Res.* **2020**, *87*, 127–133. [CrossRef] [PubMed]
14. Oancea, S. Identification of glycomacropeptide as indicator of milk and dairy drinks adulteration with whey by immunochromatographic assay. *Rom. Biotechnol. Lett.* **2009**, *14*, 4146–4151.
15. Neelima; Sharma, R.; Rajput, Y.S.; Mann, B. Chemical and functional properties of glycomacropeptide (GMP) and its role in the detection of cheese whey adulteration in milk: A review. *Dairy Sci. Technol.* **2013**, *93*, 21–43. [CrossRef]
16. Chandrajith, V.G.G.; Karunasena, G.A.D.V. Applications of Whey as A Valuable Ingredient in Food Industry. *J. Dairy Vet. Sci.* **2018**, *6*, 555698. [CrossRef]
17. Hammam, A.; Tammam, A.; Elderwy, Y.; Hassan, A. Functional Peptides in Milk Whey: An Overview. *Assiut J. Agric. Sci.* **2017**, *48*, 77–91. [CrossRef]
18. Romero, P.A.; Calderón, R.A.; Rodríguez, R.V. Evaluación de la calidad de leches crudas en tres subregiones del departamento de Sucre, Colombia. *Rev. Colomb. Cienc. Anim. RECIA* **2018**, *10*, 43–50. [CrossRef]
19. Furlanetti, A.M.; Prata, L.F. Free and total GMP (glycomacropeptide) contents of milk during bovine lactation. *Ciên. Tecnol. Aliment.* **2003**, *23*, 121–125. [CrossRef]
20. Mikkelsen, T.L.; Frøkiær, H.; Topp, C.; Bonomi, F.; Iametti, S.; Picariello, G.; Ferranti, P.; Barkholt, V. Caseinomacropeptide Self-Association is Dependent on Whether the Peptide is Free or Restricted in κ-Casein. *J. Dairy Sci.* **2005**, *88*, 4228–4238. [CrossRef]
21. Vreeman, H.J.; Visser, S.; Slangen, C.J.; Van Riel, J.A. Characterization of bovine kappa-casein fractions and the kinetics of chymosin-induced macropeptide release from carbohydrate-free and carbohydrate-containing fractions determined by high-performance gel-permeation chromatography. *Biochem. J.* **1986**, *240*, 87–97. [CrossRef] [PubMed]
22. Farías, M.E.; Martinez, M.J.; Pilosof, A.M.R. Casein glycomacropeptide pH-dependent self-assembly and cold gelation. *Int. Dairy J.* **2010**, *20*, 79–88. [CrossRef]
23. Mollé, D.; Léonil, J. Heterogeneity of the bovine κ-casein caseinomacropeptide, resolved by liquid chromatography on-line with electrospray ionization mass spectrometry. *J. Chromatogr. A* **1995**, *708*, 223–230. [CrossRef]
24. OPERON S.A. Stick cGMP. Available online: https://operondx.com/es/stick-cgmp/ (accessed on 1 August 2022).
25. Martín-Hernández, C.; Muñoz, M.; Daury, C.; Weymuth, H.; Kemmers-Voncken, A.E.M.; Corbatón, V.; Toribio, T.; Bremer, M.G.E.G. Immunochromatographic lateral-flow test strip for the rapid detection of added bovine rennet whey in milk and milk powder. *Int. Dairy J.* **2009**, *19*, 205–208. [CrossRef]

26. Raymundo, N.K.L.; Daguer, H.; Osaki, S.C.; Bersot, L.S. Correlating mesophilic counts to the pseudo-CMP content of raw milk. *Arq. Bras. Med. Vet. Zootec.* **2018**, *70*, 1660–1664. [CrossRef]
27. Cousin, M.A. Presence and Activity of Psychrotrophic Microorganisms in Milk and Dairy Products: A Review1. *J. Food Prot.* **1982**, *45*, 172–207. [CrossRef] [PubMed]
28. de Oliveira, G.B.; Favarin, L.; Luchese, R.H.; McIntosh, D. Psychrotrophic bacteria in milk: How much do we really know? *Braz. J. Microbiol.* **2015**, *46*, 313–321. [CrossRef] [PubMed]
29. Punch, J.D.; Olson, J.C., Jr.; Thomas, E.L. Preliminary observations on population levels of pure cultures of psychrophilic bacteria necessary to induce flavor or physical change in pasteurized milk. *J. Dairy Sci* **1961**, *44*, 1160–1161.
30. Tolle, A.; Suhren, G.; Otte, I.; Zur, W.H. Bakteriologie und Sensorik der pasteurisierten Trinkmilch. *Milchwissenschaft* **1979**, *34*, 406–410.
31. Comisi, L.A.; Adoptado, H.A.; Presente, E.L.; Delegado, R.; Europeo, P.; Europeo, P. *REGLAMENTO DE EJECUCIÓN (UE) 2018/150 DE LA COMISIÓN de 30 de enero de 2018*; Diario oficial de la unión Europea; Unión Europea: Brussels, Belgium, 2018; pp. 14–47. Available online: https://eur-lex.europa.eu/legal-content/ES/TXT/PDF/?uri=CELEX:32018R0150&from=EN (accessed on 2 August 2022).
32. Presidencia, M. *De Real Decreto 2021/1993, de 19 de noviembre, por el que se aprueba un método oficial de análisis de leche y productos lácteos. Última modificación: Sin modificaciones*; Agencia Estatal Boletín Oficial del Estado: España, Madrid, 2021; pp. 1–7. Available online: https://www.boe.es/eli/es/rd/1993/11/19/2021 (accessed on 2 August 2022).
33. Brasil. *Instrução Normativa no68: Métodos Analíticos Oficiais Físico-Químicos, para Controle de Leite e Produtos Lácteos*; Ministério Da Agricultura, Pecuária e Abastecimento: Brasil, Brasilia, 2006; pp. 1–49. Available online: https://wp.ufpel.edu.br/inspleite/files/2016/03/Instru%C3%A7%C3%A3o-normativa-n%C2%B0-68-de-12-dezembro-de-2006.pdf (accessed on 2 August 2022).

foods

MDPI

Article

The Longer the Storage Time, the Higher the Price, the Better the Quality? A ^{1}H-NMR Based Metabolomic Investigation of Aged Ya'an Tibetan Tea (*Camellia sinensis*)

Chenglin Zhu [1], Zhibo Yang [1], Li He [2], Xuan Lu [1], Junni Tang [1,*] and Luca Laghi [3,*]

1 College of Food Science and Technology, Southwest Minzu University, Chengdu 610041, China
2 College of Food Science, Sichuan Agricultural University, Ya'an 625014, China
3 Department of Agricultural and Food Sciences, University of Bologna, 47521 Cesena, Italy
* Correspondence: junneytang@aliyun.com (J.T.); l.laghi@unibo.it (L.L.); Tel.: +86-028-85928478 (J.T.); +39-0547-338106 (L.L.)

Abstract: As an essential beverage beneficial for Tibetan people, Ya'an Tibetan tea has received scarce attention, particularly from the point of view of the characterization of its metabolome. The aim of the study is to systematically characterize the metabolome of Tibetan tea by means of untargeted ^{1}H-NMR. Moreover, the variations of its metabolome along ageing time are evaluated by taking advantage of univariate and multivariate analyses. A total of 45 molecules are unambiguously identified and quantified, comprising amino acids, peptides and analogues, carbohydrates and derivates, organic acids and derivates, nucleosides, nucleotides and catechins. The concentrations of amino acids, organic acids, carbohydrates and catechins are mainly determined by ageing time. The present study would serve as a reference guide for further work on the Ya'an Tibetan tea metabolome, therefore contributing to the related industries.

Keywords: Ya'an Tibetan tea; metabolome; proton nuclear magnetic resonance spectroscopy; ageing time

Citation: Zhu, C.; Yang, Z.; He, L.; Lu, X.; Tang, J.; Laghi, L. The Longer the Storage Time, the Higher the Price, the Better the Quality? A ^{1}H-NMR Based Metabolomic Investigation of Aged Ya'an Tibetan Tea (*Camellia sinensis*). *Foods* **2022**, *11*, 2986. https://doi.org/10.3390/foods11192986

Academic Editor: Charis M. Galanakis

Received: 14 August 2022
Accepted: 22 September 2022
Published: 25 September 2022

1. Introduction

Tea is regarded as one of the most popular and widely consumed beverages throughout the world [1]. The consumption of tea has increased yearly, not only due to the distinct flavor and pleasant taste, but also to the important physiological state and potential health benefits, granted by the presence of various compounds, for instance, carbohydrates, polyphenols, caffeine, amino acids, vitamins and purine alkaloids [2]. There are five main marketed varieties of tea, differentiated by their fermentation process. In detail, green tea is unfermented, white tea is lightly fermented, oolong tea is partially fermented, black tea is fully fermented and dark tea is post-fermented. Among them, dark tea is a unique post-fermented tea produced by pile fermentation attributed to microbial fermentation [3], whose history could be dated back to the Ming Dynasty around 1500 A.D. [4]. In the dark tea family, it is worth mentioning that Ya'an Tibetan tea was initially produced in Southwestern China and then carried via the mountains to Tibet [5], where it has become an essential beverage benefitting millions of Tibetan people.

The Tibetan Plateau is not well suited for cultivating vegetables, fruit and trees, due to its altitude between 3000 and 5000 m. Thus, highly caloric foods are typically consumed, with low fiber intake, by Tibetan people in daily life [6]. These high-protein, high-lipid diets can effectively help them overcome the harsh environment, while they may also increase the risk of cardiovascular and indigestion diseases. Interestingly, they generally drink Tibetan tea along with meals based on high fat milk and red meat, therefore balancing cholesterol and fat absorption. Till now, numerous studies have demonstrated that Ya'an Tibetan tea exhibits antioxidant, cytoprotective [7] and antiradiation effects [8] by in vitro and in vivo experiments. Moreover, Li et al. reported that Ya'an Tibetan tea can effectively lower blood pressure, remove blood lipids and reduce the generation of atherosclerosis [9], which could be linked to its good inhibitory effects on lipase [10]. Intake of Ya'an Tibetan tea is also confirmed to have an anti-inflammatory effect through regulating gut microbiota and altering inflammation and immune system pathways expression in mice models [11].

Untargeted metabolomics, which could provide holistic information about a biofluid, is regarded as the most comprehensive representation of an organism's phenotype [12]. This approach attempts to provide qualitative and quantitative information of low weight metabolites (<900 Da) from biological samples. Until now, metabolomics has been widely applied to investigate tea metabolome profiles altered by internal and external factors, such as fermentation process [13,14], shade treatment [15] and seasonal variation [16]. As one of the mostly applied techniques for metabolomics investigations, ^{1}H-NMR spectroscopy has been applied in related fields thanks to its non-invasive nature, highly reproducible molecules' quantification and effectiveness in analyzing a diverse range of compounds. Lee et al. evaluated strong inter-country and inter-city relationships in the quantities of theanine and catechin derivatives found in green and white teas by means of ^{1}H-NMR [17]. Ohno et al. found that growing tea at higher altitudes leads to a high amount of theanine and caffeine and to low levels of thearubigins, especially thearubigin 3,3′-digallate [18].

Commonly, consumers perceive a direct connection between storage time and quality, accepting therefore higher prices for more aged teas. However, there have been no studies on the change of comprehensive metabolomic profiles during storage for Ya'an Tibetan tea, except for one paper published recently which just referred to concentrations of polyphenols and catechin compounds affected by storage time [19]. Moreover, there is limited knowledge of expected concentrations of metabolites from Ya'an Tibetan tea. To fill these gaps, the present study aims to systematically characterize the metabolomic profiles of Ya'an Tibetan tea by means of ^{1}H-NMR. Furthermore, the evolution of the metabolomic profile of Ya'an Tibetan tea along storage time was evaluated. This study could offer guidance for consumers to select Ya'an Tibetan tea products and act as a reference for the related industries to produce high-quality products.

2. Materials and Methods

2.1. Sampling

As described by Xie et al., five years could be considered as the milestone from the perspective of chemical constituents of Pu-erh tea, a fully fermented tea similar to Tibetan teas [20]. For this reason, we collected Ya'an Tibetan tea samples spanning 10 years of ageing. All the Ya'an Tibetan tea samples were purchased from Sichuan Ya'an Tea Factory Co., Ltd. (Sichuan, China). The samples comprised 30 Tibetan tea samples, sorted into three groups, namely 1 year, 5 years and 10 years. Each group included ten samples.

2.2. Metabolome Analysis

By adapting the procedure described by Ohno et al. [18], one gram of each Ya'an Tibetan tea was added to 30 mL of boiling bi-distilled water. The mixture was kept for 4 min at 95 °C and then left at room temperature for 10 min. After vortex mixing for 1 min and centrifuging for 15 min at 14,000 rpm and 4 °C, 0.5 mL of supernatant were taken to a new Eppendorf tube, and then 0.2 mL of a D_2O solution of 3-(trimethylsilyl)-propionic-2,2,3,3-d_4 acid sodium salt (TSP) 10 mM was added, used as NMR chemical-shift reference, buffered at pH 7.00 $\pm$ 0.02 using 1 mol/L phosphate buffer. In order to avoid microbial proliferation, 10 µL of NaN_3 2 mmol/L was also added. Finally, each sample was centrifuged again at the above conditions. The workflow of sample preparation procedure is shown in Figure S9a.

^{1}H-NMR spectra were performed at 298 K equipment with an AVANCE III spectrometer (Bruker, Milan, Italy) operating at a frequency of 600.13 MHz. Taking advantage of presaturation, the HOD residual signal was suppressed. This was done by employing the noesygppr1d sequence, part of the standard pulse sequence library. Each spectrum was acquired by summing up 256 transients by means of 32 K data points over a 7184 Hz spectral window, with an acquisition time of 2.28 s and a recycle delay of 5 s. The workflow of spectra processing is shown in Figure S9b. In detail, ^{1}H-NMR spectra baseline-adjusted through peak detection in accordance with the "rolling ball" principle [21] implemented in the baseline R package [22]. Differences in water and fiber content among samples were calculated by means of probabilistic quotient normalization (PQN) [23] applied to the entire spectra array. The signals were assigned by comparing their chemical shift and multiplicity to Chenomx software library (Chenomx Inc., Edmonton, AB, Canada, v.8.4) and authentic material or published data [18]. Integration of the signals was obtained for each molecule using rectangular integration.

2.3. *Statistical Analysis*

Statistical analysis was performed in R computational language [24] and online metabolomic data analysis platform MetaboAnalyst (https://www.metaboanalyst.ca, Montreal, QC, Canada, v.5.0, accessed on 25 July 2022). Prior to univariate analysis, concentrations of molecules in each group that were not-normally distributed were transformed in accordance with Box and Cox [25]. And then, to figure out perturbations caused to single molecules by the effects considered, *t*-tests were performed with a cut-off p value below 0.05.

3. Results

3.1. *^{1}H-NMR Spectra of Ya'an Tibetan Tea Samples*

Represented spectra from ^{1}H-NMR were assigned as pictorially described in Figure 1, while the entire concentrations for each sample are reported in the supporting materials. An important step for signal assignments performed by ^{1}H-NMR is the comparison with references of the fine structure of the signals visually [26]; there are supplemental material reports, for each characterized molecule, and superimpositions of spectra registered and simulated for pure compounds (Supplementary Materials, Figures S1–S8). In addition, to increase the reproducibility of our results, the functional groups and ppm for each identified metabolite are reported in Table 1.

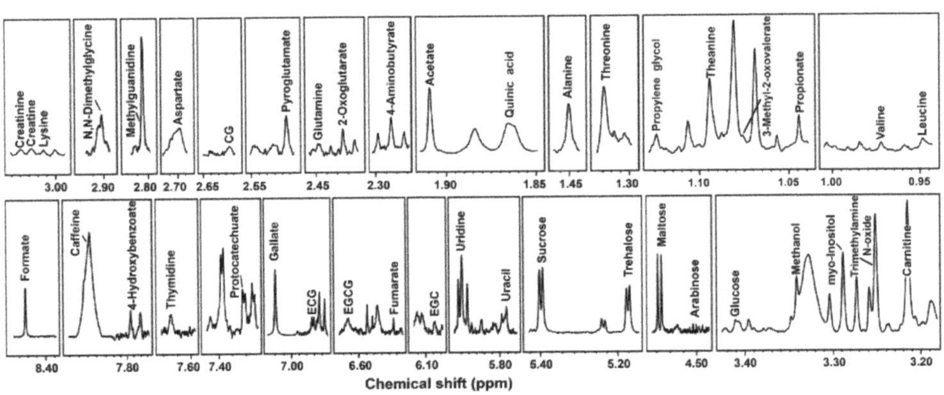

Figure 1. Portions of ^{1}H-NMR spectra from typical Ya'an Tibetan tea samples. Name of the molecules appears on the signals used for their quantification. The vertical scale of each portion is conveniently set to ease the signals observation.

Table 1. Information for molecules identification by means of ^{1}H-NMR.

	ppm	Functional Group	Multiplicity *
	Amino acids, Peptides and Analogues		
4-Aminobutyrate	2.2854	CH_2-2	t
Alanine	1.4754	CH_3	d
Aspartate	2.7009	CH_2	dd
Carnitine	3.2146	CH_3	s
Creatine	3.0270	CH_3	s
Creatinine	3.0405	CH_3	s
Glutamine	2.4492	CH_2-2	m
Isoleucine	0.9906	CH_3-4	d
Leucine	0.9479	CH_3	t
Lysine	3.0130	CH_2	t
N,N-Dimethylglycine	2.8999	CH_3	s
N-Acetylglutamate	2.2318	CH_2-2	t
Pyroglutamate	2.5275	CH_2-5	m
Threonine	1.3117	CH_3	d
Theanine	1.0936	CH_2	m
Valine	0.9718	CH_3	d
	Carbohydrates		
Arabinose	4.5082	CH_2	d
Fucose	1.2313	CH_3	d
Glucose	3.4074	CH-3	t
Maltose	4.6291	CH	d
Sucrose	5.3954	CH	d
Trehalose	5.1807	CH	d
	Organic Acids		
2-Oxoglutarate	2.4246	CH_2-2	t
3-Methyl-2-oxovalerate	1.1004	CH_3-4	d
4-Hydroxybenzoate	7.7896	CH_2-3	d
Acetate	1.9082	CH_3	s
Formate	8.4454	CH	s
Fumarate	6.5080	CH	s
Gallate	7.0203	CH	s
Propionate	1.0438	CH_3	t
Protocatechuate	7.3737	CH	dd
Quinic acid	1.8642	CH_2	d
	Nucleosides, Nucleotides and Analogues		
Thymidine	7.6287	CH-7	s
Uracil	5.7969	CH-6	d
Uridine	5.8970	CH	s
	Catechins		
Catechin gallate (CG)	2.6328	CH_2	dd
Epicatechin gallate (ECG)	6.9439	CH	d
Epigallocatechin gallate (EGCG)	6.6304	CH	d
	Miscellaneous		
Caffeine	7.8612	CH-9	s
Dimethylamine	2.7132	CH_3	s
Methanol	3.3495	CH_3	s
Methylguanidine	2.8057	CH_3	s
myo-Inositol	3.2878	CH	t
Propylene glycol	1.1248	CH_3	d
Trimethylamine N-oxide	3.2494	CH_3	s

* s stands for singlet, d stands for doublet, t stands for triplet, and m stands for multiplicity.

3.2. *Ya'an Tibetan Tea Metabolome Variations along Storage Time*

As can be seen from Table 1, a total of 45 molecules were identified and quantified, mainly pertaining to the chemical groups of amino acids, peptides and analogues (16), carbohydrates (6), organic acids (10), nucleosides, nucleotides and analogues (3), catechins

(3) and miscellaneous (7). As shown in Figure 2, the concentration of total amino acids, peptides and analogues was significantly reduced from 5 years to 10 years. Moreover, the content of total carbohydrates was reduced in the early storage period, and then the content increased markedly as storage was prolonged. As for organic acids, nucleosides, nucleotides and analogues, and miscellaneous, their trends were similar, namely, increasing first until five years and decreasing afterwards. The amount of catechins was significantly decreased after 5 years. However, it is worthy to note that several represented molecules did not exactly follow the same trend as their groups. Therefore, volcano plots, reported in Figure 3, evidence the main differences between each couple of time points. Moreover, concentrations of molecules showing a fold change above 2 in each of the two groups are shown as boxplot, in Figure 4.

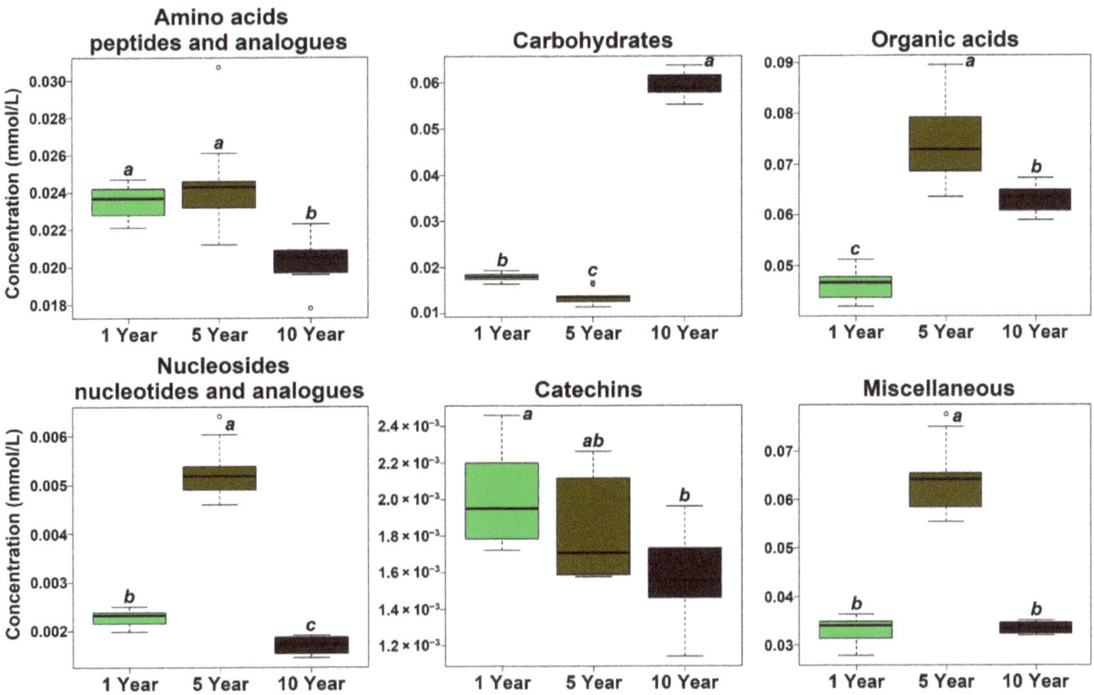

Figure 2. Concentrations of the main classes of molecules quantified by ^{1}H-NMR among the three groups. The italic lowercase letters above each box indicated the significances of the comparisons among the three groups, where a common superscript is not significantly different.

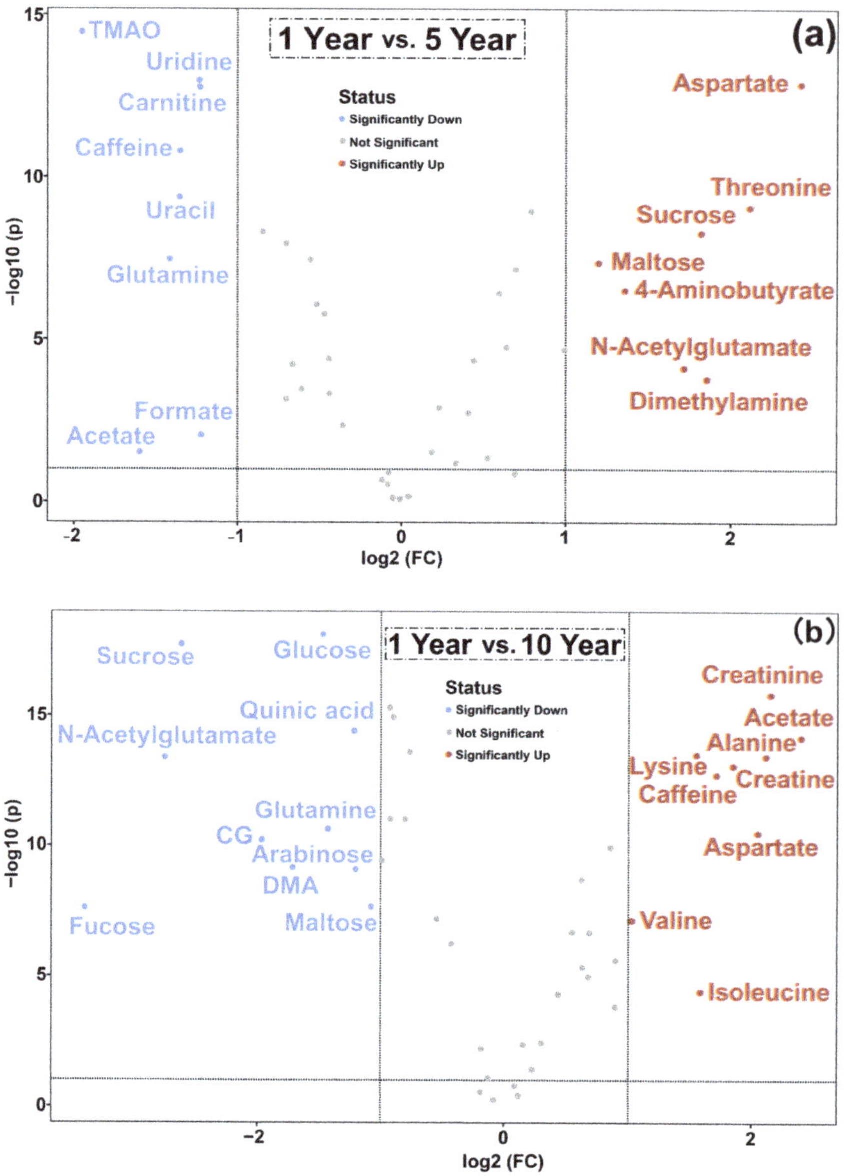

Figure 3. *Cont.*

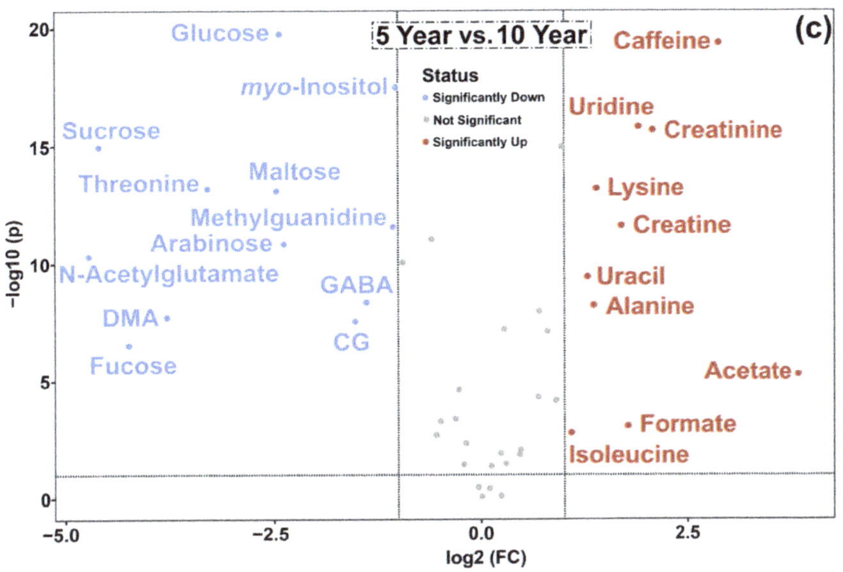

Figure 3. Volcano plot built on the concentration of molecules in each of the two groups. (**a**) indicates 1 Year vs. 5 Year, (**b**) indicates 1 Year vs. 10 Year and (**c**) indicates 5 Year vs. 10 Year.

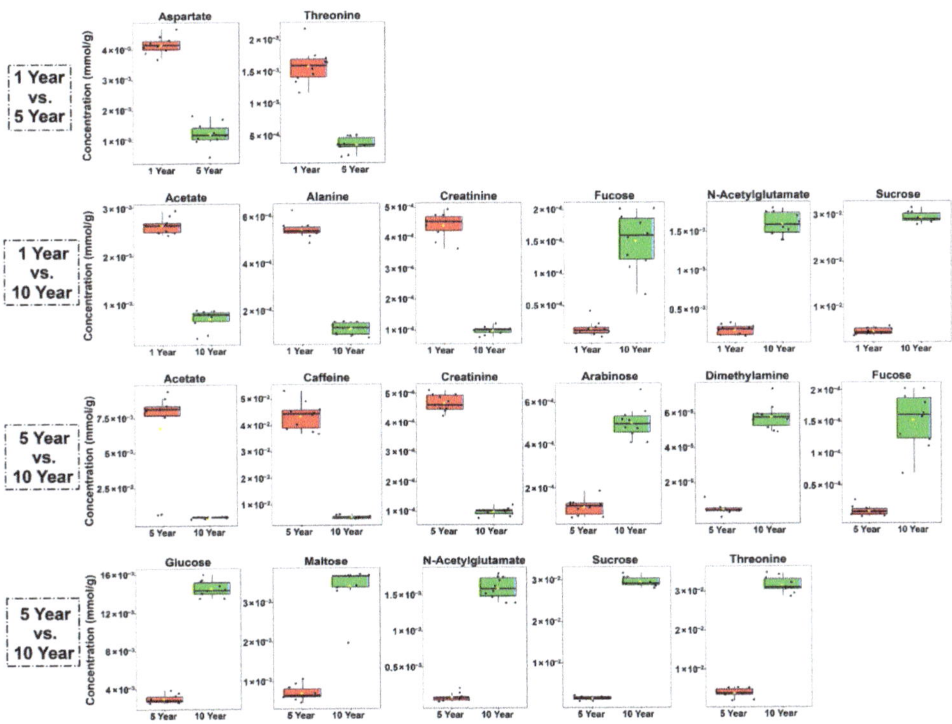

Figure 4. Boxplot based on the concentration of molecules whose fold change is above 2 in each of the two groups. Asterix indicates the mean value of the group in each box.

Comparing the first five years with the subsequent five, the concentration of several molecules exhibited opposite trends, namely maltose, n-acetylglutamate, dimethylamine, 4-aminobutyrate, caffeine, uridine, uracil, formate and acetate, as shown in Figure 3a,c and Figure 4. However, widening the view to comprise the entire period evaluated (Figure 3b), the amounts of creatinine, alanine, lysine, acetate, caffeine and isoleucine were significantly increased along storage time, while the levels of sucrose, glucose, n-acettylglutamate, CG, dimethylamine, fucose, arabinose and maltose appeared as significantly decreased.

To obtain deeper details into which metabolic pathway could undergo the widest modifications with storage time, an enrichment analysis was performed by means of the MetaboAnalyst platform. The pathways evidenced as potentially altered by storage time (p-value < 0.05) were glutamate metabolism, urea cycle and glucose-alanine cycle, as shown in Figure 5.

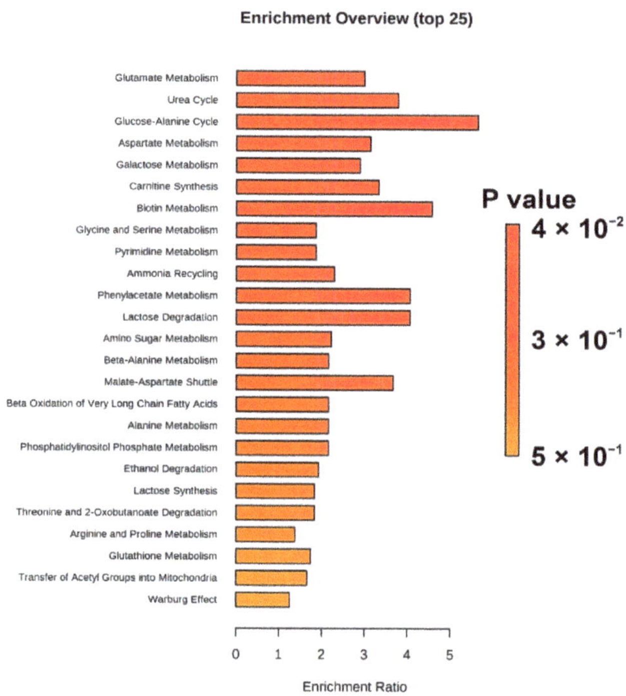

Figure 5. Enrichment analysis built on the concentration of molecules significantly varied along storage time.

4. Discussion

As one of the essential beverages for Tibetan people, most of the works dealing with Tibetan tea have been focused on its beneficial properties for human health and on safety risk assessments. For instance, Li et al., found that high doses (400 mg/Kg/d) of Tibetan tea supplementation reduced bodyweight gains and markedly attenuated serum lipid profiles and atherosclerosis index in mice model [9]. Xie et al. evaluated that Tibetan tea has antioxidative or cytoprotective properties linked to phenolic compounds, such as gallic acid and four catechins (catechin, CG, ECG and EGCG) [7]. Ye et al. assessed ten mycotoxins in Tibetan tea samples, ruling out potential risks for consumers [27]. In contrast, less attention has been devoted to Tibetan tea itself, with only a few papers having attempted to investigate the volatile [28] and phenolic compounds [7] in Tibetan tea by means of metabolomic approaches. To the best of our knowledge, there are no complete reports about quantitative information for each molecule that can be characterized by ^{1}H-NMR. Moreover, there seem to be no reports about the variations of Ya'an Tibetan tea metabolomic profiles with ageing time. To gain more information about these aspects, the present work attempts, for the first time, to provide reference quantitative values for the molecules mostly present in the Ya'an Tibetan tea metabolome, as observable by ^{1}H-NMR. A total of 45 metabolites were unambiguously characterized, a number much higher than those previously obtained based on the same platform [15,17,18]. The quantified molecules mainly pertained to the categories of amino acids, peptides and analogues, carbohydrates and derivates, organic acids and derivates, nucleosides, nucleotides and analogues, catechins and miscellaneous. The most important chemical constituents that influence the taste and flavor of tea infusions are sugars, organic acids, amino acids, polyphenols, caffeine, flavonols and volatile flavor compounds [29]. In the present work, we found that the concentrations of 12 amino acids, 3 organic acids, 6 sugars, 2 nucleosides, nucleotides and analogues, and 5 miscellaneous in total were significantly altered with ageing time by means of volcano plot. According to the above observations, we could infer that ageing time would eminently affect the taste and flavor of Ya'an Tibetan tea.

In terms of amino acids, several remarkable works have indicated that there is a relationship between the quality of tea and the amino acid contents [30], with consequences on fresh and brisk tastes of tea infusion and aroma substances [31,32]. Moreover, Alcázar et al. observed a clear relation between the amino acids content and the elaboration process of teas. In detail, unfermented or lightly fermented teas exhibit higher levels of free amino acids than fully fermented or post fermented ones [1]. Focusing on the total amino acids content, the present work noticed no significant variations but a slightly increase in the first five years of ageing, while a significant decrease occurs in the following five years. Such a phenomenon could be linked to the degradation of proteins into amino acids during the early stage of pile-fermentation process [33], and then part of amino acids could evolve into volatile compounds along storage time [34]. Among the amino acids quantified, it is worth noticing that theanine, a unique amino acid that is found almost exclusively in tea, could contribute to the brothy sweet umami taste of tea [35,36]. Cheng et al. found that theanine content was reduced by 93.51% during Qingzhuan tea processing [37]. Our results were in line with such trends, with the fermentation procedure reducing the contents of theanine, even if to lesser extents. This could be due to the distinct fermenting conditions and, in turn, to the different microbial community.

As one of the primary inhibitory neurotransmitters, 4-aminobutyrate plays an important role in the vertebrate central nervous system and has antianxiety and antihypertensive effects [38]. 4-aminobutyrate is mainly biosynthesized through the irreversible α-decarboxylation of Glutamate to 4-aminobutyrate, which is catalyzed by pyridoxal 5′-phosphate (PLP)-dependent glutamate decarboxylase (GAD) in plants [39]. Even if glutamate could not be quantified in the present study, this biosynthetic route for 4-aminobutyrate may be confirmed by the trend we highlighted for glutamine. In fact, glutamine is synthesized through glutamine synthetase from glutamate, and its concentration shows a trend opposite to that of 4-aminobutyrate. Such a pathway was highlighted by enrichment analysis, too, further indicating that it plays an important role during the pile-fermentation process of Ya'an Tibetan tea.

Organic acids are considered as major detrimental contributors to overall taste of dark teas. In terms of total organic acids, their content was significantly increased in the first five years, while decreased during the following five years. Therefore, we could speculate that five years could be regarded as the line of demarcation during pile-fermentation [20]. Among the characterized organic acids, acetate is produced by acetic acid bacteria from glucose. In the present study, the concentration of acetate increased in the first five years, followed by a decrease in the next five years. Such result was in line with previous studies on the topic [20,40]. Gallate is another important compound widely found in tea leaves, which could be regarded as a precursor for catechin catabolism. Gallate is derived from the hydrolysis of procyanidins and gallyolated catechins and degraded into methoxy phenolic compounds during dark tea processing [41]. The trend of gallate we found is in agreement with Qingzhuan tea process [37], but opposite that found for Pu-erh tea [42]. This discrepancy is probably due to the degradation of gallate, which exceeds the hydrolysis of gallyolated catechins during Ya'an Tibetan tea processing.

Catechins, which account for 60−80% of tea polyphenols, are the main components contributing to the antioxidant activity of tea [4]. Catechins, together with caffeine [43] and volatile components [44], have also been used to differentiate tea categories. The contents of catechins in dark teas share the same trends with amino acids, with a concentration much lower than the one characterizing unfermented and semi-fermented teas [4]. In the present study, the overall trend of catechins was in agreement with the previous work, while a smaller number of catechins were quantified, which could be linked to the discrepancy of detection sensitivity between different metabolomic approaches, namely ^{1}H-NMR and UPLC-QqQ-MS/MS [19]. Such an observation could be explained considering that post fermentation process highly decreases the contents of catechins and form pigments such as theabrownines, which have been suggested as linked to the oxidation and condensation of catechins during post fermentation by microorganisms [42,45]. As catechins contribute to the astringency taste of tea, their decrease with storage could lead to a decline in the astringency taste and could deepen the color of Ya'an Tibetan tea infusion. Together with a significant increase in the total content of carbohydrates, prolonging storage time may have beneficial effects on the improvement of tea infusions' sensory evaluation.

5. Conclusions

To the best of our knowledge, the present study, for the first time, has been devoted to obtaining a holistic metabolomic representation of Ya'an Tibetan tea, by providing quantitative information of Ya'an Tibetan tea metabolome through untargeted ^{1}H-NMR. A remarkably higher number of metabolites than previously reported was characterized by a single platform. The contents of amino acids, organic acids, carbohydrates and catechins are mainly determined by ageing time, which would eminently affect the taste and flavor of Ya'an Tibetan tea. As we purchased Ya'an Tibetan tea samples from the same factory, the conditions of fermentation could be considered as identical across the samples analyzed, but there are still several factors (such as the variations in raw tea leaves collected each year) that should be taken into consideration for further investigations. The present study could serve as a reference guide for further Ya'an Tibetan tea metabolome studies.

Supplementary Materials: The following are available online at https://www.mdpi.com/article/10.3390/foods11192986/s1, Figure S1–S8: Molecules assignment and quantification; Figure S9: Sample preparation and spectra processing procedures for ^{1}H-NMR.

Author Contributions: Conceptualization, C.Z., L.L. and J.T.; methodology, C.Z. and L.L.; formal analysis, C.Z., Z.Y. and X.L.; investigation, C.Z. and Z.Y.; resources, L.H.; writing—original draft preparation, C.Z. and L.L.; writing—review and editing, C.Z., Z.Y., L.H., X.L., J.T. and L.L.; supervision, J.T. and L.L.; funding acquisition, C.Z. and J.T. All authors have read and agreed to the published version of the manuscript.

Funding: This work was funded by the Natural Science Foundation of Sichuan Province (2022NSFSC1652) and the Fundamental Research Funds for the Central Universities Southwest Minzu University (2020NTD04).

Institutional Review Board Statement: Not applicable.

Informed Consent Statement: Not applicable.

Data Availability Statement: Data is contained within the article or supplementary material.

Conflicts of Interest: The authors declare no conflict of interest.

References

1. Alcázar, A.; Ballesteros, O.; Jurado, J.M.; Pablos, F.; Martín, M.J.; Vilches, J.L.; Navalón, A. Differentiation of Green, White, Black, Oolong, and Pu-Erh Teas According to Their Free Amino Acids Content. *J. Agric. Food Chem.* **2007**, *55*, 5960–5965. [CrossRef]
2. Horie, H.; Kohata, K. Analysis of Tea Components by High-Performance Liquid Chromatography and High-Performance Capillary Electrophoresis. *J. Chromatogr. A* **2000**, *881*, 425–438. [CrossRef]
3. Feng, Z.; Li, Y.; Li, M.; Wang, Y.; Zhang, L.; Wan, X.; Yang, X. Tea Aroma Formation from Six Model Manufacturing Processes. *Food Chem.* **2019**, *285*, 347–354. [CrossRef]
4. Zhang, L.; Zhang, Z.-z.; Zhou, Y.-b.; Ling, T.-j.; Wan, X.-c. Chinese Dark Teas: Post fermentation, Chemistry and Biological Activities. *Food Res. Int.* **2013**, *53*, 600–607. [CrossRef]
5. Zheng, Q.; Li, W.; Zhang, H.; Gao, X.; Tan, S. Optimizing Synchronous Extraction and Antioxidant Activity Evaluation of Polyphenols and Polysaccharides from Ya'an Tibetan Tea (*Camellia sinensis*). *Food Sci. Nutr.* **2020**, *8*, 489–499. [CrossRef]
6. Dang, S.; Yan, H.; Yamamoto, S.; Wang, X.; Zeng, L. Feeding Practice among Younger Tibetan Children Living at High Altitudes. *Eur. J. Clin. Nutr.* **2005**, *59*, 1022–1029. [CrossRef]
7. Xie, H.; Li, X.; Ren, Z.; Qiu, W.; Chen, J.; Jiang, Q.; Chen, B.; Chen, D. Antioxidant and Cytoprotective Effects of Tibetan Tea and Its Phenolic Components. *Molecules* **2018**, *23*, 179. [CrossRef]
8. Xu, J.; Liang, X.; Wang, W.; Chen, S.; Tang, Q. Study on Anti-Radiation Effect of Ya'an Tibetan Tea. In Proceedings of the Journal of Physics: Conference Series, Montréal, Canada, 10–14 May 2020; Volume 1549.
9. Li, L.; Shi, M.; Salerno, S.; Tang, M.; Guo, F.; Liu, J.; Feng, Y.; Fu, M.; Huang, Q.; Ma, L.; et al. Microbial and Metabolomic Remodeling by a Formula of Sichuan Dark Tea Improves Hyperlipidemia in ApoE-Deficient Mice. *PLoS ONE* **2019**, *14*. [CrossRef]
10. Xu, J.; Wang, W.; Du, M.; He, C.; Bian, J.; Du, X. A Comparative Analysis of Inhibitory Effect of Different Levels of Ya'an Tibetan Tea on Lipase. In Proceedings of the Journal of Physics: Conference Series, Montréal, Canada, 10–14 May 2020; Volume 1549.
11. Wang, N.; Wu, T.; Du, D.; Mei, J.; Luo, H.; Liu, Z.; Saleemi, M.K.; Zhang, R.; Chang, C.; Mehmood, M.A.; et al. Transcriptome and Gut Microbiota Profiling Revealed the Protective Effect of Tibetan Tea on Ulcerative Colitis in Mice. *Front. Microbiol.* **2022**, *12*, 165–176. [CrossRef]
12. Laghi, L.; Picone, G.; Capozzi, F. Nuclear Magnetic Resonance for Foodomics beyond Food Analysis. *TrAC Trends Anal. Chem.* **2014**, *59*, 93–102. [CrossRef]
13. Zhang, H.; Wang, J.; Zhang, D.; Zeng, L.; Liu, Y.; Zhu, W.; Lei, G.; Huang, Y. Aged Fragrance Formed during the Post-Fermentation Process of Dark Tea at an Industrial Scale. *Food Chem.* **2021**, *342*, 128175. [CrossRef]
14. Zhang, W.; Cao, J.; Li, Z.; Li, Q.; Lai, X.; Sun, L.; Chen, R.; Wen, S.; Sun, S.; Lai, Z. HS-SPME and GC/MS Volatile Component Analysis of Yinghong No. 9 Dark Tea during the Pile Fermentation Process. *Food Chem.* **2021**, 129654. [CrossRef]
15. Lee, L.S.; Choi, J.H.; Son, N.; Kim, S.H.; Park, J.D.; Jang, D.J.; Jeong, Y.; Kim, H.J. Metabolomic Analysis of the Effect of Shade Treatment on the Nutritional and Sensory Qualities of Green Tea. *J. Agric. Food Chem.* **2013**, *61*, 332–338. [CrossRef]
16. Yao, L.; Caffin, N.; D'Arcy, B.; Jiang, Y.; Shi, J.; Singanusong, R.; Liu, X.; Datta, N.; Kakuda, Y.; Xu, Y. Seasonal Variations of Phenolic Compounds in Australia-Grown Tea (*Camellia sinensis*). *J. Agric. Food Chem.* **2005**, *53*, 6477–6483. [CrossRef]
17. Lee, J.E.; Lee, B.J.; Chung, J.O.; Kim, H.N.; Kim, E.H.; Jung, S.; Lee, H.; Lee, S.J.; Hong, Y.S. Metabolomic Unveiling of a Diverse Range of Green Tea (*Camellia sinensis*) Metabolites Dependent on Geography. *Food Chem.* **2015**, *174*, 452–459. [CrossRef]
18. Ohno, A.; Oka, K.; Sakuma, C.; Okuda, H.; Fukuhara, K. Characterization of Tea Cultivated at Four Different Altitudes Using ^{1}H NMR Analysis Coupled with Multivariate Statistics. *J. Agric. Food Chem.* **2011**, *59*, 5181–5187. [CrossRef]
19. Zheng, Q.; Li, W.; Gao, X. The Effect of Storage Time on Tea Polyphenols, Catechin Compounds, Total Flavones and the Biological Activity of Ya'an Tibetan Tea (*Camellia sinensis*). *J. Food Process. Preserv.* **2021**, *45*, e16004. [CrossRef]
20. Xie, G.; Ye, M.; Wang, Y.; Ni, Y.; Su, M.; Huang, H.; Qiu, M.; Zhao, A.; Zheng, X.; Chen, T.; et al. Characterization of Pu-Erh Tea Using Chemical and Metabolic Profiling Approaches. *J. Agric. Food Chem.* **2009**, *57*, 3046–3054. [CrossRef]
21. Kneen, M.A.; Annegarn, H.J. Algorithm for Fitting XRF, SEM and PIXE X-ray Spectra Backgrounds. *Nucl. Instrum. Methods Phys. Res. Sect. B Beam Interact. Mater. At.* **1996**, *109–110*, 209–213. [CrossRef]
22. Liland, K.H.; Almøy, T.; Mevik, B.H. Optimal Choice of Baseline Correction for Multivariate Calibration of Spectra. *Appl. Spectrosc.* **2010**, *64*, 1007–1016. [CrossRef]
23. Dieterle, F.; Ross, A.; Schlotterbeck, G.; Senn, H. Probabilistic Quotient Normalization as Robust Method to Account for Dilution of Complex Biological Mixtures. Application In ^{1}H NMR Metabonomics. *Anal. Chem.* **2006**, *78*, 4281–4290. [CrossRef]
24. *R: A Language and Environment for Statistical Computing*; R Development Core Team R: Vienna, Austria, 2011; Volume 1.
25. Box, G.E.P.; Cox, D.R. An Analysis of Transformations. *J. R. Stat. Soc. Ser. B Methodol.* **2018**, *26*, 211–243. [CrossRef]
26. Zhu, C.; Laghi, L.; Zhang, Z.; He, Y.; Wu, D.; Zhang, H.; Huang, Y.; Li, C.; Zou, L. First Steps toward the Giant Panda Metabolome Database: Untargeted Metabolomics of Feces, Urine, Serum, and Saliva by ^{1}H-NMR. *J. Proteome Res.* **2020**, *19*, 1052–1059. [CrossRef]
27. Ye, Z.; Wang, X.; Fu, R.; Yan, H.; Han, S.; Gerelt, K.; Cui, P.; Chen, J.; Qi, K.; Zhou, Y. Determination of Six Groups of Mycotoxins in Chinese Dark Tea and the Associated Risk Assessment. *Environ. Pollut.* **2020**, *261*, 114180. [CrossRef]
28. Cao, L.; Guo, X.; Liu, G.; Song, Y.; Ho, C.T.; Hou, R.; Zhang, L.; Wan, X. A Comparative Analysis for the Volatile Compounds of Various Chinese Dark Teas Using Combinatory Metabolomics and Fungal Solid-State Fermentation. *J. Food Drug Anal.* **2018**, *26*, 112–123. [CrossRef]

29. Seetohul, L.N.; Islam, M.; O'Hare, W.T.; Ali, Z. Discrimination of Teas Based on Total Luminescence Spectroscopy and Pattern Recognition. *J. Sci. Food Agric.* **2006**, *86*, 2092–2098. [CrossRef]
30. Ding, Y.; Yu, H.; Mou, S. Direct Determination of Free Amino Acids and Sugars in Green Tea by Anion-Exchange Chromatography with Integrated Pulsed Amperometric Detection. *J. Chromatogr. A* **2002**, *982*, 237–244. [CrossRef]
31. Zhu, M.-z.; Li, N.; Zhou, F.; Ouyang, J.; Lu, D.-m.; Xu, W.; Li, J.; Lin, H.-y.; Zhang, Z.; Xiao, J.-b.; et al. Microbial Bioconversion of the Chemical Components in Dark Tea. *Food Chem.* **2020**, *312*, 126043. [CrossRef]
32. Zhang, L.; Cao, Q.Q.; Granato, D.; Xu, Y.Q.; Ho, C.T. Association between Chemistry and Taste of Tea: A Review. *Trends Food Sci. Technol.* **2020**, *101*, 139–149. [CrossRef]
33. Xue, J.; Liu, P.; Guo, G.; Wang, W.; Zhang, J.; Wang, W.; Le, T.; Yin, J.; Ni, D.; Jiang, H. Profiling of Dynamic Changes in Non-Volatile Metabolites of Shaken Black Tea during the Manufacturing Process Using Targeted and Non-Targeted Metabolomics Analysis. *LWT* **2022**, *156*, 113010. [CrossRef]
34. Guo, X.; Ho, C.T.; Schwab, W.; Song, C.; Wan, X. Aroma Compositions of Large-Leaf Yellow Tea and Potential Effect of Theanine on Volatile Formation in Tea. *Food Chem.* **2019**, *280*, 73–82. [CrossRef]
35. Thippeswamy, R.; Gouda, K.G.M.; Rao, D.H.; Martin, A.; Gowda, L.R. Determination of Theanine in Commercial Tea by Liquid Chromatography with Fluorescence and Diode Array Ultraviolet Detection. *J. Agric. Food Chem.* **2006**, *54*, 7014–7019. [CrossRef]
36. Yang, C.; Hu, Z.; Lu, M.; Li, P.; Tan, J.; Chen, M.; Lv, H.; Zhu, Y.; Zhang, Y.; Guo, L.; et al. Application of Metabolomics Profiling in the Analysis of Metabolites and Taste Quality in Different Subtypes of White Tea. *Food Res. Int.* **2018**, *106*, 909–919. [CrossRef]
37. Cheng, L.; Yang, Q.; Chen, Z.; Zhang, J.; Chen, Q.; Wang, Y.; Wei, X. Distinct Changes of Metabolic Profile and Sensory Quality during Qingzhuan Tea Processing Revealed by LC-MS-Based Metabolomics. *J. Agric. Food Chem.* **2020**, *68*, 4955–4965. [CrossRef]
38. Shimada, M.; Hasegawa, T.; Nishimura, C.; Kan, H.; Kanno, T.; Nakamura, T.; Matsubayashi, T. Anti-Hypertensive Effect of γ-Aminobutyric Acid (GABA)-Rich Chlorella on High-Normal Blood Pressure and Borderline Hypertension in Placebo-Controlled Double Blind Study. *Clin. Exp. Hypertens* **2009**, *31*, 342–354. [CrossRef]
39. Huang, J.; Fang, H.; Gai, Z.C.; Mei, J.Q.; Li, J.N.; Hu, S.; Lv, C.J.; Zhao, W.R.; Mei, L.H. Lactobacillus Brevis CGMCC 1306 Glutamate Decarboxylase: Crystal Structure and Functional Analysis. *Biochem. Biophys. Res. Commun.* **2018**, *503*. [CrossRef]
40. Jayabalan, R.; Marimuthu, S.; Swaminathan, K. Changes in Content of Organic Acids and Tea Polyphenols during Kombucha Tea Fermentation. *Food Chem.* **2007**, *102*, 392–398. [CrossRef]
41. Lv, H.P.; Zhong, Q.S.; Lin, Z.; Wang, L.; Tan, J.F.; Guo, L. Aroma Characterisation of Pu-Erh Tea Using Headspace-Solid Phase Microextraction Combined with GC/MS and GC-Olfactometry. *Food Chem.* **2012**, *130*, 1074–1081. [CrossRef]
42. Zhang, L.; Li, N.; Ma, Z.Z.; Tu, P.F. Comparison of the Chemical Constituents of Aged Pu-Erh Tea, Ripened Pu-Erh Tea, and Other Teas Using HPLC-DAD-ESI-MS n. *J. Agric. Food Chem.* **2011**, *59*, 8754–8760. [CrossRef]
43. Fernández, P.L.; Martin, M.J.; González, A.G.; Pablos, F. HPLC Determination of Catechins and Caffeine in Tea. Differentiation of Green, Black and Instant Teas. *Analyst* **2000**, *125*, 421–425. [CrossRef]
44. Togari, N.; Kobayashi, A.; Aishima, T. Pattern Recognition Applied to Gas Chromatographic Profiles of Volatile Components in Three Tea Categories. *Food Res. Int.* **1995**, *28*, 495–502. [CrossRef]
45. Gong, J.S.; Tang, C.; Peng, C.X. Characterization of the Chemical Differences between Solvent Extracts from Pu-Erh Tea and Dian Hong Black Tea by CP-Py-GC/MS. *J. Anal. Appl. Pyrolysis* **2012**, *95*, 189–197. [CrossRef]

foods

MDPI

Article

Characterization of Flavor Profile of "Nanx Wudl" Sour Meat Fermented from Goose and Pork Using Gas Chromatography–Ion Mobility Spectrometry (GC–IMS) Combined with Electronic Nose and Tongue

Xin Zhao [1], Jianying Feng [1], Luca Laghi [2], Jing Deng [3], Xiaofang Dao [1], Junni Tang [1], Lili Ji [4], Chenglin Zhu [1,*] and Gianfranco Picone [2]

1 College of Food Science and Technology, Southwest Minzu University, Chengdu 610041, China; zhaoxinhh@outlook.com (X.Z.); fengjianying@stu.swun.edu.cn (J.F.)
2 Department of Agricultural and Food Sciences, University of Bologna, 47521 Cesena, Italy
3 Cuisine Science Key Laboratory of Sichuan Province, Sichuan Tourism University, Chengdu 610100, China
4 Meat Processing Key Lab of Sichuan Province, Chengdu University, Chengdu 610106, China
* Correspondence: chenglin.zhu@swun.edu.cn; Tel.: +86-028-85928478

Abstract: Sour meat is a highly appreciated traditional fermented product, mainly from the Guizhou, Yunnan, and Hunan provinces. The flavor profiles of sour meat from goose and pork were evaluated using gas chromatography–ion mobility spectrometry (GC–IMS) combined with an electronic nose (E-nose) and tongue (E-tongue). A total of 94 volatile compounds were characterized in fermented sour meat from both pork and goose using GC–IMS. A data-mining protocol based on univariate and multivariate analyses revealed that the source of the raw meat plays a crucial role in the formation of flavor compounds during the fermentation process. In detail, sour meat from pork contained higher levels of hexyl acetate, sotolon, heptyl acetate, butyl propanoate, hexanal, and 2-acetylpyrrole than sour goose meat. In parallel, sour meat from goose showed higher levels of 4-methyl-3-penten-2-one, n-butyl lactate, 2-butanol, (E)-2-nonenal, and decalin than sour pork. In terms of the odor and taste response values obtained by the E-nose and E-tongue, a robust principal component model (RPCA) could effectively differentiate sour meat from the two sources. The present work could provide references to investigate the flavor profiles of traditional sour meat products fermented from different raw meats and offer opportunities for a rapid identification method based on flavor profiles.

Keywords: fermented meat; volatile compounds; GC–IMS; intelligent sensory; chemometrics

Citation: Zhao, X.; Feng, J.; Laghi, L.; Deng, J.; Dao, X.; Tang, J.; Ji, L.; Zhu, C.; Picone, G. Characterization of Flavor Profile of "Nanx Wudl" Sour Meat Fermented from Goose and Pork Using Gas Chromatography–Ion Mobility Spectrometry (GC–IMS) Combined with Electronic Nose and Tongue. *Foods* **2023**, *12*, 2194. https://doi.org/10.3390/foods12112194

Academic Editors: Oscar Núñez and Vladimir Tomović

Received: 3 April 2023
Revised: 18 May 2023
Accepted: 29 May 2023
Published: 30 May 2023

1. Introduction

Sour meat is a traditional fermented product, mainly from the Guizhou, Yunnan, and Hunan provinces, where it is known as Nanx Wudl [1]. Its manufacture is usually carried out at an artisanal level based on non-standardized production protocols, where salt, rice, and seasonings, such as Chinese prickly ash and chili, are added to the meat. Fermentation is mainly carried out by taking advantage of naturally occurring lactic acid bacteria (LAB) and requires approximately 1–2 months under anaerobic conditions [2]. The meat most often used is pork, but others may be employed, such as the more expensive goose, which plays a special role from the consumer's perspective because of its strong associations with the traditions of local Chinese communities [3]. From this point of view, it is worth noting that China accounts for as much as 70% of all the goose meat produced and consumed in the world [4].

Independent of the meat source in these fermented products, sour meat has attracted increasing attention outside the typical production areas— mainly due to its unique sensorial characteristics, but also due to its positive compositional traits, such as the richness in probiotics or the absence of nitrites. This brought about an increasing number of scientific works investigating various aspects of its production and characteristics. For instance, Lv et al. investigated the effect of fermentation temperature on the quality, bacterial community, and metabolites of sour meat. Their results showed that reduction in pH, thiobarbituric acid reactive substances (TBARS), and water content and an increase in lactic acid, free amino acids, and the number and amount of volatile compounds occurred as the fermentation temperature and time increased [5,6]. Lv et al. found that sour meat samples inoculated with *S. cerevisiae* LXPSC1 had better sensory characteristics than their naturally fermented counterparts, together with higher levels of pH, ethanol, free amino acids, and volatile organic compounds [7]. Zhang et al. found that the double-starter culture (*Lactobacillus curvatus* LAB26 and *Pediococcus pentosaceus* SWU73571) increased the L* and a* values, amino nitrogen content, and free amino acid content of sour meat significantly while also lowering the b* value; lowering the levels of nitrite, biogenic amines, total volatile basic nitrogen, and malondialdehyde; and restraining the coliform count [8]. Wang et al. found that low-salt fermentation can accelerate sourmeat maturation and facilitate the oxidation and decomposition of protein and fat and that it is more conducive to sour meat fermentation and to distinct fermented flavor production [9].

The main sensory characteristic that has been found to drive consumer preference and acceptance of fermented meat products is flavor [10]. Gas chromatography coupled with mass spectrometry (GC–MS) is considered the technique of choice for the qualitative and quantitative detection of volatile compounds in foods. In recent years, gas chromatography–ion mobility spectrometry (GC–IMS) has been increasingly used for flavor characterization in the food industry, because it can effectively distinguish the differences in flavor between products [11]. GC–IMS is an analytical technique that uses the difference in the migration rate of gas-phase ions in an electric field to characterize chemical substances [12]. It combines the excellent separation capacity of GC with the high sensitivity and fast response of IMS, granting a high accuracy of analysis [13]. To have a comprehensive view of the sensory characteristics of a food, it is ideal to couple this technique with an electronic nose and tongue. These devices are designed to mimic human olfactory and gustatory perception, respectively, without subjective judgements. They consist of a series of sensors designed to gain an overall fingerprint of the molecular profiles that give rise to complex odors and flavors [14]. Their application offers numerous advantages, among them rapidity of response, ease of use, reliability, and accuracy [15].

Despite the importance of aroma in determining consumer preferences for fermented meat foods and the potential of the E-nose, E-tongue, and GC–IMS for the purpose, few studies have evaluated their combination in this area. Moreover, most of the studies have considered products specifically manufactured at a laboratory scale or obtained at the retail level, which were industrially produced [16,17]. Finally, to the best of our knowledge, no studies have investigated the effects of different meat sources on the final product's flavor profile, even though it has been demonstrated that the action of fermenting microorganisms (particularly LAB), which leads to the final flavor profile, is deeply influenced by the starting raw material [18,19].

To fill these numerous gaps, we attempted, for the first time, to discriminate traditional sour meat based on goose and pork by means of an E-nose and E-tongue, and to obtain their flavor features through a metabolomics approach based on GC–IMS. This work could provide a framework for investigating the flavor profiles of traditional sour meats fermented from pork and goose through GC–IMS and could offer opportunities for a rapid identification method based on overall flavor characteristics by means of an E-nose and E-tongue.

2. Materials and Methods

2.1. Experimental Design

In accordance with the traditional artisanal procedure for producing sour meat (Nanx Wudl) in the Yunnan province, fresh pork (Large White breed) and goose meat (Chinese Goose breed) were cut into small pieces (around 3 cm × 5 cm × 0.6 cm and 200 g each), and then mixed with 3% salt and pickled in a refrigerator at 4 °C for 24 h. Next, 1% pepper was added, along with 7% glutinous rice, fried to golden yellow and ground into coarse-grained, and 1% glutinous rice, fried to golden yellow and steamed. Finally, the ingredients were placed in ten sealed containers (five for pork and five for goose) and spontaneously fermented at room temperature (approximately 15 °C) for 60 days.

2.2. Electronic Nose Analysis

A commercial E-nose (FOX 4000, Alpha MOS, Toulouse, France), equipped with an injection system, 18 sensor chambers, a mass flow controller, and an acquisition board with a microcontroller, was used to discriminate different sour meat samples. The main pieces of information granted by each sensor are shown in Figure 1. In order to fulfill the requirement of E-nose analysis, 0.25 g of sour meat samples was put into a 10 mL headspace bottle, and then the samples were incubated at 70 °C for 5 min and manually injected. The measurement and rinsing phases took 120 s and 240 s, respectively. The observation of each sample was repeated five times, and three stable sets of data were retained. The average value for each sample was included in an RPCA plot.

(a)

Sensors	Performance description
LY2/LG	Sensitive to oxidizing gas
LY2/G	Sensitive to ammonia, carbon monoxide
LY2/AA	Sensitive to ethanol
LY2/Gh	Sensitive to ammonia/organic amines
LY2/gCT1	Sensitive to hydrogen sulfide
LY2/gCT	Sensitive to propane/butane
T30/1	Sensitive to organic solvents
P10/1	Sensitive to hydrocarbons
P10/2	Sensitive to methane
P40/1	Sensitive to fluorine
T70/2	Sensitive to aromatic compounds
PA/2	Sensitive to ethanol, ammonia/organic amines
P30/1	Sensitive to polar compounds (ethanol)
P40/2	Sensitive to heteroatom/chloride/aldehydes
P30/2	Sensitive to alcohol
T40/2	Sensitive to aldehydes
T40/1	Sensitive to chlorinated compounds
TA/2	Sensitive to air quality

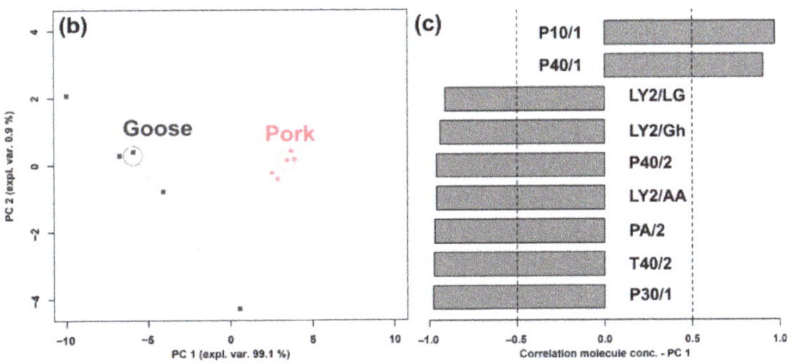

Figure 1. Performance description of the E-nose sensors (**a**); an RPCA model based on the E-nose response data, presented as a score plot (**b**); and a Pearson correlation plot of the loadings (**c**).

2.3. Electronic Tongue Analysis

E-tongue analysis was performed by the α-ASTREE (equipped with sixteen autosampler carousel positions, Alpha MOS, Toulouse, France), which provided seven sensors for sourness (AHS), saltiness (CTS), umami (NMS), sweetness (ANS), bitterness (SCS), and two reference electrodes (PKS and CPS) [20–22]. Sour meat samples (20.0 g) were mixed with 200 mL of deionized water to extract the taste substances. After the mixed solution was centrifuged 2265× *g* for 10 min at 4 °C, the water phase (100 mL) was obtained for E-tongue analysis. Each collection time was set to 120 s. The stirring rate was set to 60 rpm, and the cleaning time was 30 s. Deionized water was used as a cleaning solution. The average value measured between 100 and 120 s was taken as the output value. Following the suggestion of Li et al. [20], each sample was repeated eight times, and the data of the last 5 stable sets were selected as the original data for analysis. The average value for each sample was included in an RPCA plot.

2.4. GC–IMS Analysis

The volatile compounds in the sour meat samples were analyzed using GC–IMS (Flavorspec, G.A.S. Instrument, Munich, Germany) with an MXT-WAX capillary column (30 m × 0.53 mm × 1 μm) (Restek, Mount Ayr, USA). Without any sample pre-treatment, 0.25 g of meat samples was accurately weighed and put into a 20 mL headspace (HS) vial with a magnetic screw seal cover. Then, the samples were incubated at 50 °C for 10 min. After incubation, 100 μL of the headspace sample was automatically injected into the injector (splitless mode) via a heated syringe at 65 °C. The column was kept at 60 °C, with the drift tube temperature at 45 °C. The drift gas flow was set to a constant flow rate of 150 mL/min. Nitrogen carrier gas (99.999% purity) was used, and the GC column flow rate was programmed as follows: 2 mL/min for 5 min, 10 mL/min for 10 min, 15 mL/min for 5 min, 50 mL/min for 10 min, and 100 mL/min for 10 min. Following the suggestions of previous papers [23,24], the retention index (RI) of volatile compounds was calculated using n-ketones C4–C9 as external references. Volatile compounds were identified by comparing their RI and ions' drift time— that is, their migration time from the ionization source to the detector in the IMS chamber— with those of the standards in the GC–IMS library. In accordance with Guo et al. [25], the relevant calculation formula is as follows:

$$RI(x) = RI(x-1) + \frac{[RI(x+1) - RI(x-1)] \times [RT(x) - RT(x-1)]}{RT(x+1) - RT(x-1)}$$

$RT(x)$: The retention time of the substance/min;
$RT(n)$: The retention time of the n-ketones/min;
n: The number of carbon atoms in the n-ketones;
x: target to be carried out via qualitative and quantitative analysis;
$x-1$: The component peaking before target x;
$x+1$: The component peaking after target x.

Each sample was detected once, and the quantification of volatile compounds was based on the peak signal intensity. Using the Laboratory Analytical Viewer, Reporter, and Gallery Plot supported by the GC–IMS instrument, three-dimensional (3D) and two-dimensional (2D) topographic plots and gallery plots of the volatile compounds were constructed.

2.5. Statistical Analysis

Statistical analysis was performed in R computational language. Prior to the univariate analyses, the distribution of the data was brought to normality according to Box and Cox [26]. We used *t*-tests to look for significant differences between groups ($p < 0.05$).

Following the suggestions of previous studies [27,28], with the aim of obtaining an overall view of the data, robust principal component analysis (RPCA) models were set up based on the average values of the E-nose and E-tongue sensors and the molecules' peak signal intensities, respectively. For each RPCA model, a score plot and a Pearson correlation plot of the loadings were calculated, to highlight the structure of the data and to find out the relationships between variables and the model components.

3. Results

3.1. Electronic Nose Analysis

The E-nose analyzer, equipped with 18 sensors, was used to identify different sour meat formulations and to assess their comprehensive flavor characteristics. The main performance of the sensors is described in Figure 1a, as suggested by Wen et al. [29]. The response from nine of the sensors was found to be significantly different between the two groups ($p < 0.05$). To obtain an overview of the trends of these sensors, their response values were employed as a basis for an RPCA model, shown in Figure 1b,c.

Sour meat samples fermented from goose and pork could be clearly distinguished when observed along PC 1, with goose mainly characterized by higher response values from the sensors LY/LG, LY/Gh, P40/2, LY2/AA, PA/2, T40/2, and P30/1 and by lower response values from the sensors P10/1 and P40/1.

3.2. Electronic Tongue Analysis

The E-tongue analyzer, equipped with seven sensors, was used to identify different sour meat formulations and to assess their comprehensive taste characteristics. Six of the sensors gave a significantly different response when analyzing the two groups ($p < 0.05$). In order to obtain an overview of the trends of these sensors, their response values were employed as a basis for an RPCA model, shown in Figure 2.

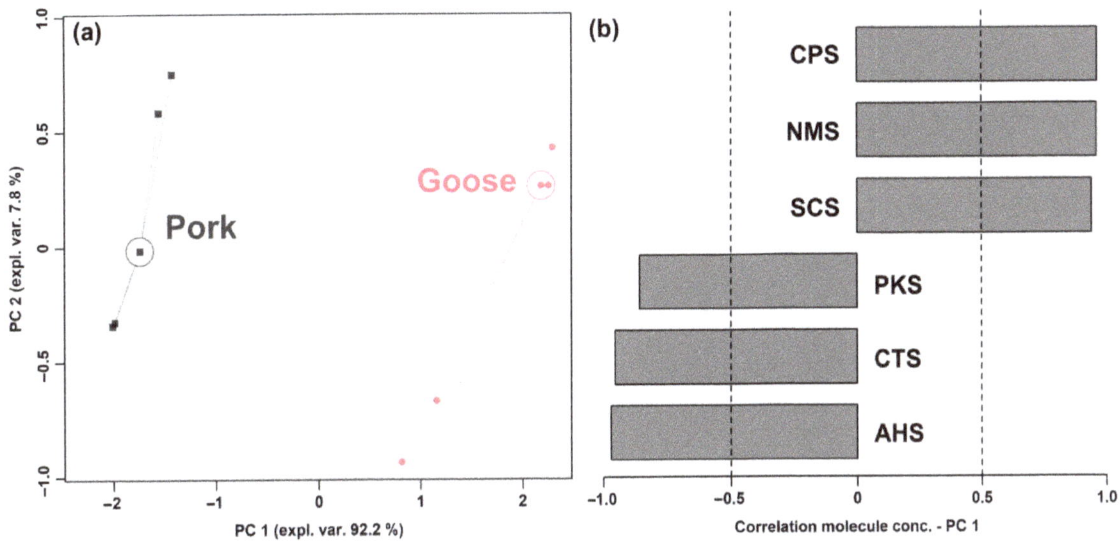

Figure 2. Score plot (**a**) and Pearson correlation plot (**b**) of the loadings of an RPCA model based on E-tongue response data.

The results showed that sour meat samples fermented from goose and pork were well distinguished along PC 1. Sour meat fermented from goose was mainly characterized by higher response values from the sensors CPS, NMS, and SCS and by lower response values from the sensors PKS, CTS, and AHS.

3.3. GC–IMS Analysis

The processing pipeline of the GC–IMS information about the volatile components in the samples from goose and pork meat is summarized by Figure 3.

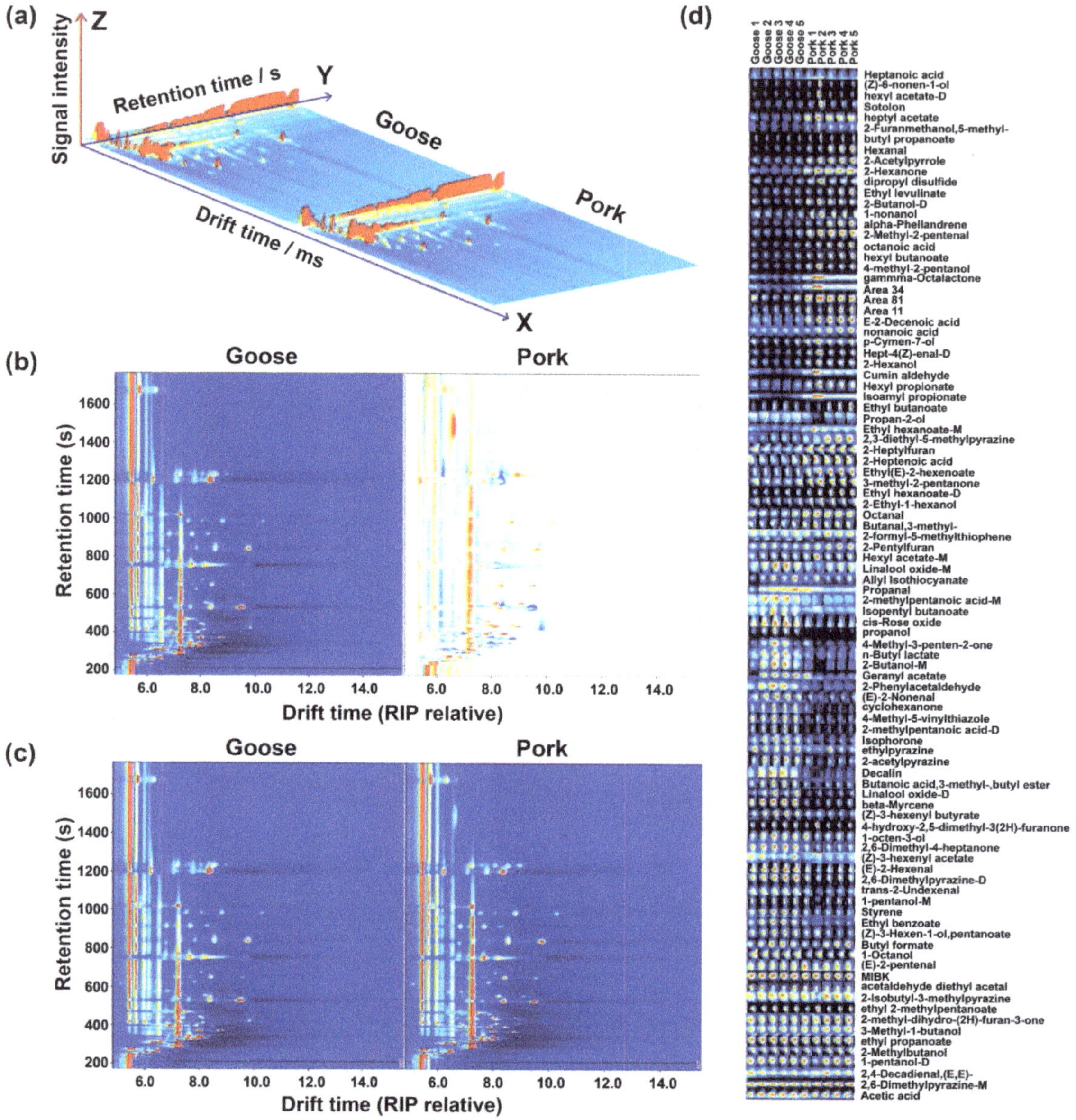

Figure 3. GC–IMS observations of sour meat fermented from goose and pork. (**a**) Their three-dimensional representation. (**b**) Their bird's-eye view representation, with spectra from goose meat employed as a reference and the corresponding spectra from pork represented as differences from goose meat. In the latter case, red and blue highlight components that were over- and under-expressed, respectively. (**c**) Their representation as ion migration spectra, where the ions are numbered and then listed in (**d**) as gallery plots, in which the color was brighter, the content was higher.

The 3D representation of Figure 3a offers an unbiased visual impression that the samples from the two meat sources differed along large portions of the GC–IMS spectrum. This allows us to establish that GC–IMS is a technique well suited for distinguishing fermented meat from the two studied sources. The point-by-point differences between the two sets of samples in Figure 3b allow us to appreciate in finer detail that most of the peculiarities regarded compounds with retention times between 200 and 1000 s, whose ions showed drift times between 6.0 and 10.0 ms. A total of 94 compounds were identified, including ketones (**10**), acids (**8**), alcohols (**19**), aldehydes (**15**), esters (**27**), and others (**10**). The relevant information about each of them is provided in Table 1.

The topographic plots of Figure 3d allow us to visually appreciate the trends distinguishing the two sets of samples, and they demonstrate that many of the volatile compounds distinguishing fermented goose meat from fermented pork were acids and alcohols. For example, hexyl acetate, sotolon, heptyl acetate, butyl propanoate, hexanal, and 2-acetylpyrrole appeared as more concentrated in sour meat from fermented pork, while levels of 4-methyl-3-penten-2-one, n-butyl lactate, 2-butanol, (E)-2-nonenal, and decalin were higher in the samples from goose meat.

In accordance with Zhu et al., to identify the molecules showing the highest differences between the two types of samples, a volcano plot was set up, which nicely combines the results of the *t*-test and fold-change analysis on a molecule-by-molecule basis [30]. Significantly different molecules with a fold change higher than 2 are shown in Figure 4a. To obtain an overview of the trends of these molecules, their signal intensities were employed as a basis for an RPCA model, as shown in Figure 4b,c.

In the score plot of Figure 4b, the PC 1 accounted for as much as 96.4% of the samples' overall variability and perfectly summarized the differences between goose and pork samples, with negative and positive PC 1 scores, respectively. The Pearson correlation plot of the loadings of Figure 4c shows that sour meat fermented from goose has higher amounts of β-myrcene, 2-butanol, 2-methylpentanoic acid, isophorone, decaline, n-butyl lactate, 4-methyl-5-vinylthiazole, and cyclohexanone and lower concentrations of 2-hexanone, hexanal, α-phellandrene, 2-methyl-2-pentenal, γ-octanoic lactone, isoamyl propionate, and cuminaldehyde.

Table 1. Molecules' peak areas (mean ± sd) characterized by GC–IMS in sour meat fermented from both goose and pork.

Count	Compound	CAS	Molecule Formula	MW *	RI	RT (s)	DT (ms)	Goose	Pork	p-Value
Ketones	2,6-Dimethyl-4-heptanone	C108838	$C_9H_{18}O$	142.2	1210.4	756.074	1.77916	$1.15 \times 10^3 \pm 43.80$	$7.19 \times 10^2 \pm 1.13 \times 10^2$	0.008
	2-Hexanone	C591786	$C_6H_{12}O$	100.2	1093.1	436.042	1.49671	$7.90 \times 10^2 \pm 66.90$	$2.23 \times 10^3 \pm 2.82 \times 10^2$	0.000
	4-Hydroxy-2,5-dimethyl-3(2 H)-furanone	C3658773	$C_6H_8O_3$	128.1	1053.3	371.81	1.61084	$3.05 \times 10^3 \pm 3.90 \times 10^2$	$2.29 \times 10^3 \pm 1.15 \times 10^3$	0.139
	4-Methyl-3-penten-2-one	C141797	$C_6H_{10}O$	98.1	1106.1	462.228	1.44482	$7.86 \times 10^2 \pm 4.19 \times 10^2$	$3.88 \times 10^2 \pm 1.79 \times 10^2$	0.093
	Isophorone	C78591	$C_9H_{14}O$	138.2	1115	480.134	1.25323	$1.01 \times 10^3 \pm 1.40 \times 10^2$	$4.97 \times 10^2 \pm 63.20$	0.000
	2-Methyl-dihydro-(2 H)-Furan-3-one	C3188009	$C_5H_8O_2$	100.1	1280.5	1019.37	1.06662	$1.74 \times 10^3 \pm 2.62 \times 10^2$	$2.07 \times 10^3 \pm 93.20$	0.056
	MIBK	C108101	$C_6H_{12}O$	100.2	1021.3	338.292	1.47542	$2.36 \times 10^2 \pm 19.10$	$3.54 \times 10^2 \pm 50.70$	0.001
	3-Methyl-2-pentanone	C565617	$C_6H_{12}O$	100.2	1056.3	374.988	1.47533	$3.54 \times 10^3 \pm 7.26 \times 10^2$	$4.16 \times 10^3 \pm 2.62 \times 10^2$	0.155
	2-Acetylpyrrole	C1072839	C_6H_7NO	109.1	1082.5	414.813	1.49823	$8.98 \times 10^2 \pm 2.03 \times 10^2$	$1.78 \times 10^3 \pm 2.21 \times 10^2$	0.003
	2-Acetylpyrazine	C22047252	$C_6H_6N_2O$	122.1	1017.1	333.894	1.20618	$8.62 \times 10^2 \pm 1.40 \times 10^2$	$4.40 \times 10^2 \pm 85.10$	0.004
Acids	Acetic acid	C64197	$C_2H_4O_2$	60.1	1447.3	1673.557	1.05277	$1.73 \times 10^4 \pm 1.07 \times 10^3$	$1.52 \times 10^4 \pm 1.77 \times 10^3$	0.101
	Octanoic acid	C124072	$C_8H_{16}O_2$	144.2	1174.4	628.473	1.44089	$9.98 \times 10^2 \pm 1.52 \times 10^2$	$1.87 \times 10^3 \pm 3.04 \times 10^2$	0.002
	2-Methylpentanoic acid (D)	C97610	$C_6H_{12}O_2$	116.2	1029.9	347.315	1.59258	$2.77 \times 10^2 \pm 37.50$	$1.15 \times 10^2 \pm 19.90$	0.000
	2-Methylpentanoic acid (M)	C97610	$C_6H_{12}O_2$	116.2	1028.9	346.226	1.26389	$4.00 \times 10^2 \pm 46.80$	$2.57 \times 10^2 \pm 24.30$	0.001
	Heptanoic acid	C111148	$C_7H_{14}O_2$	130.2	1082.4	414.564	1.3611	$3.66 \times 10^2 \pm 12.40$	$4.37 \times 10^2 \pm 54.40$	0.013
	Nonanoic acid	C112050	$C_9H_{18}O_2$	158.2	1279.5	1015.116	1.54967	$1.50 \times 10^3 \pm 2.57 \times 10^2$	$2.40 \times 10^3 \pm 1.65 \times 10^2$	0.010
	E-2-Decenoic acid	C334496	$C_{10}H_{18}O_2$	170.3	1334.9	1232.645	1.48495	$2.12 \times 10^2 \pm 40.80$	$3.27 \times 10^2 \pm 97.00$	0.054
	2-Heptenoic acid	C18999285	$C_7H_{12}O_2$	128.2	1207	744.053	1.40605	$1.26 \times 10^3 \pm 2.56 \times 10^2$	$3.10 \times 10^3 \pm 2.42 \times 10^2$	0.001
Alcohols	1-Octen-3-ol	C3391864	$C_8H_{16}O$	128.2	1447.5	1674.263	1.1622	$4.63 \times 10^3 \pm 6.10 \times 10^2$	$3.62 \times 10^3 \pm 9.00 \times 10^2$	0.096
	1-Pentanol (D)	C71410	$C_5H_{12}O$	88.1	1253.8	914.653	1.51902	$1.83 \times 10^3 \pm 1.34 \times 10^2$	$1.52 \times 10^3 \pm 4.16 \times 10^2$	0.257
	1-Pentanol (M)	C71410	$C_5H_{12}O$	88.1	1254.1	915.518	1.25162	$1.61 \times 10^3 \pm 1.27 \times 10^2$	$1.33 \times 10^3 \pm 3.01 \times 10^2$	0.199
	2-Hexanol	C626937	$C_6H_{14}O$	102.2	1235.1	843.655	1.56671	$4.13 \times 10^3 \pm 1.04 \times 10^3$	$7.17 \times 10^3 \pm 9.86 \times 10^2$	0.016
	3-Methyl-1-butanol	C123513	$C_5H_{12}O$	88.1	1207.6	746.268	1.24551	$1.15 \times 10^3 \pm 40.60$	$1.07 \times 10^3 \pm 1.03 \times 10^2$	0.235
	2-Methylbutanol	C137326	$C_5H_{12}O$	88.1	1214.5	770.435	1.24551	$2.27 \times 10^3 \pm 1.71 \times 10^2$	$1.97 \times 10^3 \pm 1.23 \times 10^2$	0.036
	p-Cymen-7-ol	C536607	$C_{10}H_{14}O$	150.2	1281.2	1021.821	1.32996	$5.51 \times 10^3 \pm 1.15 \times 10^3$	$1.03 \times 10^4 \pm 1.82 \times 10^3$	0.006
	(Z)-6-nonen-1-ol	C35854865	$C_9H_{18}O$	142.2	1181.1	652.231	1.75376	$8.75 \times 10^2 \pm 30.50$	$1.59 \times 10^3 \pm 9.61 \times 10^2$	0.016
	4-Methyl-2-pentanol	C108112	$C_6H_{14}O$	102.2	1180.5	650.067	1.54657	$2.02 \times 10^3 \pm 5.14 \times 10^2$	$4.15 \times 10^3 \pm 1.43 \times 10^3$	0.009
	1-Nonanol	C143088	$C_9H_{20}O$	144.3	1144.7	540.011	1.53779	$2.82 \times 10^3 \pm 8.05 \times 10^2$	$3.71 \times 10^3 \pm 9.29 \times 10^2$	0.105

Table 1. *Cont.*

Count	Compound	CAS	Molecule Formula	MW *	RI	RT (s)	DT (ms)	Goose	Pork	*p*-Value
Alcohols	Linalool oxide (D)	C60047178	$C_{10}H_{18}O_2$	170.3	1087.7	425.155	1.80714	$2.61 \times 10^3 \pm 7.70 \times 10^2$	$1.27 \times 10^3 \pm 3.23 \times 10^2$	0.016
	Linalool oxide (M)	C60047178	$C_{10}H_{18}O_2$	170.3	1082.4	414.503	1.25444	$4.16 \times 10^2 \pm 64.60$	$3.2 \times 610^2 \pm 68.00$	0.137
	2-Ethyl-1-hexanol	C104767	$C_8H_{18}O$	130.2	1030.4	347.834	1.4048	$3.83 \times 10^3 \pm 6.13 \times 10^2$	$4.42 \times 10^3 \pm 4.86 \times 10^2$	0.129
	2-Butanol (D)	C78922	$C_4H_{10}O$	74.1	1016.3	333.005	1.33429	$1.41 \times 10^3 \pm 3.90 \times 10^2$	$2.71 \times 10^3 \pm 3.06 \times 10^2$	0.010
	2-Butanol (M)	C78922	$C_4H_{10}O$	74.1	1016	332.708	1.14461	$6.46 \times 10^2 \pm 1.90 \times 10^2$	$2.32 \times 10^2 \pm 85.80$	0.007
	Propanol	C71238	C_3H_8O	60.1	976.8	295.222	1.24209	$6.75 \times 10^2 \pm 2.14 \times 10^2$	$1.54 \times 10^2 \pm 37.60$	0.000
	Propan-2-ol	C67630	C_3H_8O	60.1	940.1	275.184	1.08085	$7.06 \times 10^2 \pm 60.40$	$5.53 \times 10^2 \pm 1.74 \times 10^2$	0.207
	1-Octanol	C111875	$C_8H_{18}O$	130.2	1030.2	347.582	1.46611	$7.99 \times 10^3 \pm 3.00 \times 10^2$	$7.04 \times 10^3 \pm 4.65 \times 10^2$	0.037
	2-Furanmethanol, 5-methyl-	C3857258	$C_6H_8O_2$	112.1	949.8	280.488	1.26173	$1.93 \times 10^2 \pm 46.40$	$9.20 \times 10^1 \pm 9.12$	0.001
Aldehydes	Cumin aldehyde	C122032	$C_{10}H_{12}O$	148.2	1240.6	862.903	1.33504	$3.60 \times 10^3 \pm 1.00 \times 10^3$	$8.59 \times 10^3 \pm 2.16 \times 10^3$	0.005
	trans-2-Undecenal	C53448070	$C_{11}H_{20}O$	168.3	1335.1	1233.494	1.56546	$3.86 \times 10^3 \pm 2.27 \times 10^2$	$2.22 \times 10^3 \pm 5.03 \times 10^2$	0.017
	2,4-Decadienal,(E,E)-	C25152845	$C_{10}H_{16}O$	152.2	1315.6	1156.984	1.41449	$3.79 \times 10^2 \pm 16.70$	$4.62 \times 10^2 \pm 51.90$	0.005
	(Z)-4-Heptenal	C6728310	$C_7H_{12}O$	112.2	1272	985.747	1.62114	$2.59 \times 10^3 \pm 4.75 \times 10^2$	$4.85 \times 10^3 \pm 1.84 \times 10^3$	0.010
	2-Hexenal	C6728263	$C_6H_{10}O$	98.1	1210.7	757.253	1.50243	$1.28 \times 10^4 \pm 8.16 \times 10^2$	$7.70 \times 10^3 \pm 1.07 \times 10^3$	0.006
	2-Methyl-2-pentenal	C623369	$C_6H_{10}O$	98.1	1153.1	556.885	1.5028	$1.03 \times 10^3 \pm 2.27 \times 10^2$	$2.55 \times 10^3 \pm 2.73 \times 10^2$	0.001
	Hexanal	C66251	$C_6H_{12}O$	100.2	1073.1	395.761	1.55606	$5.58 \times 10^2 \pm 89.50$	$2.27 \times 10^3 \pm 9.09 \times 10^2$	0.001
	(E)-2-Nonenal	C18829566	$C_9H_{16}O$	140.2	1144.7	540.011	1.40541	$1.37 \times 10^3 \pm 1.13 \times 10^2$	$7.37 \times 10^2 \pm 58.30$	0.000
	(E)-2-Pentenal	C1576870	C_5H_8O	84.1	1094.4	438.764	1.36432	$1.01 \times 10^3 \pm 73.90$	$1.05 \times 10^3 \pm 1.34 \times 10^2$	0.975
	2-Formyl-5-methylthiophene	C13679704	C_6H_6OS	126.2	1108.5	467.078	1.17289	$3.40 \times 10^2 \pm 46.20$	$3.20 \times 10^2 \pm 26.50$	0.838
	Octanal	C124130	$C_8H_{16}O$	128.2	1006.9	323.217	1.39288	$1.13 \times 10^3 \pm 2.03 \times 10^2$	$1.54 \times 10^3 \pm 75.20$	0.033
	Butanal, 3-methyl-	C590863	$C_5H_{10}O$	86.1	909.8	258.669	1.19285	$4.51 \times 10^2 \pm 38.00$	$5.86 \times 10^2 \pm 52.20$	0.008
	Acetaldehyde diethyl acetal	C105577	$C_6H_{14}O_2$	118.2	863.4	233.347	1.13106	$5.32 \times 10^2 \pm 90.20$	$4.71 \times 10^2 \pm 21.70$	0.563
	Propanal	C123386	C_3H_6O	58.1	842	221.677	1.03933	$5.49 \times 10^2 \pm 1.50 \times 10^2$	$4.89 \times 10^2 \pm 1.33 \times 10^2$	0.863
	2-Phenylacetaldehyde	C122781	C_8H_8O	120.2	1013.7	330.292	1.25765	$1.95 \times 10^3 \pm 3.40 \times 10^2$	$1.87 \times 10^3 \pm 56.70$	0.710
Esters	Isoamyl propionate	C105680	$C_8H_{16}O_2$	144.2	1190.1	683.977	1.33833	$5.35 \times 10^3 \pm 2.37 \times 10^3$	$1.57 \times 10^4 \pm 3.81 \times 10^3$	0.005
	Ethyl-(E)-2-hexenoate	C27829727	$C_8H_{14}O_2$	142.2	1334.8	1232.212	1.32568	$3.78 \times 10^3 \pm 2.68 \times 10^2$	$6.16 \times 10^3 \pm 5.68 \times 10^2$	0.001
	Hexyl propionate	C2445763	$C_9H_{18}O_2$	158.2	1329.3	1210.43	1.43	$4.53 \times 10^3 \pm 6.92 \times 10^2$	$6.47 \times 10^3 \pm 6.47 \times 10^2$	0.021
	(Z)-3-hexenyl acetate	C3681718	$C_8H_{14}O_2$	142.2	1315.4	1155.965	1.30104	$3.60 \times 10^2 \pm 23.20$	$2.92 \times 10^2 \pm 23.80$	0.015
	Hexyl acetate (D)	C142927	$C_8H_{16}O_2$	144.2	1271.7	984.728	1.88721	$1.25 \times 10^3 \pm 91.60$	$2.58 \times 10^3 \pm 2.26 \times 10^3$	0.044
	(Z)-3-Hexen-1-ol, Pentanoate	C35852461	$C_{11}H_{20}O_2$	184.3	1272.2	986.766	1.48608	$1.53 \times 10^3 \pm 28.50$	$1.26 \times 10^3 \pm 3.50 \times 10^2$	0.122
	Ethyl hexanoate (D)	C123660	$C_8H_{16}O_2$	144.2	1234.6	841.923	1.79323	$6.16 \times 10^3 \pm 3.86 \times 10^2$	$9.92 \times 10^3 \pm 1.89 \times 10^3$	0.002
	Hexyl acetate (M)	C142927	$C_8H_{16}O_2$	144.2	1271.9	985.651	1.38787	$1.63 \times 10^3 \pm 85.90$	$1.89 \times 10^3 \pm 4.85 \times 10^2$	0.383

Table 1. *Cont.*

Count	Compound	CAS	Molecule Formula	MW *	RI	RT (s)	DT (ms)	Goose	Pork	p-Value
Eesters	Ethyl benzoate	C93890	$C_9H_{10}O_2$	150.2	1160.1	577.835	1.26386	$9.73 \times 10^2 \pm 1.61 \times 10^2$	$5.00 \times 10^2 \pm 68.00$	0.001
	γ-Octalactone	C104507	$C_8H_{14}O_2$	142.2	1255.6	921.364	1.33563	$4.68 \times 10^3 \pm 9.39 \times 10^2$	$1.06 \times 10^4 \pm 2.47 \times 10^3$	0.002
	Ethyl hexanoate (M)	C123660	$C_8H_{16}O_2$	144.2	1234.7	842.104	1.34131	$3.84 \times 10^3 \pm 4.45 \times 10^2$	$6.48 \times 10^3 \pm 1.12 \times 10^3$	0.002
	Hexyl butanoate	C2639636	$C_{10}H_{20}O_2$	172.3	1188.4	678.191	1.48375	$9.06 \times 10^2 \pm 87.10$	$1.39 \times 10^3 \pm 1.86 \times 10^2$	0.003
	(Z)-3-hexenyl butyrate	C16491364	$C_{10}H_{18}O_2$	170.3	1181.4	653.414	1.41322	$9.61 \times 10^2 \pm 58.10$	$7.61 \times 10^2 \pm 1.07 \times 10^2$	0.038
	Ethyl 2-methylpentanoate	C39255328	$C_8H_{16}O_2$	144.2	1141.7	534.023	1.74779	$1.12 \times 10^4 \pm 1.16 \times 10^2$	$9.35 \times 10^3 \pm 3.20 \times 10^3$	0.137
	Sotolon	C28664359	$C_6H_8O_3$	128.1	1097.4	444.752	1.61692	$2.79 \times 10^2 \pm 24.50$	$5.70 \times 10^2 \pm 2.84 \times 10^2$	0.007
	Ethyl levulinate	C539888	$C_7H_{12}O_3$	144.2	1082	413.724	1.64736	$4.97 \times 10^2 \pm 55.40$	$7.92 \times 10^2 \pm 2.66 \times 10^2$	0.032
	Butanoic acid, 3-methyl-, Butyl ester	C109193	$C_9H_{18}O_2$	158.2	1074.2	397.938	1.38714	$1.22 \times 10^3 \pm 84.70$	$8.69 \times 10^2 \pm 91.10$	0.006
	Isopentyl butanoate	C106274	$C_9H_{18}O_2$	158.2	1060.1	378.974	1.40381	$5.11 \times 10^2 \pm 45.20$	$3.71 \times 10^2 \pm 57.00$	0.017
	Butyl formate	C592847	$C_5H_{10}O_2$	102.1	1014	330.632	1.5051	$5.57 \times 10^2 \pm 77.30$	$4.87 \times 10^2 \pm 1.33 \times 10^2$	0.282
	Ethyl propanoate	C105373	$C_5H_{10}O_2$	102.1	1011.2	327.666	1.44949	$6.71 \times 10^3 \pm 2.54 \times 10^2$	$6.11 \times 10^3 \pm 2.84 \times 10^2$	0.021
	Butyl propanoate	C590012	$C_7H_{14}O_2$	130.2	908.2	257.789	1.27492	$5.02 \times 10^2 \pm 69.60$	$8.46 \times 10^2 \pm 1.97 \times 10^2$	0.005
	Ethyl acetate	C141786	$C_4H_8O_2$	88.1	943.6	277.07	1.33409	$3.49 \times 10^2 \pm 26.40$	$4.93 \times 10^2 \pm 59.90$	0.003
	Ethyl butanoate	C105544	$C_6H_{12}O_2$	116.2	1014.5	331.173	1.5593	$1.10 \times 10^3 \pm 1.46 \times 10^2$	$7.65 \times 10^2 \pm 1.45 \times 10^2$	0.020
	n-Butyl lactate	C34451199	$C_7H_{14}O_3$	146.2	1018.6	335.453	1.26872	$2.65 \times 10^2 \pm 51.10$	$1.33 \times 10^2 \pm 7.96$	0.000
	Heptyl acetate	C112061	$C_9H_{18}O_2$	158.2	1073.5	396.604	1.448	$5.63 \times 10^3 \pm 1.87 \times 10^3$	$4.62 \times 10^3 \pm 2.04 \times 10^3$	0.811
	Geranyl acetate	C105873	$C_{12}H_{20}O_2$	196.3	1400	1487.894	1.21911	$4.14 \times 10^2 \pm 59.60$	$1.83 \times 10^2 \pm 32.80$	0.000
	cis-Rose oxide	C3033236	$C_{10}H_{18}O$	154.3	1111.6	473.354	1.38248	$3.18 \times 10^3 \pm 3.65 \times 10^2$	$1.68 \times 10^3 \pm 2.53 \times 10^2$	0.002
Others	Dipropyl disulfide	C629196	$C_6H_{14}S_2$	150.3	1098.5	446.929	1.47388	$1.23 \times 10^3 \pm 2.11 \times 10^2$	$2.25 \times 10^3 \pm 6.31 \times 10^2$	0.006
	2,6-Dimethylpyrazine (D)	C108509	$C_6H_8N_2$	108.1	1327.6	1204.024	1.53744	$1.76 \times 10^4 \pm 7.42 \times 10^2$	$1.03 \times 10^4 \pm 1.37 \times 10^3$	0.004
	2,6-Dimethylpyrazine (M)	C108509	$C_6H_8N_2$	108.1	1331.2	1218.118	1.14507	$3.30 \times 10^3 \pm 1.67 \times 10^2$	$3.09 \times 10^3 \pm 1.49 \times 10^2$	0.059
	α-Phellandrene	C99832	$C_{10}H_{16}$	136.2	1152	554.708	1.67323	$4.86 \times 10^2 \pm 25.00$	$1.34 \times 10^3 \pm 4.68 \times 10^2$	0.000
	4-Methyl-5-vinylthiazole	C1759280	C_6H_7NS	125.2	1029.9	347.315	1.53323	$4.62 \times 10^2 \pm 32.60$	$1.99 \times 10^2 \pm 37.40$	0.001
	Decalin	C91178	$C_{10}H_{18}$	138.3	1063.2	382.237	1.23101	$5.61 \times 10^2 \pm 1.44 \times 10^2$	$1.82 \times 10^2 \pm 15.10$	0.000
	β-Myrcene	C123353	$C_{10}H_{16}$	136.2	977.3	295.442	1.28554	$2.45 \times 10^3 \pm 2.71 \times 10^2$	$1.02 \times 10^3 \pm 2.30 \times 10^2$	0.002
	Ethylpyrazine	C13925003	$C_6H_8N_2$	108.1	933.6	271.661	1.12623	$4.13 \times 10^2 \pm 81.50$	$3.21 \times 10^2 \pm 52.40$	0.051
	2,3-Diethyl-5-Methylpyrazine	C18138040	$C_9H_{14}N_2$	150.2	1154	558.675	1.27281	$9.39 \times 10^2 \pm 69.70$	$8.97 \times 10^2 \pm 65.40$	0.364
	2-Isobutyl-3-Methylpyrazine	C13925069	$C_9H_{14}N_2$	150.2	1142.5	535.449	1.30033	$1.78 \times 10^2 \pm 44.70$	$1.96 \times 10^2 \pm 93.00$	0.863
	2-Heptylfuran	C3777717	$C_{11}H_{18}O$	166.3	1212.6	763.852	1.40704	$3.13 \times 10^3 \pm 6.22 \times 10^2$	$4.40 \times 10^3 \pm 2.19 \times 10^2$	0.035
	2-Pentylfuran	C3777693	$C_9H_{14}O$	138.2	1230.8	828.288	1.25257	$8.35 \times 10^2 \pm 77.00$	$8.67 \times 10^2 \pm 63.50$	0.383

* Notes: MW—molecular mass; RI—retention index; Rt—retention time; Dt—drift time. *p*-value was calculated by *t*-test, and the cutoff value was set as below 0.05.

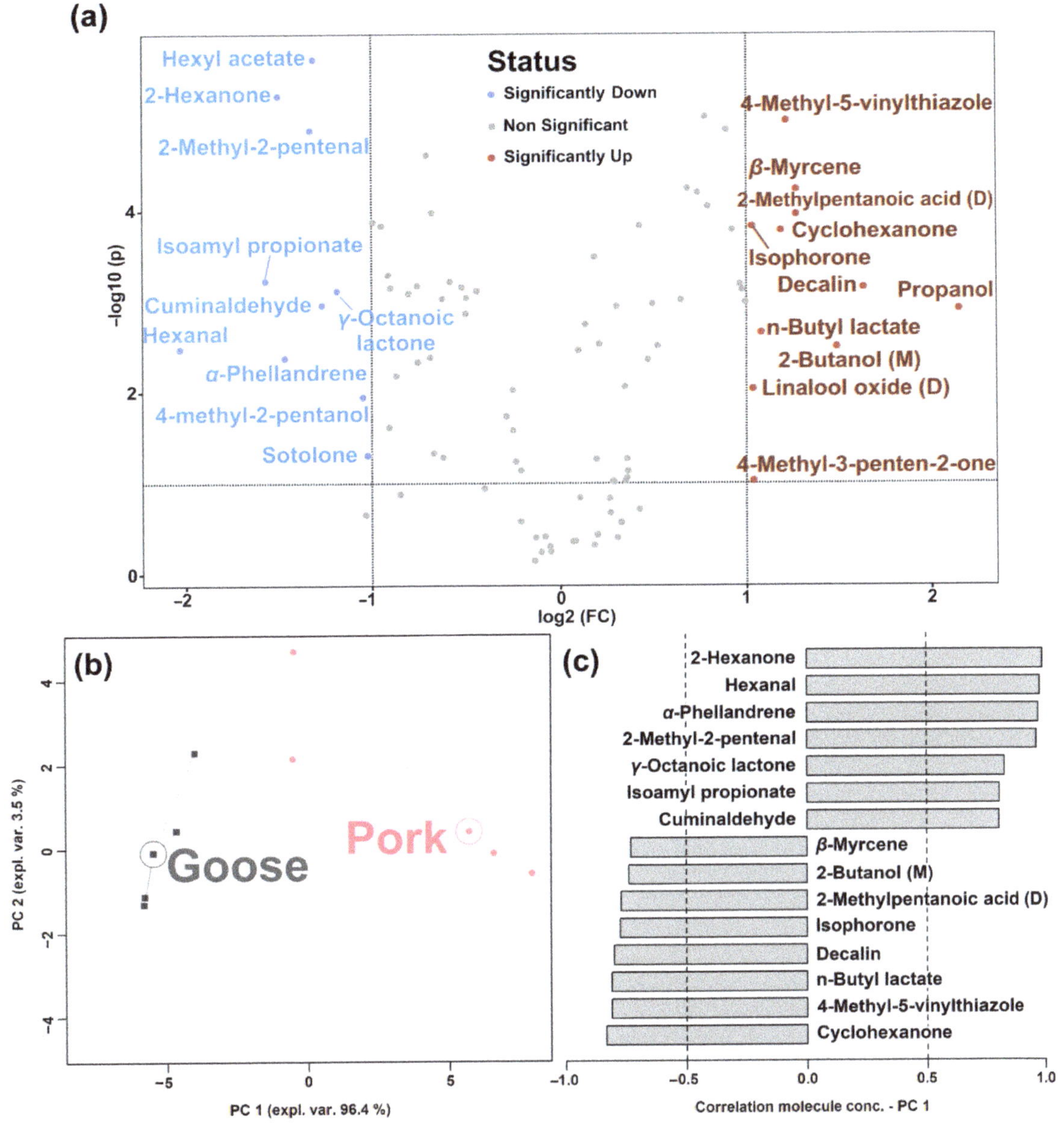

Figure 4. Volcano plot (**a**) indicating the changes in the concentrations of metabolites in sour meat samples from the two groups. The RPCA model was set up on the basis of the molecules selected by the volcano plot. In the score plot (**b**), squares and circles indicate goose and pork samples, respectively. The median of each sample group is indicated by wide, empty circles. The Pearson correlation plot of the loadings (**c**) shows the molecules with significant correlations between concentration and importance over PC 1 ($p < 0.05$).

3.4. Correlation between E-Nose and GC–IMS

The E-nose and GC–IMS can classify sour meat fermented from pork and goose meat from different points of view. For example, E-nose was able to provide overall information on the volatile compounds in each sample. In contrast, GC–IMS could provide the specific volatile profile of each sour meat. Therefore, in order to promote the overall performance of both techniques, the potential correlation between E-nose sensor responses and volatile compound levels detected by GC–IMS was analyzed, as shown in Figure 5.

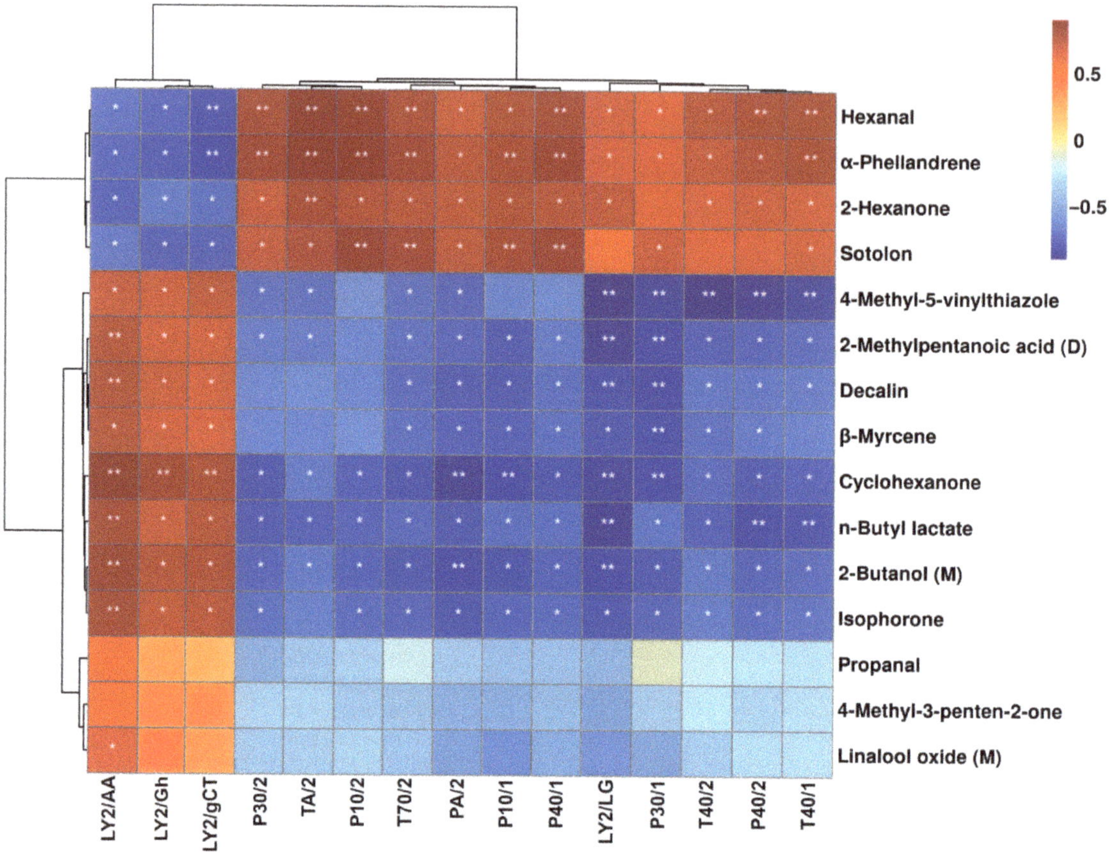

Figure 5. Spearman's correlation heatmap showing the correlation between the significantly altered volatile compound levels and the electronic nose sensor responses. Colors represent correlation coefficients, with red and blue indicating positive and negative correlation, respectively. Asterisks * and ** stand for significance at $p < 0.05$ and $p < 0.01$, respectively.

As shown in Figure 5, the LY2/AA and LY2/Gh sensors were found to positively correlate with hexanal, α-phellandrene, and 2-hexanone. These compounds were characterized by GC–IMS at high levels in the sour meat fermented from pork. In contrast, the PA/2, P10/1, P40/1, LY2/LG, P30/1, T40/2, and P40/2 sensors were found to positively correlate with 2-methylpentanoic acid (D), decaline, β-myrcene, cyclohexanone, n-butyl lactate, 2-butanol (M), and isophorone. These compounds were characterized by GC–IMS at high levels in the sour meat fermented from goose meat. These results indicate that the E-nose sensor response values and volatile compound quantifications characterized through GC–IMS can discriminate the unique flavors of sour meat fermented from pork versus goose meat.

4. Discussion

Sour meat (Nanx Wudl) is a fermented meat product traditionally manufactured by minorities (the Dong, the Miao, the Dai, the Tujia, the Maonan, etc.) but increasingly appreciated in wider parts of China due to its unique flavor, richness in nutrients, and long shelf life [2,16,31]. However, studies evaluating the effects of different raw meats on the formation of the typical complex flavor during fermentation are still lacking. To shed light on the issue, the present study attempted, for the first time, to comprehensively characterize the flavor profiles of sour meat from goose and pork by means of GC–IMS combined with an electronic nose and tongue—a perfect combination for the purpose, but still rarely employed.

Nine of the eighteen sensors on the E-nose analyzer showed a significantly different response between the samples from goose meat and pork, evidencing that this tool is extremely sensitive to peculiarities in the overall flavor profiles connected to the raw materials for fermented meat. From this point of view, it is worth underlining that this tool is not able to identify the specific volatile compounds giving rise to the overall response; a specific high-throughput technique tailored for the purpose, such as GC–IMS, should be used in parallel to obtain fine-grained information, at least in the first stages of investigation. Six of the seven sensors on the E-tongue showed a significantly different response to the two tested products, demonstrating that, in this context, this technique gives interesting complimentary information from the point of view of taste attributes.

A total of 94 compounds was characterized in each of the tested samples by GC–IMS, by comparing their RI and ion drift time to the standards in the GC–IMS library and references [32–35], pertaining to ketones (**10**), acids (**8**), alcohols (**19**), aldehydes (**15**), esters (**27**), and others (**10**). Among them, twenty-one compounds exhibited significant differences between the two types of samples, namely hexyl acetate, 2-hexanone, 2-methyl-2-pentenal, isoamyl propionate, cuminaldehyde, hexanal, γ-octanoic lactone, α-phellandrene, 4-methyl-2-pentanol, sotolon, 4-methyl-5-vinylthiazole, β-myrcene, 2-methylpentanoic acid (D), cyclohexanone, isophorone, decaline, propanol, n-butyl lactate, 2-butanol (M), linalool oxide (D), and 4-methyl-3-penten-2-one.

2-Butanol is a flavor-enriching substance with a sweet and pleasant scent [36]. The compound is derived from the reduction of 2-butanone. Moreover, it can also derive from pyruvate, similar to 1-propanol [37]. In turn, pyruvate generation could be attributed to amino acid metabolism, active in lactic acid bacteria, particularly for aspartate [38]. The distinct content of 2-butanol in sour meat fermented from goose meat and pork could be explained, at least in part, by the different amino acid profiles of goose meat versus pork [39,40].

Hexanal and 2-methyl-2-pentenal, belonging to the chemical class of aldehydes, have been found in beef meat [41] and chicken meat [42] and are considered to be useful markers of lipid oxidation [43,44]. Choi et al. found, by studying plant substrates during drying, that hexanal and 2-methyl-2-pentenal are produced from the action of residual enzymes so that, especially at low drying temperatures, their concentration is proportional to the residual humidity of the sample [45]. From these studies, it could be inferred that the different contents of hexanal and 2-methyl-2-pentenal in sour meat fermented from goose meat versus pork could be linked to different activity of residual enzymes, in turn leading to variable extents of lipid oxidation in the final product [46].

2-Hexanone, like many ketones, has an unpleasant, pungent odor [47]. By observing silver carp during chill storage, Jia et al. found that 2-hexanone was initially absent and increased gradually with storage time, showing a good correlation with microbial growth [48]. Similar to 2-hexanone, 2-methyl-2-pentenal, 2-methylpentanoic acid, and 4-methyl-3-penten-2-one were found to be linked to microbiota metabolism, though they were mainly linked to lipid oxidation, carbohydrate fermentation, and amino acid degradation [49,50].

Hexyl acetate is an ester with a pleasant fruity scent, which can usually be found in meat products. Li et al. found that hexanal was first reduced to 1-hexanol by *Lb. fermentum* and subsequently converted into hexyl acetate by *P. kluyveri* during pork fermentation [51]. Similarly, Jiang et al. found that there were positive correlations between the levels of *Leuconostoc* and *Lactobacillus* versus hexyl acetate in smoked horse meat sausages [52]. Zhang et al. found that hexyl acetate was a crucial positive contributor to the flavor profile of unsmoked bacon [53].

Overall, the trends we observed of the above-mentioned molecules seem to confirm that microorganisms' role in determining flavor profile is determined by the raw materials, particularly the meat. The trends of other molecules seem to confirm this observation. Linalool oxide can contribute to woody and floral aromas [54]. Sotolon is formed through the aldol condensation of α-ketobutyric acid, produced by threonine and acetaldehyde, which in turn result from the oxidation of ethanol generated, by the glucose metabolism of yeasts during fermentation, among other processes. Confirming this, Ohata et al. used 10% commercial koji and 10% salts to ferment a pork meat sauce for 12 months, and they found that the main odor contributors in their fermented meat sauce were sotolon and ethyl furaneol, which gave the meat sauce a sweet and caramel-like note [55]. Other compounds such as α-phellandrene, β-myrcene, and cuminaldehyde can be considered plant-derived [56–58]. The presence of these compounds in the final product is most likely due to the spices added to fresh meat, particularly pepper.

In the present study, it is worth noting that the combination of the E-nose, E-tongue, and GC–IMS could improve the overall performance of all techniques and provide a comprehensive characterization of sour meat fermented from pork and goose meat. In particular, the correlation between E-nose response values and GC–IMS molecule peak areas highlighted that lipid-oxidation-related compounds (such as hexanal and 2-hexanone) played important roles in discriminating sour meat fermented from goose meat and from pork. Moreover, a few of the E-nose sensors (such as PA/2, P10/1, P40/1, LY2/LG, T40/2, and P40/2), which exhibited significantly higher response values for the above compounds, could be considered potential candidates for developing targeted analysis methods by means of an E-nose for practical sample analysis.

5. Conclusions

In this study, for the first time, the flavor features of sour meat traditionally fermented from goose and pork were systematically characterized by means of GC–IMS, an E-nose, and an E-tongue. Taking advantage of a protocol based on univariate and multivariate analyses, we found that the raw material played a crucial role in introducing peculiarities to large portions of the flavor profile, so that sour meat from goose was readily distinguishable from that based on pork. Though the E-nose and E-tongue are able to grant an overall view of the odor and taste features of samples, it is still necessary to apply high-throughput techniques in parallel, to spot the specific volatile compounds conveying the overall response. Notably, none of the tested analyses required sample preparation, which implies that these operations are simple, fast, and nondestructive. Therefore, the present work could provide a basis for investigating the flavor profiles of traditional sour meats fermented from different meat sources and shed light on establishing more comprehensive and rapid methods for identifying flavor characteristics.

Author Contributions: Conceptualization, C.Z.; methodology, C.Z. and J.D.; formal analysis, C.Z., X.Z. and J.F.; investigation, C.Z. and X.Z.; resources, X.D.; writing—original draft preparation, C.Z., X.Z. and L.L.; writing—review and editing, X.Z., J.F., L.L., J.D., X.D., J.T., L.J., C.Z. and G.P.; supervision, C.Z.; funding acquisition, C.Z. All authors have read and agreed to the published version of the manuscript.

Funding: This work was funded by the Natural Science Foundation of Sichuan Province (2022NSFSC1652), the Fundamental Research Funds for the Central Universities, Southwest Minzu University (ZYN2022112), Cuisine Science Key Laboratory of Sichuan Province, Sichuan Tourism University (PRKX2022Z03), and Open Fund of Meat Processing Key Lab of Sichuan Province (22-R-21).

Data Availability Statement: The data presented in this study are available on request from the corresponding author.

Conflicts of Interest: The authors declare no conflict of interest.

References

1. Zhong, A.; Chen, W.; Hu, L.; Wu, Z.; Xiao, Y.; Li, K.; Li, Z.; Wang, Y.; Wang, C. Characterisation of key volatile compounds in fermented sour meat after fungi growth inhibition. *LWT* **2022**, *165*, 113662. [CrossRef]
2. Chen, C.; Chen, X.; Jiang, M.; Rui, X.; Li, W.; Dong, M. A newly discovered bacteriocin from Weissella hellenica D1501 associated with Chinese Dong fermented meat (Nanx Wudl). *Food Control* **2014**, *42*, 116–124. [CrossRef]
3. Ying, W.; Ya-Ting, J.; Jin-Xuan, C.; Yin-Ji, C.; Yang-Ying, S.; Xiao-Qun, Z.; Dao-Dong, P.; Chang-Rong, O.; Ning, G. Study on lipolysis-oxidation and volatile flavour compounds of dry-cured goose with different curing salt content during production. *Food Chem.* **2016**, *190*, 33–40. [CrossRef] [PubMed]
4. Wołoszyn, J.; Wereńska, M.; Goluch, Z.; Haraf, G.; Okruszek, A.; Teleszko, M.; Król, B. The selected goose meat quality traits in relation to various types of heat treatment. *Poult. Sci.* **2020**, *99*, 7214–7224. [CrossRef] [PubMed]
5. Lv, J.; Li, C.; Li, S.; Liang, H.; Ji, C.; Zhu, B.; Lin, X. Effects of temperature on microbial succession and quality of sour meat during fermentation. *LWT* **2019**, *114*, 108391. [CrossRef]
6. Lv, J.; Yang, Z.; Xu, W.; Li, S.; Liang, H.; Ji, C.; Yu, C.; Zhu, B.; Lin, X. Relationships between bacterial community and metabolites of sour meat at different temperature during the fermentation. *Int. J. Food Microbiol.* **2019**, *307*, 108286. [CrossRef]
7. Lv, J.; Lin, X.; Liu, M.; Yan, X.; Liang, H.; Ji, C.; Li, S.; Zhang, S.; Chen, Y.; Zhu, B. Effect of Saccharomyces cerevisiae LXPSC1 on microorganisms and metabolites of sour meat during the fermentation. *Food Chem.* **2023**, *402*, 134213. [CrossRef]
8. Zhang, Y.; Hu, P.; Xie, Y.; Wang, X. Co-fermentation with Lactobacillus curvatus LAB26 and Pediococcus pentosaceus SWU73571 for improving quality and safety of sour meat. *Meat Sci.* **2020**, *170*, 108240. [CrossRef]
9. Wang, Q.; Li, X.; Xue, B.; Wu, Y.; Song, H.; Luo, Z.; Shang, P.; Liu, Z.; Huang, Q. Low-salt fermentation improves flavor and quality of sour meat: Microbiology and metabolomics. *LWT* **2022**, *171*, 114157. [CrossRef]
10. Khan, M.I.; Jo, C.; Tariq, M.R. Meat flavor precursors and factors influencing flavor precursors—A systematic review. *Meat Sci.* **2015**, *110*, 278–284. [CrossRef]
11. Wang, Z.; Mi, S.; Wang, X.; Mao, K.; Liu, Y.; Gao, J.; Sang, Y. Characterization and discrimination of fermented sweet melon juice by different microbial strains via GC–IMS-based volatile profiling and chemometrics. *Food Sci. Hum. Wellness* **2023**, *12*, 1241–1247. [CrossRef]
12. Guo, Y.; Chen, D.; Dong, Y.; Ju, H.; Wu, C.; Lin, S. Characteristic volatiles fingerprints and changes of volatile compounds in fresh and dried Tricholoma matsutake Singer by HS-GC–IMS and HS-SPME-GC–MS. *J. Chromatogr. B Anal. Technol. Biomed. Life Sci.* **2018**, *1099*, 46–55. [CrossRef]
13. Yuan, Z.-Y.; Li, J.; Zhou, X.-J.; Wu, M.-H.; Li, L.; Pei, G.; Chen, N.-H.; Liu, K.-L.; Xie, M.-Z.; Huang, H.-Y. HS-GC–IMS-Based metabonomics study of Baihe Jizihuang Tang in a rat model of chronic unpredictable mild stress. *J. Chromatogr. B Anal. Technol. Biomed. Life Sci.* **2020**, *1148*, 122143. [CrossRef] [PubMed]
14. Chen, Q.; Hu, Y.; Wen, R.; Wang, Y.; Qin, L.; Kong, B. Characterisation of the flavour profile of dry fermented sausages with different NaCl substitutes using HS-SPME-GC-MS combined with electronic nose and electronic tongue. *Meat Sci.* **2021**, *172*, 108338. [CrossRef] [PubMed]
15. Zhu, D.; Ren, X.; Wei, L.; Cao, X.; Ge, Y.; Liu, H.; Li, J. Collaborative analysis on difference of apple fruits flavour using electronic nose and electronic tongue. *Sci. Hortic.* **2020**, *260*, 108879. [CrossRef]
16. Lv, J.; Xu, W.; Ji, C.; Liang, H.; Li, S.; Yang, Z.; Zhang, S.; Lin, X. Relationships between the bacterial diversity and metabolites of a Chinese fermented pork product, sour meat. *Int. J. Food Sci. Technol.* **2021**, *56*, 2742–2750. [CrossRef]
17. Zhong, A.; Chen, W.; Duan, Y.; Li, K.; Tang, X.; Tian, X.; Wu, Z.; Li, Z.; Wang, Y.; Wang, C. The potential correlation between microbial communities and flavors in traditional fermented sour meat. *LWT* **2021**, *149*, 111873. [CrossRef]
18. Kononiuk, A.D.; Karwowska, M. Comparison of selected parameters related to food safety of fallow deer and beef uncured fermented sausages with freeze-dried acid whey addition. *Meat Sci.* **2020**, *161*, 108015. [CrossRef]
19. Settanni, L.; Barbaccia, P.; Bonanno, A.; Ponte, M.; Di Gerlando, R.; Franciosi, E.; Di Grigoli, A.; Gaglio, R. Evolution of indigenous starter microorganisms and physicochemical parameters in spontaneously fermented beef, horse, wild boar and pork salamis produced under controlled conditions. *Food Microbiol.* **2020**, *87*, 103385. [CrossRef] [PubMed]
20. Li, X.; Yang, Y.; Zhu, Y.; Ben, A.; Qi, J. A novel strategy for discriminating different cultivation and screening odor and taste flavor compounds in Xinhui tangerine peel using E-nose, E-tongue, and chemometrics. *Food Chem.* **2022**, *384*, 132519. [CrossRef] [PubMed]
21. Yang, C.; Duan, W.; Xie, K.; Ren, C.; Zhu, C.; Chen, K.; Zhang, B. Effect of salicylic acid treatment on sensory quality, flavor-related chemicals and gene expression in peach fruit after cold storage. *Postharvest Biol. Technol.* **2020**, *161*, 111089. [CrossRef]
22. Han, D.; Zhang, C.-H.; Fauconnier, M.-L.; Jia, W.; Wang, J.-F.; Hu, F.-F.; Xie, D.-W. Characterization and comparison of flavor compounds in stewed pork with different processing methods. *LWT* **2021**, *144*, 111229. [CrossRef]
23. Wang, D.; Zhang, J.; Zhu, Z.; Lei, Y.; Huang, S.; Huang, M. Effect of ageing time on the flavour compounds in Nanjing water-boiled salted duck detected by HS-GC–IMS. *LWT* **2022**, *155*, 112870. [CrossRef]

24. Liu, H.; Yu, Y.; Zou, B.; Yu, Y.; Yang, J.; Xu, Y.; Chen, X.; Yang, F. Evaluation of Dynamic Changes and Regularity of Volatile Flavor Compounds for Different Green Plum (*Prunus mume* Sieb. et Zucc) Varieties during the Ripening Process by HS-GC–IMS with PLS-DA. *Foods* **2023**, *12*, 551. [CrossRef]
25. Guo, S.; Zhao, X.; Ma, Y.; Wang, Y.; Wang, D. Fingerprints and changes analysis of volatile compounds in fresh-cut yam during yellowing process by using HS-GC–IMS. *Food Chem.* **2022**, *369*, 130939. [CrossRef] [PubMed]
26. Box, G.E.P.; Cox, D.R. An Analysis of Transformations. *J. R. Stat. Soc. Ser. B Methodol.* **2018**, *26*, 211–243. [CrossRef]
27. Montanari, C.; Gatto, V.; Torriani, S.; Barbieri, F.; Bargossi, E.; Lanciotti, R.; Grazia, L.; Magnani, R.; Tabanelli, G.; Gardini, F. Effects of the diameter on physico-chemical, microbiological and volatile profile in dry fermented sausages produced with two different starter cultures. *Food Biosci.* **2018**, *22*, 9–18. [CrossRef]
28. Karpiński, P.; Kruszewski, B.; Stachelska, M.A.; Szabłowska, E. Development of volatile profile of Kumpiak podlaski dry-cured ham during traditional ripening. *Int. J. Food Sci. Technol.* **2020**, *55*, 3630–3638. [CrossRef]
29. Wen, R.; Kong, B.; Yin, X.; Zhang, H.; Chen, Q. Characterisation of flavour profile of beef jerky inoculated with different autochthonous lactic acid bacteria using electronic nose and gas chromatography–ion mobility spectrometry. *Meat Sci.* **2022**, *183*, 108658. [CrossRef]
30. Zhu, C.; Yang, Z.; He, L.; Lu, X.; Tang, J.; Laghi, L. The Longer the Storage Time, the Higher the Price, the Better the Quality? A ^{1}H-NMR Based Metabolomic Investigation of Aged Ya'an Tibetan Tea (*Camellia sinensis*). *Foods* **2022**, *11*, 2986. [CrossRef]
31. Sidira, M.; Kandylis, P.; Kanellaki, M.; Kourkoutas, Y. Effect of immobilized Lactobacillus casei on the evolution of flavor compounds in probiotic dry-fermented sausages during ripening. *Meat Sci.* **2015**, *100*, 41–51. [CrossRef] [PubMed]
32. Li, X.; Dong, Y.; Jiang, P.; Qi, L.; Lin, S. Identification of changes in volatile compounds in sea cucumber Apostichopus japonicus during seasonings soaking using HS-GC–IMS. *LWT* **2022**, *154*, 112695. [CrossRef]
33. Zhang, Q.; Ding, Y.; Gu, S.; Zhu, S.; Zhou, X.; Ding, Y. Identification of changes in volatile compounds in dry-cured fish during storage using HS-GC–IMS. *Food Res. Int.* **2020**, *137*, 109339. [CrossRef]
34. Wan, J.; Liu, Q.; Ma, C.; Muhoza, B.; Huang, Y.; Sun, M.; Song, S.; Ho, C.-T. Characteristic flavor fingerprint disclosure of dzo beef in Tibet by applying SAFE-GC-O-MS and HS-GC–IMS technology. *Food Res. Int.* **2023**, *166*, 112581. [CrossRef] [PubMed]
35. Nie, S.; Li, L.; Wang, Y.; Wu, Y.; Li, C.; Chen, S.; Zhao, Y.; Wang, D.; Xiang, H.; Wei, Y. Discrimination and characterization of volatile organic compound fingerprints during sea bass (*Lateolabrax japonicas*) fermentation by combining GC–IMS and GC-MS. *Food Biosci.* **2022**, *50*, 102048. [CrossRef]
36. Kim, K.-W.; Kim, H.-J.; Lee, E.-D.; Kim, D.-K.; Lee, J.; Lee, S.-S.; Jang, A.; Lee, S.-H. Comparison of Meat Quality Characteristics and Aromatic Substances of Korean Native Black Goat Ribs by Different Sex. *J. Food Nutr. Res.* **2020**, *8*, 585–590. [CrossRef]
37. Montanari, C.; Bargossi, E.; Gardini, A.; Lanciotti, R.; Magnani, R.; Gardini, F.; Tabanelli, G. Correlation between volatile profiles of Italian fermented sausages and their size and starter culture. *Food Chem.* **2016**, *192*, 736–744. [CrossRef]
38. Le Bars, D.; Yvon, M. Formation of diacetyl and acetoin by Lactococcus lactis via aspartate catabolism. *J. Appl. Microbiol.* **2008**, *104*, 171–177. [CrossRef]
39. Moya, V.-J.; Flores, M.; Aristoy, M.-C.; Toldrá, F. Pork meat quality affects peptide and amino acid profiles during the ageing process. *Meat Sci.* **2001**, *58*, 197–206. [CrossRef]
40. Geldenhuys, G.; Hoffman, L.C.; Muller, N. The fatty acid, amino acid, and mineral composition of Egyptian goose meat as affected by season, gender, and portion. *Poult. Sci.* **2015**, *94*, 1075–1087. [CrossRef]
41. Yim, D.-G.; Kim, H.J.; Kim, S.-S.; Lee, H.J.; Kim, J.-K.; Jo, C. Effects of different X-ray irradiation doses on quality traits and metabolites of marinated ground beef during storage. *Radiat. Phys. Chem.* **2023**, *202*, 110563. [CrossRef]
42. Feng, X.; Ahn, D.U. Volatile profile, lipid oxidation and protein oxidation of irradiated ready-to-eat cured turkey meat products. *Radiat. Phys. Chem.* **2016**, *127*, 27–33. [CrossRef]
43. Jo, Y.; An, K.-A.; Arshad, M.S.; Kwon, J.-H. Effects of e-beam irradiation on amino acids, fatty acids, and volatiles of smoked duck meat during storage. *Innov. Food Sci. Emerg. Technol.* **2018**, *47*, 101–109. [CrossRef]
44. Bak, K.H.; Rankin, S.A.; Richards, M.P. Hexanal as a marker of oxidation flavour in sliced and uncured deli turkey with and without phosphates using rosemary extracts. *Int. J. Food Sci. Technol.* **2020**, *55*, 3104–3110. [CrossRef]
45. Choi, S.M.; Lee, D.-J.; Kim, J.-Y.; Lim, S.-T. Volatile composition and sensory characteristics of onion powders prepared by convective drying. *Food Chem.* **2017**, *231*, 386–392. [CrossRef]
46. Xu, Y.; Shui, M.; Chen, D.; Ma, X.; Feng, T. Optimization of Jinhua Ham Classification Method Based on Volatile Flavor Substances and Determination of Key Odor Biomarkers. *Molecules* **2022**, *27*, 7087. [CrossRef]
47. Wu, W.; Wang, X.; Hu, P.; Zhang, Y.; Li, J.; Jiang, J.; Zheng, R.; Zhang, L. Research on flavor characteristics of beef cooked in tomato sour soup by gas chromatography–ion mobility spectrometry and electronic nose. *LWT* **2023**, *179*, 114646. [CrossRef]
48. Jia, S.; Li, Y.; Zhuang, S.; Sun, X.; Zhang, L.; Shi, J.; Hong, H.; Luo, Y. Biochemical changes induced by dominant bacteria in chill-stored silver carp (*Hypophthalmichthys molitrix*) and GC–IMS identification of volatile organic compounds. *Food Microbiol.* **2019**, *84*, 103248. [CrossRef]
49. Ozturk, I.; Sagdic, O.; Yetim, H. Effects of Autochthonous Yeast Cultures on Some Quality Characteristics of Traditional Turkish Fermented Sausage "Sucuk". *Food Sci. Anim. Resour.* **2021**, *41*, 196–213. [CrossRef]
50. Qiu, Y.; Wu, Y.; Li, L.; Chen, S.; Zhao, Y.; Li, C.; Xiang, H.; Wang, D.; Wei, Y.; Wang, Y. Elucidating the mechanism underlying volatile and non-volatile compound development related to microbial amino acid metabolism during golden pomfret (*Trachinotus ovatus*) fermentation. *Food Res. Int.* **2022**, *162*, 112095. [CrossRef]

51. Li, X.; Liu, S.-Q. Effect of co-inoculation and sequential inoculation of Lactobacillus fermentum and Pichia kluyveri on pork hydrolysates fermentation. *Food Biosci.* **2021**, *44*, 101400. [CrossRef]
52. Jiang, L.; Chen, Y.; Deng, L.; Liu, F.; Wang, T.; Shi, X.; Wang, B. Bacterial community diversity and its potential contributions to the flavor components of traditional smoked horsemeat sausage in Xinjiang, China. *Front. Microbiol.* **2022**, *13*, 942932. [CrossRef] [PubMed]
53. Zhang, J.; Zhang, W.; Zhou, L.; Zhang, R. Study on the influences of ultrasound on the flavor profile of unsmoked bacon and its underlying metabolic mechanism by using HS-GC–IMS. *Ultrason. Sonochem.* **2021**, *80*, 105807. [CrossRef] [PubMed]
54. Jeong, I.-S.; Park, G.-W.; Kim, Y.-S.; Park, K.-H.; Hwangbo, H.; Lee, S.-Y.; Kim, S.-G. Effects of a Novel Convection Heated Roasting Method on Aroma and Umami Taste Components of Vegetable- and Meat-Based Extracts. *J. Food Qual.* **2022**, *2022*, 7690931. [CrossRef]
55. Ohata, M.; Zhou, L.; Higuchi, K.; Nagai, T.; Kasamatsu, H.; Arihara, K. Investigation of volatile components and identification of the most potent odour-active component in fermented meat sauce. *Flavour Fragr. J.* **2017**, *32*, 171–177. [CrossRef]
56. Soncu, E.D.; Özdemir, N.; Arslan, B.; Küçükkaya, S.; Soyer, A. Contribution of surface application of chitosan–thyme and chitosan–rosemary essential oils to the volatile composition, microbial profile, and physicochemical and sensory quality of dry-fermented sausages during storage. *Meat Sci.* **2020**, *166*, 108127. [CrossRef]
57. Shao, Y.; Liu, X.; Zhang, Z.; Wang, P.; Li, K.; Li, C. Comparison and discrimination of the terpenoids in 48 species of huajiao according to variety and geographical origin by E-nose coupled with HS-SPME-GC-MS. *Food Res. Int.* **2023**, *167*, 112629. [CrossRef] [PubMed]
58. Thangaleela, S.; Sivamaruthi, B.S.; Kesika, P.; Tiyajamorn, T.; Bharathi, M.; Chaiyasut, C. A Narrative Review on the Bioactivity and Health Benefits of Alpha-Phellandrene. *Sci. Pharm.* **2022**, *90*, 57. [CrossRef]

foods

MDPI

Article

An Integrative Glycomic Approach for Quantitative Meat Species Profiling

Sean Chia [1,†], Gavin Teo [1,†], Shi Jie Tay [1], Larry Sai Weng Loo [2,3], Corrine Wan [1], Lyn Chiin Sim [1], Hanry Yu [2,3,4,5], Ian Walsh [1] and Kuin Tian Pang [1,*]

1 Bioprocessing Technology Institute, Agency for Science Technology and Research (A*STAR), Singapore 138668, Singapore; sean_chia@bti.a-star.edu.sg (S.C.); gavin_teo@bti.a-star.edu.sg (G.T.); tay_shi_jie@bti.a-star.edu.sg (S.J.T.); corrine_wan@bti.a-star.edu.sg (C.W.); sim_lyn_chiin@bti.a-star.edu.sg (L.C.S.); ian_walsh@bti.a-star.edu.sg (I.W.)
2 Institute of Bioengineering and Bioimaging, Agency for Science Technology and Research (A*STAR), Singapore 138669, Singapore; larry_loo@ibb.a-star.edu.sg (L.S.W.L.); hyu@ibb.a-star.edu.sg (H.Y.)
3 Department of Physiology, the Institute for Digital Medicine (WisDM), Yong Loo Lin School of Medicine, National University of Singapore, Singapore 117593, Singapore
4 Mechanobiology Institute, National University of Singapore, Singapore 117411, Singapore
5 CAMP, Singapore-MIT Alliance for Research and Technology, Singapore 138602, Singapore
* Correspondence: zach_pang@bti.a-star.edu.sg
† These authors contributed equally to this work.

Abstract: It is estimated that food fraud, where meat from different species is deceitfully labelled or contaminated, has cost the global food industry around USD 6.2 to USD 40 billion annually. To overcome this problem, novel and robust quantitative methods are needed to accurately characterise and profile meat samples. In this study, we use a glycomic approach for the profiling of meat from different species. This involves an O-glycan analysis using LC-MS qTOF, and an N-glycan analysis using a high-resolution non-targeted ultra-performance liquid chromatography-fluorescence-mass spectrometry (UPLC-FLR-MS) on chicken, pork, and beef meat samples. Our integrated glycomic approach reveals the distinct glycan profile of chicken, pork, and beef samples; glycosylation attributes such as fucosylation, sialylation, galactosylation, high mannose, α-galactose, Neu5Gc, and Neu5Ac are significantly different between meat from different species. The multi-attribute data consisting of the abundance of each O-glycan and N-glycan structure allows a clear separation between meat from different species through principal component analysis. Altogether, we have successfully demonstrated the use of a glycomics-based workflow to extract multi-attribute data from O-glycan and N-glycan analysis for meat profiling. This established glycoanalytical methodology could be extended to other high-value biotechnology industries for product authentication.

Keywords: O-glycan; N-glycan; glycomic; meat species

Citation: Chia, S.; Teo, G.; Tay, S.J.; Loo, L.S.W.; Wan, C.; Sim, L.C.; Yu, H.; Walsh, I.; Pang, K.T. An Integrative Glycomic Approach for Quantitative Meat Species Profiling. *Foods* **2022**, *11*, 1952. https://doi.org/10.3390/foods11131952

Academic Editors: Gianfranco Picone and Nazimah Hamid

Received: 18 May 2022
Accepted: 24 June 2022
Published: 30 June 2022

1. Introduction

With the growing human population and the increasing demand for food, food adulteration has become a global problem estimated to affect 10–20% of all food consumed in the world [1,2]. Such contamination by either additions or substitutions of meat from a different species is a significant dietary issue, particularly for individuals with allergies or those of a certain religious conviction [3,4]. It is thus prudent to develop techniques in authenticating meat products as a means of ensuring safe trade and ethics [2,3].

Currently, many methods have been developed for the means of food fraud detection, including microscopic, spectroscopic (NMR, FTIR), and DNA-based techniques (PCR) [2,5–7]. Indeed, amidst these techniques, biomarkers identification by means of omics technology allows such quantification and distinction at a molecular level [8,9]. In fact, the significant popularity and application of these technologies in resolving food compositions in general at a high resolution has developed an entire field of "foodomics" to identify molecular traits pertaining to the production and processing of these complex mixtures [10,11].

Interestingly, within the available omics technologies in the analysis of meat from different species, the use of glycomics remains relatively unknown. Considering the multiple diverse glycan structures that can arise as a result of the complex protein glycosylation pathways, it is likely that significant structural differences in glycan compositions can be detected between samples derived from different animal species [12,13]. In particular, studies have found at least six different N-glycan structural differences between duck and meat samples as a means of differentiating these two species [13]. Additionally, the types of O-glycan structures in meat samples are not presently known, and thus the discrimination patterns between different meat samples using an O-glycan analysis have also not been resolved. With recent advances in deciphering glycan structures, such as a combined fluorescence-based quantitation with an LC-MS technique (LC-FLR-MS) in resolving N-glycan structures at great sensitivity [14,15], as well as novel methods in releasing O-glycans with free reducing-end aldehydes for O-glycan analysis [16–18], a deep and total structural analysis of both N-linked and O-linked glycans can be performed to characterise samples with a degree of resolution that previously would not have been possible.

In this study, we demonstrate the use of this approach to determine the diverse O-linked and N-linked glycan profiles of three meat samples (chicken, beef, and pork). In the O-glycan analysis, we find a clear difference in the distinct structures between the different meat samples. The abundance and diversity of each O-glycan structure appears to be significantly dissimilar, suggesting well-defined O-glycosylation patterns between meat samples derived from different animal species. In the N-glycan analysis, our high-resolution measurements have identified the presence of up to 17 different N-glycan structures in the meat samples, which is an increase by a factor of two compared to the previously carried out analysis [13]. The individual glycan structures, as well as the total glycosylation attributes, are clearly distinguishable between the different meat samples. Finally, a principal component analysis (PCA) is also performed with all O-glycan and N-glycan structures within the meat samples, revealing straightforward discrimination between the samples, and thus the potential use of such integrative glycomic approaches for the high-throughput authentication of meat samples in the future [19].

2. Materials and Methods

2.1. Meat Lysis and Protein Extraction

Meat samples for each species were purchased and processed within the same day. Chicken samples were isolated from peroneus longus, whilst pork and beef samples were isolated from extensor carpis radialis. Fat and connective tissues were trimmed off from the meat. Pea-sized meat samples were snap-frozen using nitrogen and were minced to homogeneity using a pestle and mortar. Approximately 150 μg of each homogenised meat sample was lysed in 800 μL of T-PER tissue protein extraction reagent supplemented with protease inhibitor (1:100, both from ThermoFisher, UK) for 10 min on a rotary shaker. After this, the samples were centrifuged at 10,000× g for 5 min and the supernatant was collected. Proteins extracted were stored at −80 °C before being subjected to the O- and N-glycan analysis workflow.

2.2. Release and Permethylation of O-Glycans

O-glycans were released from 100 μg of meat samples by adding 200 μL of (0.5 M) sodium borohydride in 0.05 M potassium hydroxide and incubating in a 50 °C oven for 16 h. The reaction was terminated by adding glacial acetic acid dropwise followed by a clean-up using Dowex 50W-X8(H) 50–100 mesh resin chromatography. The samples were loaded onto the pre-prepared Dowex resin column and the flowthrough was collected in a glass tube. The O-glycans were eluted using 5 mL of 5% acetic acid and combined with the flowthrough. Eluted O-glycans were evaporated to dryness using a nitrogen sample concentrator. Then, 500 μL of 10% acetic acid in methanol was added and dried to remove borate (repeated five times). Sodium hydroxide dissolved in dimethyl sulfoxide and iodomethane were added to the dried glycan samples in glass tubes. The reaction was allowed to proceed under rotation at 30 rpm for about 3 h. Next, 1 mL of deionised water was added dropwise to quench the reaction. After 2 mL of chloroform was added, the mixture was mixed thoroughly. After allowing the mixture to separate into 2 layers, the upper aqueous layer was removed. Deionised water was added to the chloroform layer and this step of mixing and removal of aqueous layer was repeated several times until the chloroform layer was clear. The chloroform layer was then evaporated to dryness using a nitrogen sample concentrator.

2.3. Sep-Pak Separation of Permethylated Glycans

C18 Sep-Pak® cartridge (Water Corporation, Milford, MA, USA) was primed sequentially with 5 mL methanol, 5 mL deionised water, 5 mL acetonitrile and 5 mL deionised water. The dried permethylated sample was redissolved in 200 μL of 50% methanol and loaded to the Sep-Pak® cartridge. Elution was carried out by adding 2 mL of 15, 35, 50 and 75% acetonitrile in water (*v*/*v*). Each elution fraction was collected and evaporated to dryness using a SpeedVac.

2.4. Mass Spectrometry Analysis of O-Glycans

Permethylated O-glycans from the 35% and 50% fractions were combined and reconstituted in 100 μL of 80% methanol with 0.1% formic acid. Then, 10 μL of reconstituted released O-glycans were injected into Agilent 1290 infinity LC system coupled to an Agilent 6550 iFunel qTOF mass spectrometer (Agilent Technologies, Santa Clara, CA, USA). O-glycans were separated using an Agilent Zorbax Eclipse Plus C18 RRHD column (1.8 μm, 2.1 mm × 50 mm) at 500 μL/min, with an elution gradient of 3 to 10%, 10 to 40%, 40 to 70%, and 70 to 90% of 0.1% formic acid in acetonitrile (ACN, mobile phase B) at 0 to 10 min, 10 to 25 min, 25 to 30 min, and 30 to 38 min, respectively. For mobile phase A, 0.1% formic acid in water was used. The column was flushed with 90% mobile phase B for 12 min before re-equilibrating with 3% mobile phase B for 15 min.

Mass spectra were acquired in positive ion mode over a mass range of m/z 100–2000 with an acquisition rate of 1 Hz. The following parameters were used for the acquisition: drying gas temperature 150 °C at 12 L/min, sheath gas temperature 300 °C at 12 L/min, nebuliser pressure at 45 psi and capillary voltage at 2500 V. Mass correction was enabled using an infused calibrant solution with a reference mass of m/z 121.0873 and 922.0098.

2.5. O-Glycan Assignment

LC-MS data were processed using Molecular Feature Extractor (MFE) algorithm of MassHunter Qualitative Analysis Software (version B.06.00 Build 6.0.633.10 SP1, Agilent Technologies, Santa Clara, CA, USA). A permethylated mass list was generated based on the neutral masses of O-glycans found on the GlycoStore and Consortium for Functional Glycomics database [20,21]. This list with a mass filter of 10 ppm was used to search the LC-MS data. Mass peaks were filtered with a peak height of at least 100 counts and resolved into individual ion species. Using a Glycans Isotopic distribution model, charge state of a maximum of 3 and retention time, all ion species with singly and doubly protonated ions and their sodium adducted ions associated with a single compound were summed together. The neutral compound mass was then calculated and a list of all compound peaks in the samples and standards were generated with relative abundances depicted by chromatographic peak areas.

Targeted tandem MS was acquired in positive ion mode over a mass range of m/z 100–2000 with an acquisition time of 1.5 Hz. A targeted mass list was generated based on the desired MFE compounds found on the samples for MS/MS analysis. The precursor masses of interest, along with its charge state, retention time and peak width were indicated. The isolation width used was medium (~4 m/z) and the collision energy (CE) used for each precursor compound were automatically calculated by the acquisition software based on the following equation:

$$CE\ (eV) = 3\ (m/z \div 100) - 4.8 \quad (1)$$

The targeted tandem MS data were processed using MFE algorithm with the same settings used for searching LC-MS data and the MS/MS spectrum were extracted from each of the targeted compounds.

2.6. N-Glycan Release and Labelling

N-glycans of meat were released and labelled using GlycoWorks RapiFluor-MS (RFMS) N-glycan kit (Waters Corporation, Milford, MA, USA) according to the manufacturer's protocol. Briefly, 15 μg of dried protein extracted from meat was reconstituted in 22.8 μL of LCMS grade water and 6 μL of 5% RapiGest solution (final concentration 0.01% RapiGest, Waters Corporation, Milford, MA, USA). The solution was incubated at 95 °C for 5 min to denature the protein extracted from meat. N-glycans were released enzymatically by adding 600 U of recombinant PNGase F (New England Biolabs, Ipswich, MA, USA) followed by 10 min incubation at 55 °C. Released N-glycans were labelled with 12 μL of the RapiFluor-MS Reagent Solution (fluorescence label, 0.07 mg/μL in anhydrous dimethyformamide (DMF), Waters Corporation, Milford, MA, USA) at room temperature for 10 min. The solution was diluted in 358 μL of ACN, followed by clean-up using a GlycoWorks HILIC μElution Plate (Waters Corporation, Milford, MA, USA). Isolated released N-glycans were dried and reconstituted in 9 μL of LCMS grade water, 10 μL of DMF, and 21 μL of ACN sequentially for LC-MS analysis.

2.7. Liquid Chromatography-Mass Spectrometry Analysis of RFMS-Labelled N-Glycan

Released N-glycans were analysed as previously described [22]. First, 10 μL of reconstituted released N-glycans were injected into an ACQUITY H-Class UPLC (Waters Corporation, Milford, MA, USA) coupled to a SYNAPT XS mass spectrometer (Waters Corporation, Milford, MA, USA). Samples were separated using an ACQUITY UPLC Glycan BEH amide column (130 A, 1.7 μm, 2.1 mm × 150 mm, Waters Corporation, Milford, MA, USA) at 60 °C and 400 μL/min, with a 40 min gradient from 25 to 49% of 50 mM Ammonium Formate (mobile phase A). As mobile phase B, 100% ACN was used. RFMS-labelled glycans were excited at 265 nm and measured at 425 nm with an ACQUITY UPLC FLR detector (Waters Corporation, Milford, MA, USA). The MS1 profile scans of m/z 400–2000 were acquired using the SYNAPT XS in positive mode with an acquisition rate of 1 Hz. The electrospray ionisation capillary voltage was set at 1.8 kV, cone voltage at 30 V, desolvation gas flow at 850 L/h, and ion source temperature and desolvation temperature were kept at 120 °C and 350 °C, respectively. Leucine Enkephalin (Waters Corporation, Milford, MA, USA) was used as the LockSpray compound for real-time mass correction. RapiFluor-MS Dextran Calibration ladder (Waters Corporation, Milford, MA, USA) was also injected into LC-MS to calibrate the retention time of sample peaks. The retention times were normalised using the dextran calibration curve to Glucose Units (GU).

2.8. N-Glycan Assignment

Released N-glycans were analysed using the UNIFI Scientific Information System (Version 1.8, Waters Corporation, Milford, MA, USA). Fluorescence peaks were integrated manually using the UNIFI Scientific Information System and relative quantitation of peaks was obtained by area-under-curve measurements followed by normalisation to the total area. Glycan assignment was carried out by matching neutral mass and/or Glucose Units (GU) of each peak to the modified "N-glycan 309 mammalian no sodium" database available in the Byonic software.

2.9. Statistical Analyses

Data are presented as mean ± standard error of the mean. Statistical analyses were performed by Student's t-test or one-way ANOVA with Tukey's post hoc test using Graph-

Pad Prism 8 (GraphPAD Software Inc., San Diego, CA, USA). Multiple comparisons were performed between chicken, pork, and beef in ANOVA tests. O-glycan and N-glycan relative abundance was analysed using principal component analysis (PCA). The criterion for significance was $p < 0.05$ (* $p < 0.05$; ** $p < 0.01$; *** $p < 0.001$; **** $p < 0.0001$).

3. Results

3.1. Framework for an Integrated Glycomic Study of Meat Samples from Different Species

The overall glycomic-based workflow consists of: (i) an experimental approach in extracting and isolating glycan structures from meat samples, (ii) a quantitative glycomic analysis of the abundance of both N-glycans and O-glycans, and finally (iii) the computation of the principal component analysis (PCA) to successfully discriminate glycans belonging to the different meat samples (Figure 1).

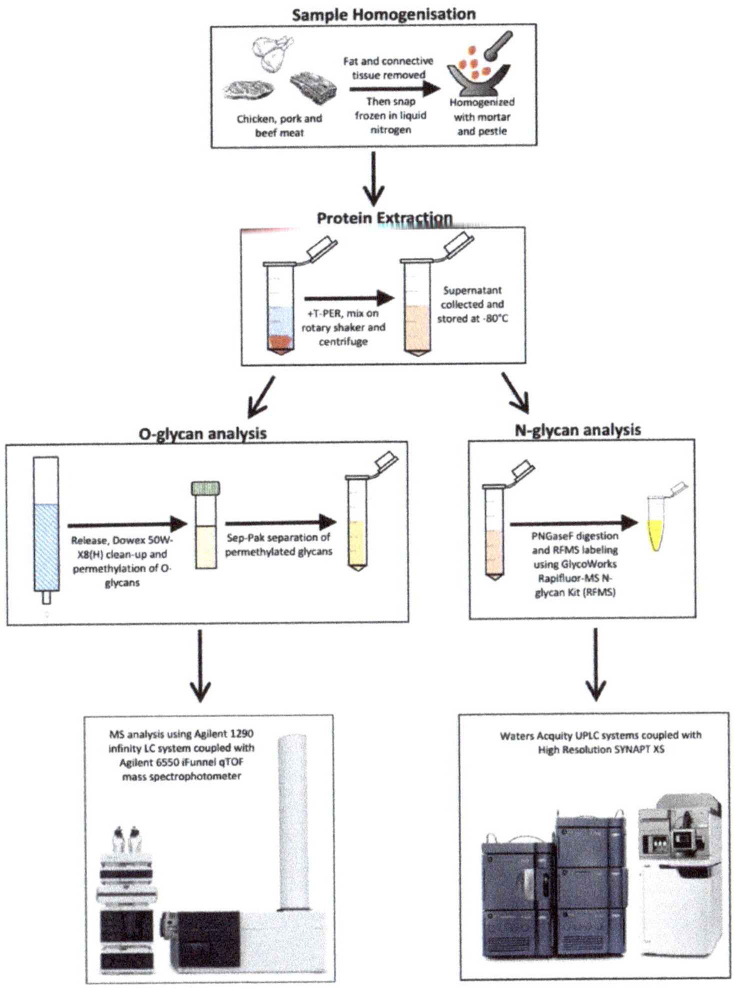

Figure 1. Glycomics-based workflow for the characterisation and profiling of meat samples. Samples are first processed for the extraction of proteins, which undergo separate procedures for the release of N-linked and O-linked glycans. These glycans are finally quantitated via a mass spectrometry-based analysis. Permethylated O-glycans are quantitated through the mass abundance of the ions whilst N-glycans are quantitated via the fluorescence of the RFMS label.

In the first step, whole meat samples are grounded and subjected to lysis using T-PER^{TM} Tissue Protein Extraction Reagent. Subsequently, a centrifuge step is employed to pellet the tissue debris, and the final supernatant product contains all proteins successfully extracted from the meat samples, including the glycoproteins of interest. In the next step, the same samples are split for N-glycan and O-glycan analysis. In the O-glycan analysis, O-glycans are released and permethylated before the LC-MS analysis, while in N-glycans, they are released and labelled with Rapifluor-MS (RFMS) before an UPLC-FLR-MS analysis. Both workflows allow the differentiation and quantification of the abundance of distinct N-glycans and O-glycans. Finally, these results are pooled together for a PCA analysis which allows the novel discrimination of the different meat samples on a molecular level (Supplementary Figures S1–S54).

3.2. O-Glycan Characterisation of Meat

We demonstrate the application of this workflow by applying the above workflow to three types of meat samples, namely chicken, pork, and beef. The extracted proteins were first analysed based on their O-glycan profiles. O-glycan structures, in particular, have been shown to exhibit a diverse glycosylation pattern in eukaryotes, owing to the many biosynthetic pathways [23]. In the case of the three meat samples, we observe the presence of four distinct O-glycan structures found through the analysis of released permethylated O-glycans. Through the combined analysis of the retention time and MS2 fragment data of the O-glycan standards, two of the most abundant glycans were identified as Gal-GalNAc and NeuAc-Gal-GalNAc (Figures 2 and 3A,B). Gal-GalNAc, in particular, was observed to be the most abundant structure in all samples, though its relative abundance ranged from 74.7 ± 0.6% in pork meat to 45.7 ± 3.3% in chicken meat (Figure 3A, Table 1). This significant abundance corroborates previous observations of Gal-GalNAc as one of the core, and thus most abundant, O-glycan structures that can be found, particularly in mammals [24,25]. On the other hand, the relative abundance of NeuAc-Gal-GalNAc was observed to range from 10–25% instead, depending on the sample. The presence of two other distinct glycans was also observed, though their exact linkage could not be ascertained. These two structures, labelled instead through their chemical compositions as Hex-HexNAc (RT~15.5 min), and Hex(NeuAc)HexNAc (RT~18.8 min), respectively, was observed to be more abundant in the chicken meat samples than the other animal samples (Figures 2 and 3C,D).

Table 1. Summary of relative abundances of O-glycans in percentage detected in chicken, pork, and beef. (±, standard error of mean; ND, not detected).

	Chicken (%)	**Pork (%)**	**Beef (%)**
Hex-HexNAc	23.6 ± 2.2	ND	13.8 ± 1.0
Gal-GalNAc	45.7 ± 3.3	74.7 ± 0.6	63.1 ± 3.6
Hex(NeuAc)HexNAc	19.3 ± 2.4	ND	ND
NeuAc-Gal-GalNAc	11.3 ± 1.9	25.3 ± 0.6	23.1 ± 3.6

Furthermore, a quantitative comparison of the relative abundances of all four O-glycans was made across all three meat samples (Figure 3A–D, Table 1). ANOVA test indicated that the relative abundances of Gal-GalNAc and Hex-HexNAc (by chemical composition) in chicken, pork, and beef meat samples are significantly different from each other (Figure 3A,C). An interesting trend was observed in the case of sialylated O-glycans within the meat samples. In particular, chicken meat samples were observed to contain a significant amount of Hex(NeuAc)HexNAc (by chemical composition) which was otherwise undetected in the beef and pork samples (Figure 3D). However, in the case of the other sialylated O-glycan NeuAc-Gal-GalNAc, chicken meat samples had the lowest relative abundance as compared to the other two samples (Figure 3B, 11.3 ± 1.9% versus 25.3 ± 0.6% and 23.1 ± 3.6%, respectively, $p < 0.05$, $p < 0.01$). Indeed, the combined quantification of all sialylated glycans (Hex(NeuAc)HexNAc and NeuAc-Gal-GalNAc) yield no statistically significant difference between species (Figure 3E). This highlights the importance of high resolution glycomics in characterising individual O-glycan structures for the proper distinction between meat samples of different animal origins.

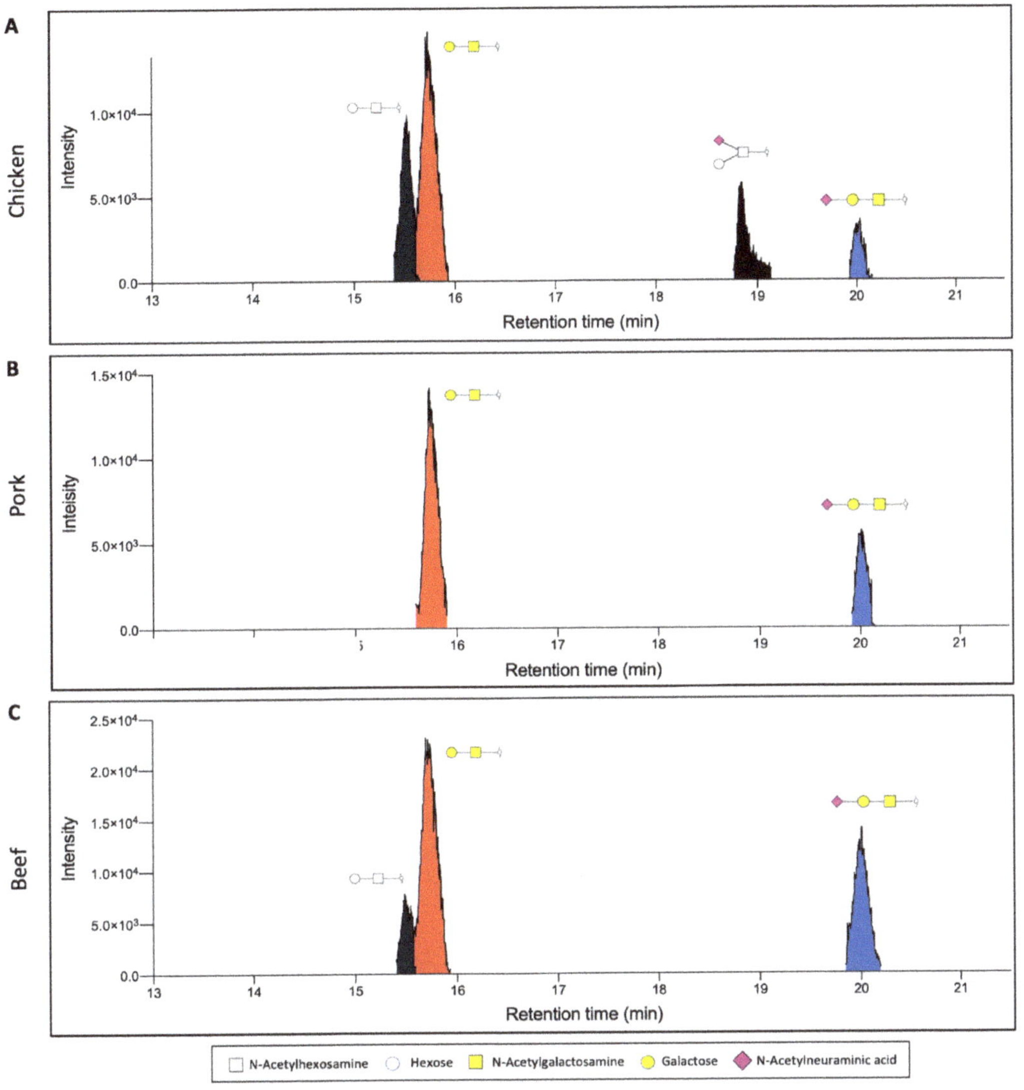

Figure 2. Representative extracted ion chromatogram of permethylated O-glycans extracted from (**A**) chicken, (**B**) pork, and (**C**) beef. Gal-GalNAc and NeuAc-Gal-GalNAc were detected in all species, whilst Hex-HexNAc was detected only in chicken and beef and Hex(NeuAc)HexNAc was detected exclusively in chicken. (White square, N-acetylhexosamine; white circle, Hexose; yellow square, N-acetylgalactosamine; yellow circle, galactose; purple diamond, N-acetylneuraminic acid).

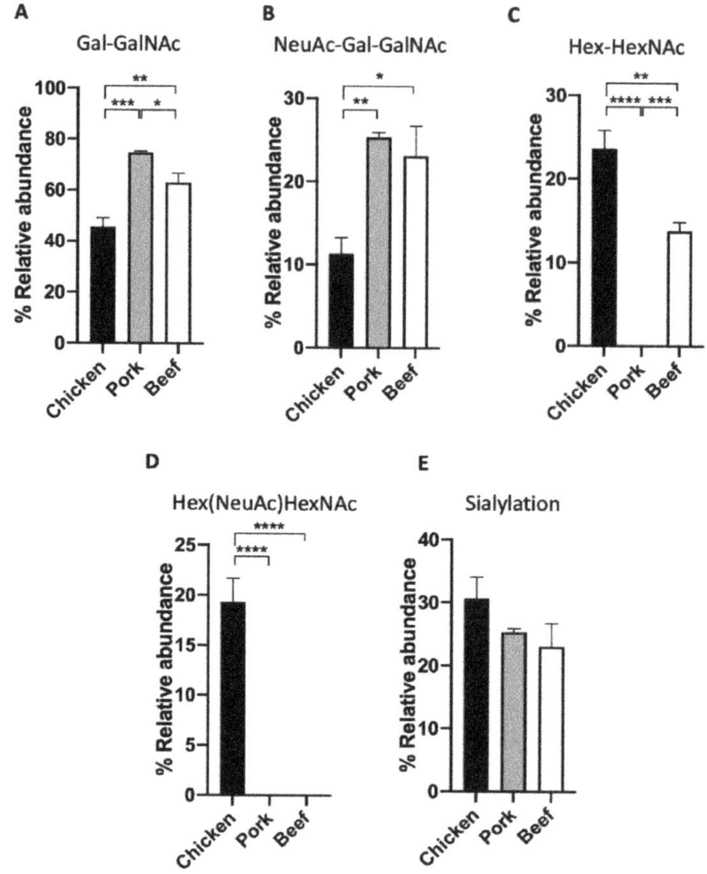

Figure 3. The relative abundance of permethylated O-glycans in meat samples. (**A**) Gal-GalNAc, (**B**) NeuAc-Gal-GalNAc, (**C**) Hex-HexNAc, (**D**) Hex(NeuAc)HexNAc, and (**E**) sialylation of chicken, pork and beef. (One-way ANOVA followed by Tukey's post hoc test, multiple comparisons were performed between chicken, pork, and beef, $n = 4$; * $p < 0.05$; ** $p < 0.01$; *** $p < 0.001$; **** $p < 0.0001$).

3.3. N-Glycan Characterisation of Meat

We also performed a full characterisation of the N-glycan structures of the extracted proteins from the three meat samples (Figure 4). We released N-glycans using the enzyme PNGaseF and labelled them with RFMS fluorescence tag before subjecting them to FLR-LC-MS workflow. Interestingly, from the FLR chromatogram, which reflects the abundance of the fluorescently labelled N-glycans, we can observe distinct overall N-glycomic signatures between the samples. For instance, in the case of the chicken meat sample, an even distribution of peaks was observed (Figure 4A), whilst distribution of peaks from pork and beef meat samples were skewed towards those with higher retention time, which also corresponds to higher neutral mass (Figure 4B,C). In particular, the overall N-glycome of the pork sample appeared to be less heterogeneous, with one major peak observed at around 20.8 min. N-glycan compositions were further confirmed with neutral mass and/or glucose unit (GU) and labelled in the representative FLR chromatogram (Figure 4). We subsequently quantified the relative abundance of each N-glycan structure between the meat samples (Table 2). In each sample, up to 17 different glycan structures could be identified, which is significantly higher than that reported from previous studies [13]. We note that other rare, low-abundant glycan structures may also be present in these samples, whose signal is masked by the relatively much more abundant glycan structures. These N-glycans of these samples were grouped based on their glycosylation attributes—fucosylation, sialylation, galactosylation, and the presence of high mannose (Figure 5). We observed that the pork meat samples contained the highest abundance of fucosylated N-glycan structures (Figure 5A, $p < 0.0001$ and $p < 0.05$, for comparison between pork and chicken or beef, respectively), followed by beef and chicken samples (Figure 5A, $p < 0.001$, for comparison between chicken and beef). Chicken meat samples contained the lowest relative abundances of sialylated and galactosylated N-glycans compared to pork and beef meat samples (sialylation: Figure 5B, $p < 0.001$ and $p < 0.01$, for comparison between chicken and pork or beef, respectively; galactosylation: Figure 5C, $p < 0.001$, for both comparison between chicken and pork or beef), with no statistically significant difference between pork and beef meat samples. In contrast, the chicken meat sample had the highest amount of N-glycans with high mannose structure (Figure 5D, $p < 0.001$, for both comparison between chicken and pork or beef), while no statistically significant difference was found between pork and beef samples.

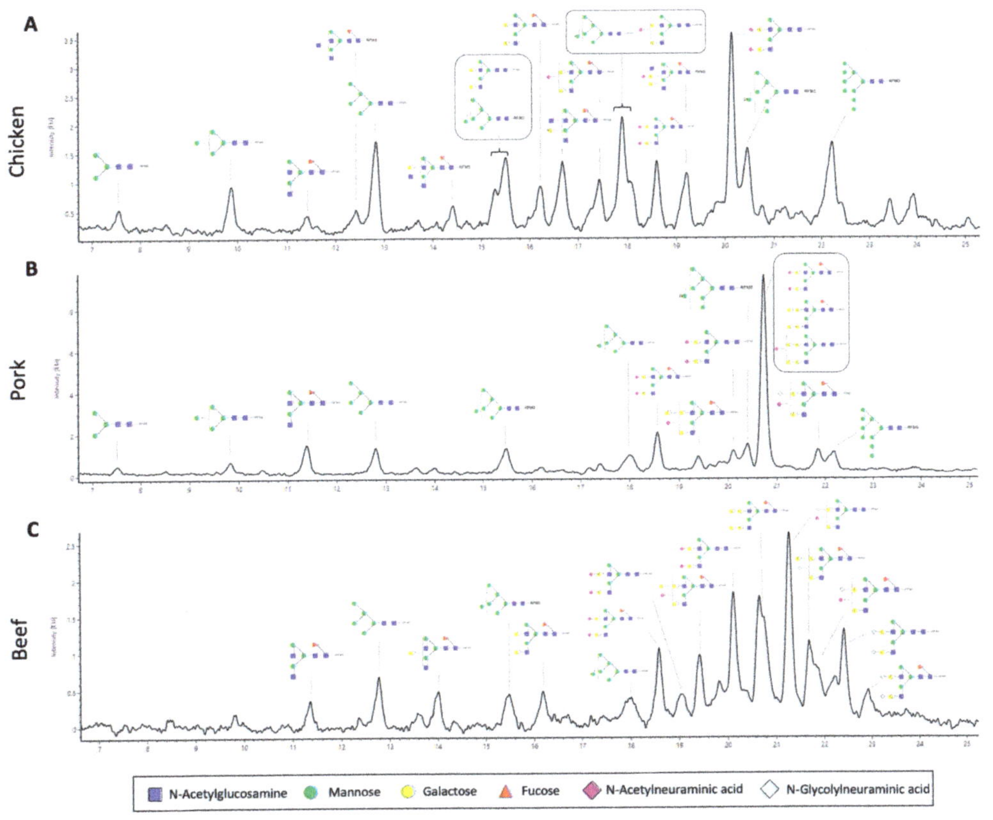

Figure 4. Representative fluorescence (FLR) chromatograms of RFMS-labelled released N-glycans from (**A**) chicken, (**B**) pork, and (**C**) beef. (Blue square, N-Acetylglycosamine; green circle, mannose; yellow circle, galactose; red triangle, fucose; purple diamond, N-acetylneuraminic acid; light blue, N-Glycolylneuraminic acid.) Glycan images represent compositions and linkage type is not determined. (Samples were separated using an ACQUITY UPLC Glycan BEH amide column).

Table 2. Summary of relative abundances of N-glycans detected in chicken, pork, and beef.

	Chicken (%)	Pork (%)	Beef (%)
HexNAc(2)Hex(3)	1.0 ± 0.4	0.3 ± 0.3	ND
HexNAc(2)Hex(4)	2.9 ± 0.8	0.5 ± 0.5	1.0 ± 1.0
HexNAc(4)Hex(3)Fuc(1)	1.1 ± 0.2	6.3 ± 0.2	0.8 ± 0.5
HexNAc(2)Hex(5)	9.3 ± 0.2	4.3 ± 0.5	5.8 ± 0.3
HexNAc(4)Hex(4)Fuc(1)	ND	1.2 ± 0.4	1.8 ± 1.4
HexNAc(5)Hex(4)Fuc(1)	1.6 ± 0.1	ND	ND
HexNAc(4)Hex(5)	2.0 ± 0.2	0.3 ± 0.1	ND
HexNAc(2)Hex(6)	8.1 ± 0.8	4.5 ± 1.0	4.1 ± 1.4
HexNAc(4)Hex(5)Fuc(1)	2.2 ± 0.4	ND	2.5 ± 0.5
HexNAc(5)Hex(5)Fuc(1)	5.9 ± 0.5	ND	ND
HexNAc(4)Hex(5)NeuAc(1)	9.4 ± 0.5	ND	ND
HexNAc(4)Hex(5)Fuc(1)NeuAc(1)	6.2 ± 0.6	0.2 ± 0.2	ND
HexNAc(2)Hex(7)	6.6 ± 0.6	3.9 ± 0.6	4.1 ± 0.9
HexNAc(4)Hex(5)Fuc(1)NeuAc(2)	4.3 ± 0.5	60.7 ± 6.3	4.6 ± 0.1
HexNAc(5)Hex(5)Fuc(1)NeuAc(1)	6.0 ± 0.3	ND	ND
HexNAc(4)Hex(6)Fuc(1)NeuAc(1)	ND	1.2 ± 0.5	8.1 ± 0.7
HexNAc(4)Hex(5)NeuAc(2)	11.9 ± 1.0	2.3 ± 0.2	9.7 ± 1.0
HexNAc(2)Hex(8)	10.8 ± 1.1	2.2 ± 1.4	0.6 ± 0.1
HexNAc(4)Hex(7)Fuc(1)	ND	3.3 ± 0.3	17.8 ± 1.3
HexNAc(4)Hex(6)Fuc(1)NeuGc(1)	ND	ND	4.2 ± 0.4
HexNAc(4)Hex(7)NeuAc(1)	ND	1.1 ± 0.1	2.3 ± 0.2
HexNAc(4)Hex(5)NeuAc(1)NeuGc(1)	ND	ND	18.5 ± 1.4
HexNAc(4)Hex(5)Fuc(1)NeuAc(1)NeuGc(1)	ND	4.9 ± 0.3	2.2 ± 1.0
HexNAc(2)Hex(9)	9.2 ± 0.7	2.3 ± 1.4	ND
HexNAc(4)Hex(5)NeuGc(2)	ND	ND	5.8 ± 0.4
HexNAc(4)Hex(5)Fuc(1)NeuGc(2)	ND	ND	2.5 ± 1.0

Hex, hexose sugars; HexNAc, N-acetylhexosamine; Fuc, fucose; NeuAc, N-acetylneuraminic acid; NeuGc, N-glycolylneuraminic acid. The number beside each sugar refers to the number of such molecules present in that structure. (±, standard error of mean; ND, not detected).

The relative abundances of two predominant sialic acids on N-glycans, Neu5Ac and Neu5Gc were also characterised, considering the prominent role of sialic acids in multiple biological functions [26]. In particular, the uptake of Neu5Gc from red meat has been shown to trigger inflammatory response and cancer development [27]. Our results show that pork samples contained significantly higher relative abundance of Neu5Ac than chicken and beef samples, with no difference between chicken and beef samples (Figure 5E, $p < 0.001$, $p < 0.01$, for comparison between pork and chicken or beef, respectively). On the other hand, no Neu5Gc was detected in chicken, with 4.8 ± 0.3% and 34.4 ± 1.9% of Neu5Gc detected in pork and beef, respectively (Figure 5F). More than half of the total sialic acid content (Neu5Ac and Neu5Gc) found in beef was Neu5Gc, compared to only a significantly small proportion of Neu5Gc detected in pork (Figure 5G, 58.2 ± 1.8% versus 6.9 ± 0.4%). The ratio of Neu5Gc to total sialic acid in all three meats agrees with previously published work that analysed free sialic acid using HPLC method [27].

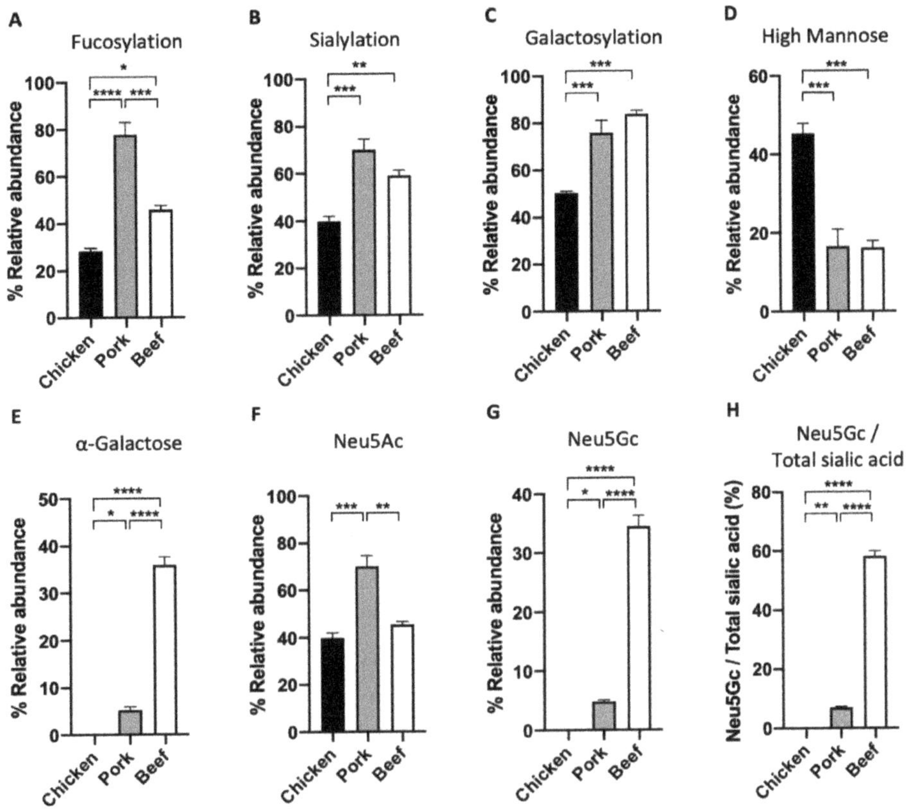

Figure 5. The relative abundance of (**A**) core fucosylation, (**B**) sialylation, (**C**) galactosylation, (**D**) high mannose, (**E**) α-galactose, (**F**) N-acetylneuraminic acid (Neu5Ac), and (**G**) N-glycolylneuraminic acid (Neu5Gc) measured in N-glycans of chicken, pork, and beef. (**H**) Ratio of Neu5Gc and total sialic acid was also calculated. (One-way ANOVA followed by Tukey's post hoc test, $n = 4$ independent biological replicates; * $p < 0.05$; ** $p < 0.01$; *** $p < 0.001$; **** $p < 0.0001$).

Our high resolution glycomic workflow also allows the full characterisation of N-glycans with α-galactose (galactose-α-1,3-galactose). It is important to characterise the abundance of α-galactose in meat samples because it is implicated in alpha-gal syndrome (also known as red meat allergy)—a potentially life-threatening allergy to red meat [28]. The highest relative abundance of α-galactose amongst the meat samples was detected in the beef meat sample, followed by a moderate amount found in the pork sample, and an undetectable amount in chicken meat samples (Figure 5H, 35.8 ± 1.8%, 5.1 ± 0.8%, and 0%, respectively). This agrees with our understanding that α-galactose is found more abundantly in red meat than in white meat [29].

3.4. PCA Analysis

Our characterisation of the N-linked and O-linked glycans of the three different meat samples shows the distinct differences in their relative abundances, suggesting a unique overall glycomic profile of each sample that can be distinguished with the integrative glycomic workflow. In order to achieve a unified glycome analysis of both N-linked and O-linked glycans, the datasets were pooled together and subjected to principal component analysis (PCA). The dataset was transformed into two principal components—PC1 and

PC2,—which explained percentages of variance, which were 57.4% and 25.8%, respectively. The reliability of this PCA analysis can be observed by more than 82% of the total variance that can be accounted for by the first two PCs. The score plot of PC1 and PC2 of each meat allowed visual discrimination of different species (Figure 6). This analysis demonstrates, with strong statistical significance, the well-defined glycomic characteristics of meat samples pertaining to different animal species, and the strength of the overall integrated glycomic approach in profiling meat samples.

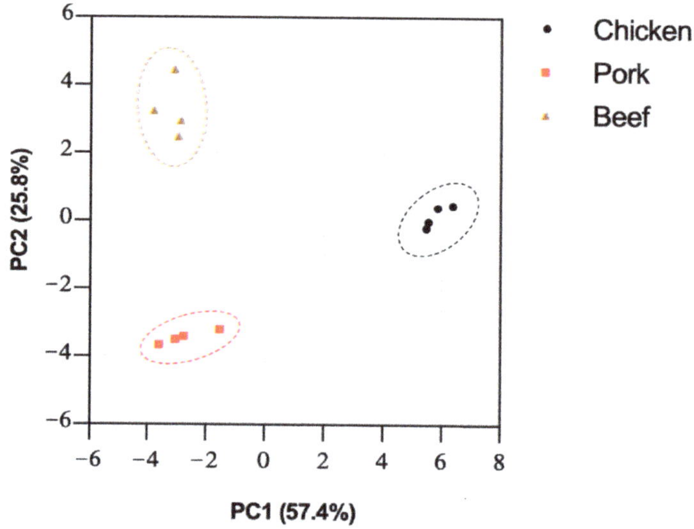

Figure 6. Principal component analysis was performed on relative abundances of O- and N-glycans of chicken, pork, and beef. A clear separation was observed between meat from different species. Each dot represents an independent biological replicate.

4. Discussion

Current glycomic approaches are extensively applied for the use as biomarkers in different areas, particularly in medical and biotechnology fields [30–32]. In this study, we have described in detail the extended use of an integrated glycomics approach to characterise meat samples in terms of their O-linked and N-linked glycan structures. The abundances of individual glycan types are significantly different both in terms of the O-linked as well as N-linked glycans. Previous studies have investigated the O-linked glycans found in food, such as bovine whey protein product and mucins from salmon and chicken [33–35]. However, to the best of our knowledge, O-linked glycan characterisation has not been employed in meat profiling. The novel discovery of the main types of O-glycan structures found in meat samples shows different molecular signatures pertaining to each species, and an orthogonal measurement in glycomics for meat differentiation purposes. Similarly, the identification of a diverse number of N-glycan structures in each sample shows the diversity of mechanisms which each species undergoes for glycosylation, resulting in their distinct structural differences. Despite the variety of N-glycan structures that can be present, it is interesting that meat samples of each animal species possess a distinct subset of these structures. This suggests a high degree of specificity, especially in the glycoenzymes involved in the synthesis of these glycans [8].

It is worth noting that meat tissues were isolated only from one region of the animal (peroneus longus from chicken samples, and extensor carpis radialis from the pork and beef samples). As glycosylation is context- and tissue-specific, we anticipate potential differences in the glycan profile of different tissues belonging to the same animal [36,37]. By systematically comparing the differences between tissue sources, and between animals, future studies can further reveal the potential of glycomics in its versatile applicability in different types of food samples. In addition, this study has not characterised the glycan profile of all animal species. It has been observed that meat species adulteration can occur between meat samples of closely related species such as horse meat and beef [1,38]. Thus, further studies investigating the glycome differences between meat samples derived from closely related species are warranted.

The meat industry presents other major challenges such as fraudulent labelling of geographical origin and production system (e.g., organic vs. non-organic) of the meat samples. Amongst the many analytical techniques available in the meat industry, there are only a few capable of performing the authentication of geographical origin and production system. For instance, DNA-based methods such as polymerase chain reactions and genomics rely on a specific DNA sequence or taxonomic marker to differentiate the species of meat, but they are unable to detect fraudulent labels of geographical origin and production system. Since dietary patterns, lifestyle and environmental changes are known to affect protein glycosylation in humans, animals would likely also undergo glycan changes. As such, such changes in glycan profile could be exploited in the authentication of geographical origin and production system. The use of glycomic techniques in meat species profiling presented in this paper serves as a proof-of-concept and its use in the authentication of other meat product features should be investigated. Importantly, glycoproteomic approaches harnessing both proteomic and glycomic potentials could also be a promising tool to further the molecular characterisation of meat samples.

Considering the traction of cultured meat products (growing cells to generate meat-like tissue structures in the laboratory) as an alternative protein source for human consumption, the glycoprofile workflow described in this study can also be used in establishing the critical quality attributes (CQA) of cultured meat products [39–41]. Given the considerable risk of cell line contamination and product adulteration in this growing industry, such techniques can help to establish a stringent quality control of these cultured meat samples in the future [42,43]. With the advent of FDA-approved genetically modified pigs with α-galactose for human consumption, our glycomic techniques can also be used to monitor the controlled manipulation of cultured meat [44]. This includes their modification in consideration of unwanted glycans such as α-galactose and Neu5Gc for health and safety reasons.

While more studies are warranted to investigate how these glycomic signatures may differ based on breed genetic compositions, parts from which the meat samples were obtained, and other extrinsic factors (e.g., feed intake, growth conditions, regional differences), it is evident that the use of O-linked and N-linked glycome profiling allows the successful differentiation between different meat samples, and as a proof-of-concept paves the way for a new high-throughput and robust approach in quantifying meat adulteration. This may be particularly advantageous over other approaches as N-glycan profiling is unaffected by the harsh processing of meat (heat-induced treatments) that can degrade DNA and affect the accuracy of genomic approaches [8]. Given its highly quantitative and efficient procedure, the adoption of such glycome profiling approaches provides a powerful future alternative technique to traditional methods in meat identification and authentication.

Supplementary Materials: The following are available online at: https://www.mdpi.com/article/10.3390/foods11131952/s1, Supplementary file containing representative of N-glycan MS spectra.

Author Contributions: Conceptualization, K.T.P.; methodology, S.C., G.T. and S.J.T.; validation, S.C., G.T., C.W., I.W. and K.T.P.; formal analysis, G.T. and K.T.P.; investigation, S.C., G.T. and K.T.P.; resources, I.W. and K.T.P.; data curation, K.T.P.; writing—original draft preparation, S.C., G.T. and K.T.P.; writing—review and editing, S.C., G.T., L.S.W.L., H.Y., I.W. and K.T.P.; visualization, L.C.S. and K.T.P.; supervision, I.W. and K.T.P.; project administration, K.T.P. All authors have read and agreed to the published version of the manuscript.

Funding: This work was supported by the Career Development Award (No. C210112057) awarded to I.W. from A*STAR. K.T.P. are supported by A*STAR scholarship program. S.C. is funded by the A*STAR Young Achiever Award. L.L.S.W. is funded by Singapore-New Zealand Bilateral Programme on Future Food Grant (No. A20D3b0073) by Biomedical Research Council (BMRC), Agency for Science, Technology and Research (A*STAR) awarded to H.Y.

Data Availability Statement: The data presented in this study are available on request from the corresponding author.

Conflicts of Interest: The authors declare no conflict of interest.

References

1. Black, C.; Chevallier, O.P.; Elliott, C.T. The current and potential applications of Ambient Mass Spectrometry in detecting food fraud. *TrAC Trends Anal. Chem.* **2016**, *82*, 268–278. [CrossRef]
2. Sajali, N.; Wong, S.C.; Abu Bakar, S.; Khairil Mokhtar, N.F.; Manaf, Y.N.; Yuswan, M.H.; Mohd Desa, M.N. Analytical approaches of meat authentication in food. *Int. J. Food Sci. Technol.* **2021**, *56*, 1535–1543. [CrossRef]
3. Ballin, N.Z.; Vogensen, F.K.; Karlsson, A.H. Species determination—Can we detect and quantify meat adulteration? *Meat Sci.* **2009**, *83*, 165–174. [CrossRef] [PubMed]
4. Bansal, S.; Singh, A.; Mangal, M.; Mangal, A.K.; Kumar, S. Food adulteration: Sources, health risks, and detection methods. *Crit. Rev. Food Sci. Nutr.* **2017**, *57*, 1174–1189. [CrossRef]
5. Hong, E.; Lee, S.Y.; Jeong, J.Y.; Park, J.M.; Kim, B.H.; Kwon, K.; Chun, H.S. Modern analytical methods for the detection of food fraud and adulteration by food category. *J. Sci. Food Agric.* **2017**, *97*, 3877–3896. [CrossRef]
6. Huck, C.W.; Pezzei, C.K.; Huck-Pezzei, V.A. An industry perspective of food fraud. *Curr. Opin. Food Sci.* **2016**, *10*, 32–37. [CrossRef]
7. Doosti, A.; Ghasemi Dehkordi, P.; Rahimi, E. Molecular assay to fraud identification of meat products. *J. Food Sci. Technol.* **2014**, *51*, 148–152. [CrossRef]
8. Dirong, G.; Nematbakhsh, S.; Selamat, J.; Chong, P.P.; Idris, L.H.; Nordin, N.; Fatchiyah, F.; Razis, A.F.A. Omics-based analytical approaches for assessing chicken species and breeds in food authentication. *Molecules* **2021**, *26*, 6502. [CrossRef]
9. Böhme, K.; Calo-Mata, P.; Barros-Velázquez, J.; Ortea, I. Recent applications of omics-based technologies to main topics in food authentication. *TrAC Trends Anal. Chem.* **2019**, *110*, 221–232. [CrossRef]
10. Capozzi, F.; Bordoni, A. Foodomics: A new comprehensive approach to food and nutrition. *Genes Nutr.* **2013**, *8*, 1–4. [CrossRef]
11. Barabási, A.L.; Menichetti, G.; Loscalzo, J. The unmapped chemical complexity of our diet. *Nat. Food* **2020**, *1*, 33–37. [CrossRef]
12. Walsh, I.; Zhao, S.; Campbell, M.; Taron, C.H.; Rudd, P.M. Quantitative profiling of glycans and glycopeptides: An informatics' perspective. *Curr. Opin. Struct. Biol.* **2016**, *40*, 70–80. [CrossRef] [PubMed]
13. Shi, Z.; Yin, B.; Li, Y.; Zhou, G.; Li, C.; Xu, X.; Luo, X.; Zhang, X.; Qi, J.; Voglmeir, J.; et al. N-Glycan Profile as a Tool in Qualitative and Quantitative Analysis of Meat Adulteration. *J. Agric. Food Chem.* **2019**, *67*, 10543–10551. [CrossRef] [PubMed]
14. Lauber, M.A.; Yu, Y.Q.; Brousmiche, D.W.; Hua, Z.; Koza, S.M.; Magnelli, P.; Guthrie, E.; Taron, C.H.; Fountain, K.J. Rapid preparation of released N -glycans for HILIC analysis using a labeling reagent that facilitates sensitive fluorescence and ESI-MS detection. *Anal. Chem.* **2015**, *87*, 5401–5409. [CrossRef]
15. Pallister, E.G.; Choo, M.S.F.; Walsh, I.; Tai, J.N.; Tay, S.J.; Yang, Y.S.; Ng, S.K.; Rudd, P.M.; Flitsch, S.L.; Nguyen-Khuong, T. Utility of Ion-Mobility Spectrometry for Deducing Branching of Multiply Charged Glycans and Glycopeptides in a High-Throughput Positive ion LC-FLR-IMS-MS Workflow. *Anal. Chem.* **2020**, *92*, 15323–15335. [CrossRef]
16. Goso, Y.; Sugaya, T.; Ishihara, K.; Kurihara, M. Comparison of Methods to Release Mucin-Type O-Glycans for Glycomic Analysis. *Anal. Chem.* **2017**, *89*, 8870–8876. [CrossRef]
17. Reinhold, V.; Zhang, H.; Hanneman, A.; Ashline, D. Toward a platform for comprehensive glycan sequencing. *Mol. Cell. Proteomics* **2013**, *12*, 866–873. [CrossRef]
18. Yusufi, F.N.K.; Lakshmanan, M.; Ho, Y.S.; Loo, B.L.W.; Ariyaratne, P.; Yang, Y.; Ng, S.K.; Tan, T.R.M.; Yeo, H.C.; Lim, H.L.; et al. Mammalian Systems Biotechnology Reveals Global Cellular Adaptations in a Recombinant CHO Cell Line. *Cell Syst.* **2017**, *4*, 530–542.e6. [CrossRef]
19. Li, Y.C.; Liu, S.Y.; Meng, F.B.; Liu, D.Y.; Zhang, Y.; Wang, W.; Zhang, J.M. Comparative review and the recent progress in detection technologies of meat product adulteration. *Compr. Rev. Food Sci. Food Saf.* **2020**, *19*, 2256–2296. [CrossRef]
20. Royle, L.; Mattu, T.S.; Hart, E.; Langridge, J.I.; Merry, A.H.; Murphy, N.; Harvey, D.J.; Dwek, R.A.; Rudd, P.M. An analytical and structural database provides a strategy for sequencing O-glycans from microgram quantities of glycoproteins. *Anal. Biochem.* **2002**, *304*, 70–90. [CrossRef]
21. Zhao, S.; Walsh, I.; Abrahams, J.L.; Royle, L.; Nguyen-Khuong, T.; Spencer, D.; Fernandes, D.L.; Packer, N.H.; Rudd, P.M.; Campbell, M.P. GlycoStore: A database of retention properties for glycan analysis. *Bioinformatics* **2018**, *34*, 3231–3232. [CrossRef] [PubMed]
22. Pang, K.T.; Tay, S.J.; Wan, C.; Walsh, I.; Choo, M.S.F.; Yang, Y.S.; Choo, A.; Ho, Y.S.; Nguyen-Khuong, T. Semi-Automated Glycoproteomic Data Analysis of LC-MS Data Using GlycopeptideGraphMS in Process Development of Monoclonal Antibody Biologics. *Front. Chem.* **2021**, *9*, 1–14. [CrossRef] [PubMed]
23. Corfield, A.P.; Berry, M. Glycan variation and evolution in the eukaryotes. *Trends Biochem. Sci.* **2015**, *40*, 351–359. [CrossRef] [PubMed]

24. Watson, M.E.; Diepeveen, L.A.; Stubbs, K.A.; Yeoh, G.C. Glycosylation-related diagnostic and therapeutic drug target markers in hepatocellular carcinoma. *J. Gastrointest. Liver Dis.* **2015**, *24*, 349–357. [CrossRef]
25. Saldova, R.; Wilkinson, H. Current methods for the characterization of o-glycans. *J. Proteome Res.* **2020**, *19*, 3890–3905. [CrossRef]
26. Li, Y.; Chen, X. Sialic acid metabolism and sialyltransferases: Natural functions and applications. *Appl. Microbiol. Biotechnol.* **2012**, *94*, 887–905. [CrossRef]
27. Samraj, A.N.; Pearce, O.M.T.; Läubli, H.; Crittenden, A.N.; Bergfeld, A.K.; Band, K.; Gregg, C.J.; Bingman, A.E.; Secrest, P.; Diaz, S.L.; et al. A red meat-derived glycan promotes inflammation and cancer progression. *Proc. Natl. Acad. Sci. USA* **2015**, *112*, 542–547. [CrossRef]
28. Commins, S.P.; Satinover, S.M.; Hosen, J.; Mozena, J.; Borish, L.; Lewis, B.D.; Woodfolk, J.A.; Platts-Mills, T.A.E. Delayed anaphylaxis, angioedema, or urticaria after consumption of red meat in patients with IgE antibodies specific for galactose-α-1,3-galactose. *J. Allergy Clin. Immunol.* **2009**, *123*, 426–433.e2. [CrossRef]
29. Wilson, J.M.; Schuyler, A.J.; Workman, L.; Gupta, M.; James, H.R.; Posthumus, J.; McGowan, E.C.; Commins, S.P.; Platts-Mills, T.A.E. Investigation into the α-Gal Syndrome: Characteristics of 261 Children and Adults Reporting Red Meat Allergy. *J. Allergy Clin. Immunol. Pract.* **2019**, *7*, 2348–2358.e4. [CrossRef]
30. Reis, C.A.; Osorio, H.; Silva, L.; Gomes, C.; David, L. Alterations in glycosylation as biomarkers for cancer detection. *J. Clin. Pathol.* **2010**, *63*, 322–329. [CrossRef]
31. Rudman, N.; Gornik, O.; Lauc, G. Altered N-glycosylation profiles as potential biomarkers and drug targets in diabetes. *FEBS Lett.* **2019**, *593*, 1598–1615. [CrossRef] [PubMed]
32. Seeling, M.; Brückner, C.; Nimmerjahn, F. Differential antibody glycosylation in autoimmunity: Sweet biomarker or modulator of disease activity? *Nat. Rev. Rheumatol.* **2017**, *13*, 621–630. [CrossRef] [PubMed]
33. Van Leeuwen, S.S.; Schoemaker, R.J.W.; Timmer, C.J.A.M.; Kamerling, J.P.; Dijkhuizen, L. N -and O -Glycosylation of a Commercial Bovine Whey Protein Product. *J. Agric. Food Chem.* **2012**, *60*, 12553–12564. [CrossRef] [PubMed]
34. Struwe, W.B.; Gough, R.; Gallagher, M.E.; Kenny, D.T.; Carrington, S.D.; Karlsson, N.G.; Rudd, P.M. Identification of O-glycan structures from chicken intestinal mucins provides insight into Campylobactor jejuni pathogenicity. *Mol. Cell. Proteom.* **2015**, *14*, 1464–1477. [CrossRef]
35. Jin, C.; Padra, J.T.; Sundell, K.; Sundh, H.; Karlsson, N.G.; Lindén, S.K. Atlantic Salmon Carries a Range of Novel O-Glycan Structures Differentially Localized on Skin and Intestinal Mucins. *J. Proteome Res.* **2015**, *14*, 3239–3251. [CrossRef]
36. Suzuki, N.; Abe, T.; Natsuka, S. Structural analysis of N-glycans in chicken trachea and lung reveals potential receptors of chicken influenza viruses. *Sci. Rep.* **2022**, *12*, 2081. [CrossRef]
37. Jahan, M.; Thomson, P.C.; Wynn, P.C.; Wang, B. The non-human glycan, N-glycolylneuraminic acid (Neu5Gc), is not expressed in all organs and skeletal muscles of nine animal species. *Food Chem.* **2021**, *343*, 128439. [CrossRef]
38. O'Mahony, P.J. Finding horse meat in beef products-a global problem. *QJM* **2013**, *106*, 595–597. [CrossRef]
39. Post, M.J. Cultured beef: Medical technology to produce food. *J. Sci. Food Agric.* **2014**, *94*, 1039–1041. [CrossRef]
40. Arshad, M.S.; Javed, M.; Sohaib, M.; Saeed, F.; Imran, A.; Amjad, Z. Tissue engineering approaches to develop cultured meat from cells: A mini review. *Cogent Food Agric.* **2017**, *3*, 1320814. [CrossRef]
41. Datar, I.; Betti, M. Possibilities for an in vitro meat production system. *Innov. Food Sci. Emerg. Technol.* **2010**, *11*, 13–22. [CrossRef]
42. Alston-Roberts, C.; Barallon, R.; Bauer, S.R.; Butler, J.; Capes-Davis, A.; Dirks, W.G.; Elmore, E.; Furtado, M.; Kerrigan, L.; Kline, M.C.; et al. Cell line misidentification: The beginning of the end. *Nat. Rev. Cancer* **2010**, *10*, 441–448. [CrossRef]
43. Horbach, S.P.J.M.; Halffman, W. The ghosts of HeLa: How cell line misidentification contaminates the scientific literature. *PLoS ONE* **2017**, *12*, e0186281. [CrossRef] [PubMed]
44. US Food and Drug Administration. FDA Approves First-of-Its-Kind Intentional Genomic Alteration in Line of Domestic Pigs for Both Human Food, Potential Therapeutic Uses. 2020. Available online: https://www.fda.gov/news-events/press-announcements/fda-approves-first-its-kind-intentional-genomic-alteration-line-domestic-pigs-both-human-food (accessed on 23 November 2021).

MDPI
St. Alban-Anlage 66
4052 Basel
Switzerland
www.mdpi.com

Foods Editorial Office
E-mail: foods@mdpi.com
www.mdpi.com/journal/foods

www.ingramcontent.com/pod-product-compliance
Lightning Source LLC
Chambersburg PA
CBHW061521210926
51850CB00008B/1995